CAMBRIDGE LIBRARY COLLECTION

Books of enduring scholarly value

Cambridge

The city of Cambridge received its royal charter in 1201, having already been home to Britons, Romans and Anglo-Saxons for many centuries. Cambridge University was founded soon afterwards and celebrates its octocentenary in 2009. This series explores the history and influence of Cambridge as a centre of science, learning, and discovery, its contributions to national and global politics and culture, and its inevitable controversies and scandals.

A Flora of Cambridgeshire

This flora, published in 1964, was the first comprehensive account of Cambridgeshire's plants since Babington's of 1860. Based on records to the end of 1962, it details 1509 species. These comprise 27 pteridophytes, 3 gymnosperms, 1223 angiosperms and 256 bryophytes. The following information is provided for each of the species: scientific name; well known vernacular name, if any; first known record of the plant in the county; synonyms; habitat; notes on rare, difficult or interesting species; distribution by OS grid reference numbers. The introduction examines local topography, climate, the main geological areas and vegetation types, together with a history of botanical investigation in the county. Important localities are noted, highlighting key species that could then be found. Botanists, conservationists and naturalists will find this historic flora provides a valuable baseline for contemporary studies, including those focusing on biodiversity, extinction or the effects of climate change.

Cambridge University Press has long been a pioneer in the reissuing of out-of-print titles from its own backlist, producing digital reprints of books that are still sought after by scholars and students but could not be reprinted economically using traditional technology. The Cambridge Library Collection extends this activity to a wider range of books which are still of importance to researchers and professionals, either for the source material they contain, or as landmarks in the history of their academic discipline.

Drawing from the world-renowned collections in the Cambridge University Library, and guided by the advice of experts in each subject area, Cambridge University Press is using state-of-the-art scanning machines in its own Printing House to capture the content of each book selected for inclusion. The files are processed to give a consistently clear, crisp image, and the books finished to the high quality standard for which the Press is recognised around the world. The latest print-on-demand technology ensures that the books will remain available indefinitely, and that orders for single or multiple copies can quickly be supplied.

The Cambridge Library Collection will bring back to life books of enduring scholarly value across a wide range of disciplines in the humanities and social sciences and in science and technology.

A Flora of Cambridgeshire

FRANKLYN HUGH PERRING
PETER D. SELL
STUART MAX WALTERS
HAROLD LESLIE KERR WHITEHOUSE

CAMBRIDGE UNIVERSITY PRESS

Cambridge New York Melbourne Madrid Cape Town Singapore São Paolo Delhi

Published in the United States of America by Cambridge University Press, New York

www.cambridge.org
Information on this title: www.cambridge.org/9781108002400

This edition first published 1964
This digitally printed version 2009

ISBN 978-1-108-00240-0

A FLORA OF
CAMBRIDGESHIRE

Ioannes Rajus.
Societatis Regiæ Socius.

A FLORA OF
CAMBRIDGESHIRE

BY

F. H. PERRING, P. D. SELL

AND

S. M. WALTERS

with a section on Bryophyta by
H. L. K. WHITEHOUSE

CAMBRIDGE

Published in association with the Cambridge Natural History Society

AT THE UNIVERSITY PRESS

1964

PUBLISHED BY
THE SYNDICS OF THE CAMBRIDGE UNIVERSITY PRESS
Bentley House, 200 Euston Road, London, N.W.1
American Branch: 32 East 57th Street, New York 22, N.Y.
West African Office: P.O. Box 33, Ibadan, Nigeria

©

CAMBRIDGE UNIVERSITY PRESS

1964

Printed in Great Britain at the University Printing House, Cambridge
(Brooke Crutchley, University Printer)

TO

THE MEMORY OF

JOHN RAY

RICHARD RELHAN

CHARLES C. BABINGTON

'We would urge men of University standing to spare a brief
interval from other pursuits for the study of nature and of
the vast library of creation so that they can gain wisdom in it
at first hand and learn to read the leaves of plants and the
characters impressed on flowers and seeds.... We are sure
that the pursuit of plants can appeal to the young; for we
have seen many sons of Trinity College finding in it both
bodily exercise and mental satisfaction. Of course there are
people entirely indifferent to the sight of flowers or of
meadows in spring, or if not indifferent at least pre-occupied
elsewhere. They devote themselves to ball-games, to drinking,
gambling, money-making, popularity-hunting. For these
our subject is meaningless.'

JOHN RAY: Preface to *Catalogus Plantarum circa
Cantabrigiam nascentium,* 1660

CONTENTS

ILLUSTRATIONS

PLATES

DISTRIBUTION MAPS

FOREWORD

To produce a new Flora of Cambridgeshire is to be responsible for accepting a great tradition and adding to what may well claim to be the most famous sequence of local taxonomic records in the world. When the present volume was planned it was hoped to publish it in the tercentenary year of John Ray's *Catalogus Plantarum circa Cantabrigiam nascentium* published in 1660 and marking not only its author's entry into the study of natural history but the first scientific treatment of our native plants. Ray's book records a large number of species new to botany and gives first lists of plants found in particular types of localities, the lanes of Chesterton and Ditton, the chalk of the Gog Magogs and Cherry Hinton, the woods of Madingley and Kingston, Newmarket Heath and the Devil's Dyke and the fens at Teversham and Stretham ferry. How thoroughly he observed is witnessed by the fact that several species are still only found in the place where he discovered them, and one, *Veronica spicata*, which he recorded as 'in a close near the beacon on the left hand of the way from Cambridge to Newmarket in great plenty', was lost from his day till after much research Dr W. H. Mills rediscovered it in the thirties.

Since Ray's time the Cambridgeshire plants have been resurveyed by Charles Cardale Babington, Professor of Botany in 1860—a careful and accurate record giving for the first time lists of some of the 'difficult' genera—a worthy successor to the *Catalogus*; and by A. H. Evans, a keen field-naturalist who brought the records up to date in 1939. It is obvious from the present volume how much has been done both in the addition of species, some of them now well established, but one or two apparently old inhabitants, and in the discovery and identification of taxonomically important races. To these as to the whole question of plant distribution much attention largely centred in Cambridge has been given in the past decade.

But the tradition in the University does not begin with Ray. It goes back to William Turner, the 'father' of English natural history, Fellow of Pembroke in the second and third decades of the sixteenth century, author of the first bird-book of the modern world, and of the first English folio herbal; to his successor Thomas Penny of Trinity whose own work was mainly on insects but who sent brilliantly exact descriptions of a number of British plants, including *Arnoseris*, to his

FOREWORD

friend Charles de l'Ecluse (Clusius), and to Peter Turner, William's son and himself a botanist with an international reputation. There was a barren half-century between them and Ray: but since then the succession has not failed; and if taxonomy still plays a large part in its activities, pollen-analysis and quaternary research prove that it has also expanded its frontiers and deepened its researches.

This new Flora is the outcome of some ten years of shared study in ecological, genetic and cytological research as well as of field-work and experimental cultivation. It had been stimulated both by the enthusiasm of Mr H. Gilbert Carter of the Botanic Garden who gathered a body of young explorers round him and by the preparation of the *Flora of the British Isles* and the 'B.S.B.I. Maps Scheme' which has united professionals and amateurs in a large-ranging effort. Taxonomy has thus spread beyond the herbaria into the research stations, the laboratories and gardens, and is opening up problems which may be expected to throw light upon the basic conditions of plant physiology and evolution. 'Name this plant' is the proper beginning of botanical knowledge: this book will help to show how wide and deep are the issues which must be settled before a worthwhile answer can be given.

November 1961   CHARLES E. RAVEN

x

PREFACE

John Ray published the first Flora of Cambridgeshire just over three hundred years ago, and since then the study of our local plants has been almost continuous. Relhan (1785) and Babington (1860) published further Floras at approximately hundred-year intervals, and this Flora was first planned by the Cambridge Natural History Society to appear in 1960, the tercentenary of Ray's work. Unfortunately we underestimated the amount of work involved, and regretfully decided to abandon the original date.

This work, based upon the Natural History Society's records, departs from the standard practice of most local Floras in that it does not give long lists of localities under each species. In their place is a concise list of Ordnance Survey Grid References which give an adequate picture of the main features of distribution of the species. The space saved allowed us to include taxonomic notes, comments and keys, which we believe will prove to be more valuable than the traditional distribution data.

1519 species are listed, of which 27 are Pteridophyta, 3 Gymnospermae, 1228 Angiospermae and 261 Bryophyta. Of the 1231 seed plants, 968 are native (65 of these being extinct), 19 are doubtfully native, 81 are naturalized, 31 are planted trees, and 195 casuals, garden escapes or relics of cultivation. Of the 27 Pteridophyta one is introduced and nine extinct. Of the Bryophyta 222 are mosses (20 extinct) and 39 liverworts (4 extinct).

The area covered by the Ordnance Survey Grid Square 52/45, which includes the City of Cambridge, must be one of the best known areas botanically in the world. In it have been recorded during the last 300 years 1174 species, of which 1002 are vascular plants and 172 Bryophyta.

We are particularly pleased that it has been possible to include in this volume Dr H. L. K. Whitehouse's excellent up-to-date account of the Bryophyta. Interest in the mosses and liverworts of the county has been steadily increasing in recent years, largely through Dr Whitehouse's own interest and enthusiasm, and we can confidently expect future generations of students of the Cambridgeshire flora to include the Bryophyta in their studies.

The number of people who have contributed to this Flora is so large

PREFACE

that it is impossible to thank them all individually, but we have tried to give a complete list of those who have recorded plants in the county (see p. xiii). The following have either determined specimens or helped with the accounts of critical genera: A. O. Chater (*Callitriche* and *Carex*), C. D. K. Cook (*Ranunculus* subgen. *Batrachium*, and *Sparganium*), D. E. Coombe (*Impatiens* and *Trifolium*), A. C. Crundwell (Bryophyta), R. A. Graham (*Mentha*), J. Heslop Harrison (*Dactylorchis*), W. H. Mills (*Rosa* and *Rubus*), Mrs J. A. Paton (Bryophyta), R. H. Richens (*Ulmus*), T. G. Tutin (*Bromus* and Cyperaceae), D. H. Valentine (*Primula* and *Viola*), E. F. Warburg (Bryophyta), P. F. Yeo (*Euphrasia*) and D. P. Young (*Epipactis*).

Professor H. Godwin has read the account of the Fens and made many useful suggestions. In addition, K. Albon has extracted temperature and rainfall readings from the Botanic Garden records, B. A. Golding has drawn the Map, M. C. F. Proctor and J. C. Faulkner supplied the photographs, W. Stigwood helped with the documentation, and Mrs A. Wright has twice performed the exacting task of typing the manuscript. Information on the crop plants was supplied by officers of the National Agricultural Advisory Service for Cambridgeshire and for the Isle of Ely.

The distribution maps are reproduced from the *Atlas of the British Flora* by kind permission of Thomas Nelson and Sons Ltd, and the Botanical Society of the British Isles.

Special thanks must be given to the University Press for publishing the book and to the Royal Society for a generous grant of £300 towards its publication. Lastly, our thanks are especially due to H. Gilbert Carter, who inspired several generations of Cambridge botanists to do better things.

We hope that this book will be a guide to the local flora, not only for the successive generations of University students, but also for those people who like to spend their spare time looking for wild flowers as a recreation. This work shows only too clearly how many of our species have become extinct or extremely rare. We hope that in a more enlightened age, the next authors of a Cambridgeshire Flora, in perhaps another hundred years, will not have to record a further dismal list of extinctions. The activity of the recently formed Cambridgeshire and Isle of Ely Naturalists' Trust gives us grounds for this hope.

F. H. PERRING
P. D. SELL
December 1962   S. M. WALTERS

xii

LIST OF CONTRIBUTORS
TO THE FLORA

The following list gives the names of all those who have in some way contributed records to this account. They have sent in records for the Maps Scheme, entered records in the Natural History Society's Card Index, published records, or made herbarium sheets. Botanists who have put many specimens from the county in the University Herbarium (CGE) are marked with an asterisk, while those marked with a dagger have written papers on Cambridgeshire plants.

B. A. Abeywickrama
†R. S. Adamson
D. E. Allen
M. C. Anderson
J. Andrews
E. Armitage
†*C. C. Babington
J. Balding
S. Balkwill
J. Ball
A. M. Barnard
†A. Bennett
S. H. Bickham
F. J. Bingley
K. Blades
P. J. Bourne
†W. T. Bree
D. Britten
†J. Britten
F. Y. Brocas
F. T. Brooks
J. Brown
P. D. Brown
J. N. Bullock
†I. H. Burkill
P. A. Buxton
J. Carpenter
J. Carter
Mrs Casbourne
D. G. Catcheside

D. F. Chamberlain
P. J. Chamberlain
J. H. Chandler
G. W. Chapman
V. J. Chapman
A. O. Chater
C. B. Clarke
J. Clarke
W. H. Coleman
R. H. Compton
A. P. Conolly
D. E. Coombe
S. Corbyn
E. J. H. Corner
J. W. Cowan
G. J. Crawford
†G. Crompton
J. L. Crosby
A. J. Crosfield
W. J. Cross
S. Dale
M. J. d'Alton
†P. Dent
J. H. Dickson
P. S. Digby
†H. N. Dixon
†F. D. Dobbs
J. G. Dony
J. Downes
†G. C. Druce

D. Dupree
P. Duval
T. A. Dymes
F. W. Edwards
W. N. Edwards
F. L. Engledow
†A. H. Evans
J. C. Faulkner
J. Fisher
T. J. Foggitt
H. Fordham
†A. Fryer
E. A. George
J. Gerarde
E. J. Gibbons
P. J. Gibbs
†G. S. Gibson
J. L. Gilbert
*H. Gilbert Carter
J. S. L. Gilmour
H. C. Gilson
†H. Godwin
†G. Goode
K. M. Goodway
V. K. Gotobed
W. B. Gourlay
P. G. Greaves
D. W. P. Greenham
P. Grieg-Smith
H. and J. Groves

LIST OF CONTRIBUTORS

P. J. Grubb
D. Guymer
P. M. Hall
G. Halliday
J. Harding
R. E. Hardy
T. M. Harris
G. D. Haviland
J. Hempstead
†*J. S. Henslow
†P. Hiern
S. Hiley
R. Hill
P. C. Hodgson
J. Holme
†A. Hosking
W. O. Howarth
R. C. L. Howitt
A. P. Hughes
B. Ing
†A. B. Jackson
W. Jackson
E. G. Jefferys
C. Jeffrey
D. H. Jennings
†L. Jenyns
G. B. Jermyn
A. P. D. Jones
E. W. Jones
†M. Kassas
R. Lancaster
T. Lawson
†F. A. Lees
W. A. Leighton
C. M. Lemann
R. A. Lewin
A. Ley
J. J. Lister
J. E. Little
J. M. Lock
R. H. Lock
H. E. Lowe
K. E. Luck
P. F. Lumley
F. G. H. Lupton

†I. Lyons
†E. S. Marshall
†W. Marshall
†J. Martyn
†T. Martyn
W. Mathews jun.
G. N. and N. Maynard
H. T. Mayo
D. N. McVean
†J. C. Melvill
C. Miller
†S. H. Miller
J. N. Mills
†W. H. Mills
*C. E. Moss
H. Mumford
M. T. Myres
D. M. Neal
W. W. Newbould
J. Newlands
Mr Newton
P. H. Oswald
†F. A. Paley
†S. Palmer
W. H. Palmer
R. E. Parker
A. Peckover
†*F. H. Perring
C. P. Petch
W. R. Philipson
*C. D. Pigott
M. E. D. Poore
†G. T. Porrit
P. M. Priestley
†M. C. F. Proctor
†R. A. Pryor
W. Pulling
J. D. Radcliffe
†C. E. Raven
†J. Ray
B. Reeve
†R. Relhan
†B. Reynolds
P. M. G. Rhodes
†P. W. Richards

†R. H. Richens
†H. J. Riddelsdell
†J. Rishbeth
F. Rose
R. Ross
E. Rosser
F. Roythorne
C. A. Rylands
J. Rylands
†C. E. Salmon
Mr Sare
W. Sargent
L. J. Sedgewick
†*P. D. Sell
A. C. Seward
C. Shepherd
J. Sherard
*A. S. Shrubbs
N. W. Simmonds
E. Skipper
W. Skrimshire
R. B. Smart
A. M. Smith
B. W. Sparrow
K. R. Sporne
M. Stanier
W. T. Stearn
†H. C. Stuart
J. B. Syme
†A. G. Tansley
F. R. Tennant
H. H. Thomas
B. Tilly
C. C. Townsend
H. M. Treen
A. C. Trueblood
J. G. Turner
W. Turner
T. G. Tutin
D. H. Valentine
Dr Venn
W. Vernon
C. H. Waddle
†A. Wallis
O. E. Wallis

LIST OF CONTRIBUTORS

†*S. M. Walters
S. W. Wanton
E. F. Warburg
S. J. P. Waters
J. Watson
W. Watson
A. S. Watt

†D. Welch
G. S. West
†W. West
F. White
†*H. L. K. Whitehouse
E. C. Wilkinson
†J. C. Willis

A. J. Wilmott
R. S. Winteringham
S. Woodham
†R. H. Yapp
T. York

NOTE ADDED IN PROOF

In 1962 E. F. Warburg found *Pottia caespitosa* (Bruch ex Brid.) C. Müll. on the Devil's Dyke, and in 1963 J. M. Lock found *Hookeria lucens* (Hedw.) Sm., *Polytrichum commune* Hedw., *Sphagnum fimbriatum* Wils. and *Tetraphis pellucida* Hedw. at Wicken Fen. These interesting records were received too late to be included in the general text.

INTRODUCTION

HISTORY OF THE STUDY OF THE FLORA

For three hundred years Cambridgeshire has been one of the best known counties, botanically, in the British Isles. The first list of plants made in the county was by Samuel Corbyn (1656), although a few records date from the sixteenth and early seventeenth centuries, made by men like Turner and How, who were beginning to study the flora of the country as a whole. However, the first work of real importance was that of the illustrious John Ray (cf. Raven, 1942), who in 1660 published a 12mo volume of 182 pages entitled *Catalogus Plantarum circa Cantabrigiam nascentium*. This has long been celebrated as the first comprehensive local British Flora. It was the result of nine years' work, and consists of an alphabetical list of plants found in the Cambridge area. It gives localities of plants, which in several cases can still be found there today, for example, *Geranium sanguineum*, 'Found on Newmarket heath in the Devils ditch, also in a wood adjoining to the highway betwixt Stitchworth (Stetchworth) and Chidley (Cheveley)'.

In 1663 Ray published a 13-page appendix to the *Cambridge Catalogue*, and after this in 1685 appeared a second appendix consisting of 30 pages, edited by Peter Dent, a Cambridge apothecary. There was no second edition of the *Cambridge Catalogue*, but in 1670 Ray published his *Catalogus Plantarum Angliae* (ed. 2, 1677), in which all plants occurring in Cambridgeshire were marked with the letter C. This latter work, however, contains very few additions to the county list, which is not surprising as Ray left Cambridge in 1662.

John Martyn, who became the second Professor of Botany at Cambridge in 1733, had published, in 1727, his *Methodus Plantarum circa Cantabrigiam nascentium*. In this he included all the plants of Ray and Dent, but added no new records.

Thomas Martyn, who succeeded his father in the chair of Botany in 1761, produced an 8vo work entitled *Plantae Cantabrigienses* (1763). This was arranged according to the Linnaean system and nomenclature. He published at the same time the *Herbationes Cantabrigienses* which consists of an account of thirteen botanical excursions to localities in the Cambridge area. Some of these, such as Newmarket Heath and Gamlingay, are still visited today. In the same year, but three months later, Israel Lyons published his *Fasciculus Plantarum* in which he lists an

INTRODUCTION

additional 105 species found growing around Cambridge since the time of Ray.

A more thorough knowledge of the flora of our county became available in 1785 with the publication of the first of three editions of Richard Relhan's *Flora Cantabrigiensis*. This work contains the first full account of the Bryophyta, Algae, Lichens and Fungi occurring in the county. Much of our knowledge of the flora of the northern Fenlands dates from this period. This is chiefly due to the work of W. Skrimshire who was a correspondent of Relhan. Details of Skrimshire's activities are contained in a manuscript to be found in the Wisbech Museum labelled *Catalogue of Plants contained in Mr Skrimshire's Hortus Siccus*. This was transcribed by an unknown hand and dated 12 September 1829. Relhan brought out three supplements to his *Flora* dated 1786, 1788 and 1793, while two further editions of the whole work appeared in 1802 and 1820.

The fourth Cambridge Professor of Botany, J. S. Henslow, who was elected in 1825, added much to our knowledge of local plants. He drew up in 1829 *A Catalogue of British Plants* in which he italicized all plants not found in the county. This was followed in 1835 by a second edition in which the letter C was appended to all the Cambridgeshire plants. To Henslow, who was a great lover of field-botany, must go the credit for being the first person to make a comprehensive herbarium collection of the plants of the county. His specimens, which are to be found in the University Herbarium, are still in constant use today.

One of Henslow's pupils, C. C. Babington, was destined to make the most important contribution to the knowledge of the flora of our county. The *Flora of Cambridgeshire*, which was published in 1860, the year before Babington became the fifth Professor of Botany, was the result of his own detailed researches and the work of his many correspondents. Besides including under each species a full list of localities, in which old and new records are distinguished, Babington divided the county into eight botanical regions. These are as follows: 1 and 2 include the main mass of the chalk; 3 the clayey drifts with the Gamlingay Greensand; 4 and 5 contain the country bordering the Fenlands (the Breckland sands were also included in no. 5). The remaining three districts comprise the Fenlands (including the silts). Although this provided a basis for a more even study of our flora, it is clear from a close inspection of the records that the Fenlands still remained under-recorded. For example, *Veronica chamaedrys*, which is now known to occur throughout the county, was omitted from district 7 by Babington.

A number of species are recorded from the county for the first time by

2

INTRODUCTION

Babington, mainly in the critical genera. This reflects his detailed knowledge of the flora, not only of the British Isles but also that of Europe. He summarized his investigations in an Appendix which deals with *Thalictrum, Papaver, Viola, Arenaria, Rubus, Bromus* and *Agropyron*.

Finally, Babington gives a list of species which may be found growing at Wicken Fen. Unfortunately, despite the care which has been taken in protecting the Fen, a number of species, including *Viola stagnina, Stellaria palustris, Potentilla palustris, Senecio paludosus* and *Stratiotes aloides*, have disappeared during the last hundred years. Babington also lists sixty-one species which were certainly or probably extinct in the county as a whole. It is, however, pleasant to record that over twenty of these have been subsequently refound, and still occur. They are as follows: *Aristolochia clematitis, Asperugo procumbens, Geranium rotundifolium, Centranthus ruber, Lactuca saligna, Lathyrus nissolia, Lysimachia nemorum, Myosurus minimus, Myrica gale, Oenanthe silaifolia, Phleum arenarium, Polygonum minus, Prunus cerasus, Ribes nigrum, Salix purpurea, Sedum album, S. telephium, Senecio viscosus, Setaria viridis, Sorbus torminalis, Thlaspi arvense, Veronica spicata.*

Babington lived for another thirty-five years after the publication of his *Flora*, remaining in Cambridge all that time. Further records of his exist in an annotated copy of his *Flora*, and in his notebooks and papers. His interest in critical genera seems to have influenced other workers in the county. Of these, the most important was Alfred Fryer. Known to the country as a whole for his work in conjunction with Arthur Bennett on the genus *Potamogeton*, he made a very useful contribution to our knowledge of the botany of the western Fenlands. He lived at Chatteris and was thus able to reach easily a part of the county which had hardly been investigated before. Although he published no general papers on the botany of the fens, many manuscript documents are available, and have been extracted for this *Flora*.

It was fortunate that Bennett, one of the leading amateurs of his day, should have taken so much interest in Cambridgeshire. He was an excellent critical botanist, and the study of a number of difficult groups in the county was begun by him. His main contribution was *Notes on Cambridgeshire Plants* (1899), but many other shorter notes appeared in the *Journal of Botany* from time to time. Other useful contributions were made about this time by R. A. Pryor, the Hertfordshire botanist, in 1874 and by W. West Jnr. in 1898. Pryor provided a list of plants from the Kirtling area, a little-known parish near the Suffolk border. An interesting account of the local flora was provided by A. Wallis (1904).

3

INTRODUCTION

The knowledge which accumulated in the last decade of the nineteenth century seemed to demand some outlet. The task fell to A. H. Evans of Clare College, who published *A Short Flora of Cambridgeshire* in 1911. This paper, which must have had a limited circulation, is nevertheless a most helpful one in every respect except for distribution data. Evans disliked Babington's eight regional divisions, preferring to record on which of the six main geological formations the species occur. Evans's 'Short Flora' traces the course of botany and botanists in the county to the beginning of this century, and gives detailed notes on thirty-eight of the rarer species of the county, many of which were already extinct by that time. Finally, the work is notable as, with the exception of a series of papers by G. S. West on the Algae published in the *Journal of Botany* in 1899, it contains the first lists of the lower plants of the county published since Relhan's *Flora*. The Rev. P. G. M. Rhodes prepared the Bryophyta and Lichens, G. S. West was responsible for the Algae, and F. T. Brooks, later to become eighth Professor of Botany, compiled the account of the Fungi.

After the death of Babington in 1895, the chair of botany was no longer occupied by an ardent taxonomist or field botanist. A new approach to the study of the flora, however, developed at this time under the direction of C. E. Moss, the Curator of the Herbarium. Moss was a careful and competent taxonomist whose major work, the *Cambridge British Flora*, unfortunately remained unfinished. He was also one of the pioneer ecologists of this country. Although his most important works on this subject are devoted to Derbyshire and Somerset, his influence in the encouragement of others to follow the 'new science' must have been considerable. Evans recalls that about this time the botanical excursions took on a new lease of life. The swing to ecology was further accentuated by the presence in Cambridge of A. G. Tansley at the same time as C. E. Moss. Tansley edited in the same year as the appearance of the 'Short Flora' his *Types of British Vegetation* which includes some of the first descriptions of the vegetation of the county, for example, the chalk grassland of the Fleam Dyke. 1911 is thus a most important landmark in the history of Cambridgeshire botany. It saw the production of a summary of the traditional study of the flora during the preceding half century, and the beginnings of the study of ecology, a study which was to overshadow and almost eliminate interest in classical taxonomy for the next twenty years. This is reflected in the very few records which were made between 1910 and 1930, and the lack of papers on taxonomy. During this time ecological information about the county began to accumulate, and the greatest attention

4

INTRODUCTION

was paid to the Fens and to Wicken Fen in particular. This culminated in the publication in 1932 of *The Natural History of Wicken Fen*, a collection of fifty-six papers on all aspects of the subject, including accounts of the flora by A. H. Evans and the vegetation by H. Godwin and A. G. Tansley. The work of Professor Godwin and his colleagues, both here and in other parts of the Fens, has given us a remarkable insight not only into the structure and interrelations of our vegetation of the present day, but a very clear idea of its origin and history.

Brief accounts of the vegetation of the county as a whole were published by Professor Godwin in 1938, first in *A Scientific Survey of the Cambridge District* prepared for the Cambridge meeting of the British Association, and, secondly, in the account of the botany of the county in vol. I of the *Victoria County History*. In each account four main types of vegetation are discussed: the fens, the boulder-clay woods, the acid sands and the chalk grasslands. These four types fit very conveniently into the four lecturing weeks of the University Easter Term, but should not, for this reason, be thought of as the only types of vegetation in the county.

During the 1920's interest in field botany in the county was kept alive mainly by those who had no direct connection with teaching in the Botany School. Foremost among these was Evans, though he was finding it increasingly difficult to get about owing to illness. This short history would be incomplete without mention of A. S. Shrubbs, an assistant in the Botany School from 1870 to 1922, who added much material to the Herbarium and whose delightful personality endeared him to all those who came in contact with him.

At first it seemed difficult to fit into this account of field botany in the county *A Flora of Cambridgeshire* by A. H. Evans published in 1939, but on second thoughts this is not perhaps so surprising. Ray was the pioneer, Relhan the first to include the Fenlands in any detail and record groups other than flowering plants, and Babington the first to turn attention to critical genera. Evans's *Flora* of 1939 came at a time when an interest in species *per se* had been almost dead for a quarter of a century, and before the effect of the revival of interest, which began about 1930, could be fully felt. This *Flora* unfortunately contained very little which was new and a great deal which was erroneous; old records were given without comment though the plants were extinct by that time, and statements on frequency were often misleading.

Though the Apocrypha records that the revival began after T. G. Tutin and J. S. L. Gilmour shared a bed at Foul Anchor, it is certain that these two, in collaboration with W. T. Stearn, and inspired by the

5

INTRODUCTION

Director of the Botanic Garden, H. Gilbert Carter, began a series of Exsiccatae based on critical material collected in the county. Many of their records were the first to be incorporated in the Cambridge Natural History Society's Card Index which was begun by E. A. George, D. H. Valentine and E. F. Warburg, in about 1938. The basis of this work was a card for each species with a cutting from Babington's *Flora* pasted on the top left-hand corner. The first task was to extract data from all books and papers which had been published since 1860, and to add notes and localities from all the annotated Floras and manuscripts which had accumulated in the library of the Botany School. At the same time part of the county was divided by a grid system, and a small group of workers started to collect records of common species from these areas, and enter them on to the appropriate index cards. The Second World War intervened, and no great progress was made until the latter half of the 1940's. By then it began to be felt that an attempt should be made to collect distribution data as evenly as possible from the whole county. Duplicated sheets were circulated listing about 100 common species to be looked for. From this in 1952 developed the 8 in. × 5 in. field record card listing nearly all the species known to occur in the county. On this card recorders were asked to mark those which they found in a particular locality or kilometre grid square of the National Grid. This type of card has become familiar to botanists throughout the country since the inception of the Botanical Society of the British Isles Distribution Maps Scheme in 1954. The Cambridge Natural History Society can take credit for pioneering this method in this country. At one time it was hoped that lists might be obtained from all the kilometre squares in the county, but the task proved to be too great, and we have had to be content with the 10 kilometre square as the recording unit. Forty of these, many only partly in Cambridgeshire, cover the county, and every effort has been made to investigate each area with equal thoroughness. Of course, with active botanists mainly living in Cambridge, the southern squares have received more attention, and it is still true that the Fenland squares are relatively not so well known. This part of the county has suffered even more than the south from human interference, and the long lists of old records for the Wisbech and Chatteris areas, for example, reflect not only the somewhat inadequate attention these areas have received in recent years, but a real decrease in the variety of habitat and species remaining there today.

Over the last twenty-five years there has also been a re-awakening of interest in the lower plants. P. W. Richards began the card index of Bryophyta and this has since been maintained by M. C. F. Proctor and

INTRODUCTION

H. L. K. Whitehouse. Proctor used this as a basis for a *Bryophyte Flora of Cambridgeshire* (1956). Since then the bryologists have become 'square-minded' and have collected systematic information on the distribution of the commoner species.

No account of botanical activities in Cambridgeshire would be complete without a reference to the growing concern for nature conservation locally, a concern which led to the inauguration in 1957 of the Cambridgeshire and Isle of Ely Naturalists' Trust. In the Trust biologists and naturalists, professional and amateur, can make a concerted attempt to carry out a reasoned policy for the protection of natural interest and beauty throughout the county. Enormously increased land values, together with revolutionary new techniques in agriculture, are threatening to reduce the countryside to dull uniformity. If we are to have anything left to study and enjoy of the rich heritage of nature which has survived to the present day, urgent conservation action is necessary. The main National Nature Reserves in Britain, such as Wicken Fen, owned by the National Trust, are legally protected; but the many smaller sites throughout the county require local concern to protect them. Public authorities and private owners are usually co-operative if the naturalists' concern is put to them reasonably. The agreement with the Cambridge City Council over Lime Kiln Close, Cherry Hinton, by which this interesting site is preserved as a nature reserve with public access, is an excellent example of such co-operation.

TOPOGRAPHY

Cambridgeshire (including the Isle of Ely) is one of the larger British counties, being about fifty miles in length and about thirty at its greatest breadth. It covers an area of 555,118 acres. No fewer than eight counties touch its borders: Lincs (v.c. 53), Norfolk (v.c. 28), Northants (v.c. 32), Hunts (v.c. 31), Beds (v.c. 30), Suffolk (v.c. 26), Essex (v.c. 19) and Herts (v.c. 20). It approaches to within five miles of the sea north of Wisbech, and the River Nene is tidal for some miles south-west of that town. The northern part of the county, consisting of the former Great Level of the Fens, is monotonously flat. The southern part is occupied by a range of low chalk hills rising to over 130 m. (400 ft.) near Great Chishill. In the west is a wide plateau ending in an outcrop of greensand at Gamlingay where formerly existed some large bogs. The county is watered by the Ouse, the Cam and the Nene. Besides the rivers, we find in the fen country a network of artificial lodes and dykes, discharging into them. The largest of these artificial waterways

7

INTRODUCTION

are the two 'Bedford Rivers' running from Earith to Denver. There is only one large town, Cambridge, and several smaller ones, Chatteris, Ely, March, Soham, Whittlesey and Wisbech, while the towns of Newmarket, Peterborough and Royston are only just over our borders. Large villages such as Bassingbourn, Burwell, Elm, Gamlingay, Histon, Linton, Sawston and Sutton are a distinguishing feature of the county. The occupation of the people is mainly agriculture and much of the land is under arable cultivation. The main crops are wheat, barley, potatoes, brussels sprouts, beet and oats. Around Cottenham, Histon, Haddenham, Wilburton and Wisbech are large orchards, and the Wisbech area grows many acres of bulb plants such as tulips and daffodils, and also asparagus and tomatoes. Most of the woodland is on the clays, but there are a number of plantations on the chalk.

CLIMATE

The most important feature of the climate of the Cambridge area is its resemblance to that of the main part of continental Europe. This feature is reflected both in temperature and rainfall.

Temperature

The mean monthly temperatures are shown in Table 1, and as in most parts of the British Isles, the lowest mean temperatures occur in January, and the highest in July. The range of mean monthly temperatures (22·4° F.) is about average for south-east England. The most significant feature of the temperature of the region is the low summer minima (Table 1 b). These indicate the frequency of frosts. Monthly minima below 32° F. are usual from October to May and quite serious frosts have been recorded at the beginning of June. Winter frosts are often severe and the damage they do is greater than might be expected, for snow gives protection on only a few days in the year. In an average year snow lies in the morning on twelve days only.

Table 1. *Temperatures*

	J.	F.	M.	A.	M.	J.	Jl.	A.	S.	O.	N.	D.	Mean annual	Period
(a) Mean monthly and mean annual temperatures (°F.)														
	39·3	39·7	42·3	46·3	53·5	58·0	61·7	61·3	58·9	50·3	42·9	39·9	49·3	1906–35
(b) Mean monthly extreme temperatures (°F.)														
Max.	54·0	55·9	63·0	69·1	75·0	81·0	83·8	82·9	78·1	68·0	59·0	55·0	—	1906–35
Min.	19·9	21·0	23·0	26·2	30·0	37·9	43·0	42·1	36·0	28·9	24·1	21·1		

INTRODUCTION

Rainfall

In the amount of rainfall and its distribution throughout the year the Cambridge climate shows affinities to the continental type. The area has a low annual total varying from 20·6 inches at Upwell to 24·7 inches at Conington. The values for Cambridge are given in Table 2. Only a few areas in Essex have a smaller annual total and, whereas in most parts of the British Isles the rainfall of the winter half of the year is greater than the summer half, in Cambridge only 48 % of the total falls between 1 October and 31 March. This is a continental feature which is shown by only a small area in east and central England; for example, in the East Riding of Yorkshire 49 % falls during this period but in north Dorset the figure is over 55 %.

Table 2. *Mean monthly and mean annual rainfall (inches)*

J.	F.	M.	A.	M.	J.	Jl.	A.	S.	O.	N.	D.	Mean annual	Period
1·92	1·32	1·18	1·64	1·91	1·48	2·32	1·93	1·95	2·04	2·18	1·54	20·72	1921–50

Relative humidity

Compared with other parts of Britain the humidity of the Cambridge area is relatively low. This is particularly marked at the beginning of the growing season in April and May, when values may be as much as 8 % below those for north Dorset to the west and the East Riding of Yorkshire to the north. It is only during these spring months that the mean values fall below 70 % (see Table 3). During the summer, however, these differences from other parts of England are reduced as the average humidity steadily rises, and during the mid-winter period of December and January humidity in Cambridge is at least as high as elsewhere in England.

Table 3. *Mean relative humidity, 09.00 hours*

J.	F.	M.	A.	M.	J.	Jl.	A.	S.	O.	N.	D.	Mean annual	Period
90·4	86·0	80·2	69·4	69·5	71·1	73·3	74·7	79·0	82·7	87·7	90·1	77·9	1947–50

Bright sunshine

The Cambridge area is in an intermediate position in the amount of bright sunshine it receives compared with other parts of the British Isles. The average annual total of about 1550 hours is 200 hours less than that

INTRODUCTION

experienced on the south coast, but it is about 200 hours greater than
that received by areas in North Scotland and the Outer Hebrides. This
results mainly from differences during the summer months; during the
period November to January variation throughout the country is
negligible. The mean daily hours of sunshine for Cambridge are shown
month by month in Table 4.

Table 4. *Mean no. of hours of bright sunshine per day*

J.	F.	M.	A.	M.	J.	Jl.	A.	S.	O.	N.	D.	Mean annual total	Period
1·7	2·5	3·7	5·2	6·6	6·8	6·3	6·0	5·0	3·5	2·1	1·3	1545	1901–30

General considerations

The data discussed above indicate that the climate of the Cambridge
area is an extreme one in relation to the British Isles as a whole, and this
is particularly marked in the spring months when rainfall and relative
humidity are low, and there is a high probability of late frosts. This
latter factor perhaps accounts for the absence from Cambridgeshire of a
number of oceanic species (e.g. *Ulex gallii* (see map, p. 11), *Oenanthe
crocata*, and *Corydalis claviculata*). Drought seems to be a particularly
potent factor and must surely explain the paucity of ferns, mosses and
particularly liverworts. The present-day fern flora of Cambs contains
only thirteen out of a total of forty-eight species in the British Isles,
whereas Sussex, a maritime county on the same longitude, has twenty-
six species. The only woodland ferns which occur with any frequency
are *Dryopteris filix-mas* and *D. dilatata*, and it is probably significant
that the latter has increased in recent years as old woodlands have
become more densely overgrown. The wall-ferns are almost confined to
north aspects and are best looked for on that side of old parish churches,
especially if the churchyard is surrounded by trees to give extra
shelter. In addition, there is a group of flowering plants which appear
to be unable to tolerate this 'dry centre' of Britain and are rare or
local in Cambridgeshire. These include *Silene dioica* (see map, p. 12),
Geum rivale, *Vicia sepium*, *Lysimachia nemorum*, *Veronica montana* and
Stellaria holostea, to mention only a few of the more conspicuous species.
In contrast a number of species appear to be well adapted to these
climatic conditions of the Cambridge area and are frequent with us,
whereas they are rare or absent elsewhere in Britain except in adjacent
counties. This group of species is mainly of continental distribution in

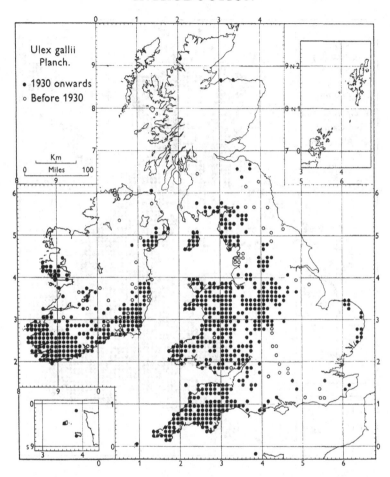

Europe and includes *Primula elatior*, *Trifolium ochroleucon* (see map, p. 13) and *Melampyrum cristatum*. The species peculiar to the Breckland should also be included in this group. Only a small percentage of that interesting flora is found in Cambridgeshire because suitable soils are very infrequent. However, those which do occur include *Silene*

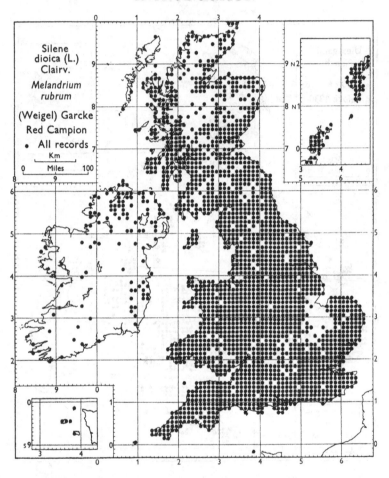

Silene
dioica (L.)
Clairv.

*Melandrium
rubrum*

(Weigel) Garcke
Red Campion
• All records
Km
0   Miles   100

conica, *S. otites*, *Herniaria glabra*, *Phleum phleoides*, *Medicago minima* and *Apera interrupta*.

The late frosts and the lack of snow cover no doubt explain why there is a small northern element in the flora including species like *Carex ericetorum* and *Astragalus danicus*, the former reaching its southern

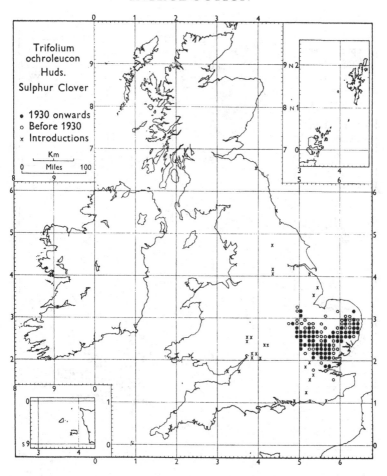

Trifolium
ochroleucon
Huds.

Sulphur Clover

• 1930 onwards
○ Before 1930
× Introductions

Km

0   Miles   100

limit in the Cambridge area. In contrast a number of species, perhaps demanding high summer temperatures, which are mainly found in central and southern Europe, are at or near their northern limit in our area (e.g. *Cephalanthera damasonium* (see map, p. 14) and *Lathyrus aphaca*).

13

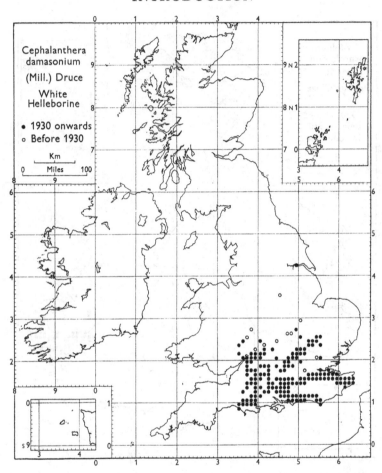

Cephalanthera damasonium (Mill.) Druce
White Helleborine
● 1930 onwards
○ Before 1930
Km
0 Miles 100

The nature of our climate is also reflected in agricultural and garden-ing practice. Apart from the obvious difficulties of raising frost-sensitive plants or growing crops of early vegetables, including potatoes, the dryness of the climate is important. Among the planted conifers around Cambridge only the xerophytic pines do at all well. Spruces and

INTRODUCTION

silver firs do badly. It is generally unwise to move a herbaceous border after the end of March, and spring-sown grass seed is unlikely to produce a satisfactory lawn. However, the reasonable amount of sunshine and the higher rainfall of the summer months do ensure the production of good crops of wheat.

THE FLORA OF CAMBRIDGESHIRE

Cambridgeshire and the Isle of Ely form an area which probably has the most distinct county flora when compared with any other part of the British Isles. There are three factors of importance which cause this to be so. First, as we have seen, the area lies in the driest part of the country, which accounts for the absence of many species demanding high humidities, particularly ferns and bryophytes. Secondly, the area is dominated by the chalk or similar parent materials giving rise to alkaline or neutral soils. Acid sands are very infrequent, and acid bogs practically non-existent—hence the calcifuge flora is extremely poor. Thirdly, the fens, once a wilderness of marsh-loving species, are now the home of arable weeds; natural vegetation has almost disappeared and there are no original woodlands. Hence our Fenlands form part of an area, thirty miles wide and sixty miles long, containing parts of Norfolk, Lincolnshire, Huntingdonshire and the Soke of Peterborough, from which exclusively woodland species are almost entirely absent.

Thus, although our county will always attract botanists for the many rare and local species which the flora contains, it is an area which lacks variety, and the total number of species must be less than neighbouring counties like Bedfordshire, Norfolk and Suffolk, all of which have a more humid climate and greater geological diversity.

GEOLOGICAL AREAS AND THEIR TYPES OF VEGETATION

For the purpose of studying the different types of vegetation, the county may be divided into six areas, based on their soil types. These are shown on the map facing p. 330.

(1) Fenlands

The Fenlands form the largest division, covering almost the whole of the northern half of the county from Wisbech south to Cambridge, and extending up the river valleys to form isolated pockets as at Fulbourn and Fowlmere.

Of the original vegetation very little trace remains at the present day. Much drainage was begun by the Romans, and it continued in medieval

15

times but not upon a very large scale until the seventeenth century. At the present time almost the whole area has been brought under cultivation, although traces of the original vegetation can still be recognized at once within the deep lodes and dykes often full of reeds (*Phragmites communis*). The dykes communicate with one another and converge upon larger drains, from which the water is lifted into the river-system by means of pumps. This extensive drainage has caused a very great lowering of the surface level of the peat, so that ground level is below Ordnance Datum over large areas. Here naturally no trace of the original fen vegetation remains.

Peat has been forming more or less continually in the deeper valleys of the fens for upwards of 10,000 years, but peat formation became general only about 5000 years ago. Three major factors have influenced the development of this: (i) the natural processes of vegetational succession, bound up with peat accumulation; (ii) the rainfall, which not only operates through that which falls on the fens themselves, but through the volume of river-water discharged into the Fens; and (iii) the relative movement of land and sea, causing a lessened drainage gradient, backing-up of fresh water during periods of marine transgression, and increased drainage during relative land-elevation.

Recent work in Professor Godwin's Sub-department of Quaternary Research (Godwin & Willis, 1961) has provided a date accurate to within two or three centuries for the impressive layer of bog-oak lying under the Wicken peat. Radiocarbon measurements on samples of root from giant oaks found rooted in the gault clay below the peat on Adventurers' Fen indicate an age of 4380 ($\pm$140) years. Thus the main development of Wicken peat began some 4500 years ago, when fresh water gradually backed up on the primeval forests of oak growing on the clay.

During periods of increased wetness we should expect the normal vegetational successions to be retarded or set back so that areas bearing fen might revert to reed swamp or open water. During periods of dryness, on the other hand, we should expect all the fen successions to be accelerated, fen would rapidly give place to carr, and this would be succeeded marginally by the development of fen woods. Such effects are often recognizable in the fen deposits.

The effect of marine transgression is the formation of estuarine or marine clays and silts. At any given time, the seaward deposits will have been silts and clays, and the landward deposits peat. During submergences the silts and clays will have overlain previously formed peats, and during emergences peat will have extended seawards

16

over the silts and clays. Thus landwards, as at Wicken, which lies towards the inner limit of the fen area, there is but one peat bed; further seaward there are two, separated by the semi-marine ' buttery clay'; and at St German's, near King's Lynn, there were established four peat beds, alternating with thick marine or semi-marine deposits. A natural outcome of this double origin from fresh and from brackish or salt water is the division of the fens at the present day into the 'silt fens' and the 'peat fens'. The seaward part is that with silt at the surface (though peat beds may occur below) and it forms a great concentric zone round the Wash, but also protrudes a great southerly arm into the Fenland, from Wisbech along the Old Croft river as far as Littleport. This is the natural bed of the Great Ouse which formerly reached the sea at Wisbech, but which has been deflected by an artificial cut, possibly Roman in age, to the sea from Littleport to Denver and so to King's Lynn. This silt-land, including the raised silt banks of the rivers, lies some feet higher in level than the peat-land, partly as a result of the conditions of formation, and partly as a result of peat wastage following drainage.

The silt-land includes the country round Wisbech, which is now extensively exploited for fruit-growing, flower-growing and market gardening, and is one of the most fertile soils in Britain. Thus though the plant life of the Fenlands as a whole is very different from that of the rest of the county there are also distinctions, though less obvious ones, between the two main types of fenland, the silts and the peats.

In the peat-fens very little original vegetation remains, but two areas, namely Wicken Fen and Chippenham Fen, are so famous botanically that they are here discussed in some detail.

(i) *Wicken Fen*

Wicken Sedge Fen lies about ten miles to the north-east of Cambridge, on the very margin of the Fens. Its lodes or main drainage channels converge at Upware, and then communicate with the Cam through sluice gates, which maintain a water level of about 6 ft. o.d. in the fen dykes. The peat of the sedge-fen is about 1 ft. higher. The surrounding fenland which has been drained and cultivated is at a much lower level, and has its own drainage system. The sedge-fen owes its preservation to its use as a catchwater, into which flood-water from the uplands could be discharged, so as to save the cultivated lowlands. Thus a specific use is responsible for the preservation of this patch of fenland, and, as we shall see, the hand of man has greatly influenced its vegetation, although most of the original plant species are still present.

INTRODUCTION

The fen contains several types of habitat, which can be conveniently grouped as follows: the lodes, ditches and brick-pits containing the aquatic and reed swamp vegetation; the fen 'droves' consolidated by trampling and kept clear by mowing once or twice every year; and the fenland communities proper dominated by sedge (*Cladium mariscus*), by the grass *Molinia*, by several species of which *Juncus subnodulosus* is the most important, or by the shrubs *Frangula alnus*, *Rhamnus catharticus* and *Salix cinerea* ('carr').

The bottoms of the lodes, which are nowhere deep, are thickly covered with *Chara*, which also forms almost the exclusive filling of many ponds and abandoned peat-cuttings. This *Chara* bed is thickly sown with abundant phanerogams, mostly rooted and with submerged vegetative parts. There are pond-weeds (*Potamogeton lucens, praelongus, crispus, perfoliatus* and *Groenlandia densa*), and large patches of *Oenanthe fluviatilis* with its finely dissected leaves, *Sparganium emersum* with its trailing pale strap-like leaves and the grass-like leaves of *Eleocharis acicularis*. In slightly shallower water the shaggy stems of *Hippuris vulgaris* grow abundantly. *Myriophyllum verticillatum* occurs in the lodes, and abundantly in the deep water of the brick-pits. The floating-leaf community consists of *Nymphaea alba*, *Nuphar lutea* and *Polygonum amphibium*. Nearer to the side are to be found groups of *Scirpus lacustris*, *Sagittaria sagittifolia*, *Alisma plantago-aquatica*, *A. lanceolatum* and *Baldellia ranunculoides*. *Hydrocharis morsus-ranae* and *Lemna* species are limited to the ponds, shallow ditches and margins of the lodes, while *Utricularia vulgaris* flowers abundantly in the brick-pits. *U. minor* was found by the edge of the lodes and in old peat-cuttings, but has not recently been seen. Still nearer the lode-bank occurs the reed-swamp, which at Wicken is nearly always a well-developed *Phragmitetum*. Since the excavation of the Mere on Adventurers' Fen in 1954–55, *Phragmites* reed-swamp has established itself around the open water, growing originally from rhizomes surviving in the peat which had been under arable cultivation for some years. The speed with which a dense reed-swamp up to 8 or 9 feet tall could establish itself here was very remarkable indeed. By 1958 the community was closed over the main available area, and the natural edge to the deeper water was quite evident. A large crop of reed was cut for sale from Adventurers' Fen in the winter of 1959–60. Other reed-swamp dominants occur locally at Wicken: e.g. in the brick-pits *Typha angustifolia* and *Scirpus lacustris*, and at the village end of Wicken Lode are to be found *Typha latifolia*, *Glyceria maxima* and *Phalaris arundinacea* with a fringe of *Glyceria fluitans*. Where repeated clearings prevent the formation of the mar-

18

INTRODUCTION

ginal reed-swamp, we find a fringe of fen sedges. These sedges include *Carex elata, acuta, otrubae, lepidocarpa, disticha, riparia* and *acutiformis*, and there are also groups of *Juncus* spp. and *Eleocharis palustris.*

An extreme case of man's interference is presented by the fen droves which are cut twice or thrice a year. Even the youngest of these show rapid deviation from the communities from which they originated, and the main drove (Sedge Fen Drove), which is at least 200 or 300 years old, is quite different from any other fen-community. It is striking that these droves are extremely rich in species. Repeated cutting prevents dominance of tall species, and many dwarf herbaceous plants occur, including numerous ruderal invaders. Some of the characteristic species of these droves are *Cirsium dissectum, Triglochin palustris, Eleocharis quinqueflora* and *uniglumis, Taraxacum palustre* and *spectabile, Carex distans, hostiana* and *panicea* and *Valeriana dioica.*

The main area of the Wicken reserve called the 'Sedge Fen' was formerly dominated by *Cladium mariscus*, which was cut as a crop for thatching every four years. Under this traditional cropping regime the Sedge Fen was stabilized; but with the decline of demand for sedge, and increase in costs of labour, the traditional pattern was gradually abandoned. In the period immediately preceding the first acquisition of parts of the Fen by the National Trust in 1896, a rapid spread of carr must have taken place; and it was indeed partly this which impressed the entomologists with the urgency of acquiring the Fen as a Nature Reserve. Under the present management policy, the National Trust endeavours to preserve (and recreate when necessary) some 40–50 acres of 'sedge field' from which a *Cladium* crop can still be taken in the traditional manner.

There is evidence at Wicken, as on the Norfolk Broads, that in the natural succession *Cladietum* would follow the *Phragmites* reed-swamp as the peat accumulated. The present distribution of the patches of sedge, however, does not show this natural succession, because of the complex history of man's interference with the Fen.

More frequent cutting of *Cladium* weakens the plant at the expense of the coarse grass *Molinia coerulea*, and 'mixed sedge', or 'litter' communities, with abundant or dominant *Molinia*, can become established in this way. Such communities were widespread in the 1920's when Professor Godwin made his detailed studies of the Fen vegetation, but are nowadays quite small in extent. *Molinietum* is apparently stabilized by annual cutting.

In pure sedge there are a few other fen species in relatively small quantity; in the mixed sedge communities, however, plants such as

2-2

INTRODUCTION

Phragmites, Eupatorium cannabinum, Angelica sylvestris, Peucedanum palustre, Lysimachia vulgaris and *Salix repens* var. *fusca* occur frequently. Species common in the annually cut 'litter' include *Juncus subnodulosus, Succisa pratensis, Thalictrum flavum, Cirsium dissectum* and *Filipendula ulmaria*. Such species are also common in the 'mixed fen' communities often dominated by *Juncus subnodulosus* which arise after clearing of young carr. In recent years several acres of 'mixed fen' have been established in this way at the entrance to the Sedge Fen, and this area now contains most of the characteristic herbaceous perennials. The two grasses, *Calamagrostis epigeios* and *C. canescens*, are more common after clearing carr, especially on peat areas which are slightly higher than the rest.

When cutting or other interference is discontinued, bush invasion is very rapid. By far the most abundant species is *Frangula alnus* which dominates the young stages, but *Salix cinerea, Viburnum opulus* and *Rhamnus catharticus* are abundant, while *Crataegus monogyna, Prunus spinosa* and *Ligustrum vulgare* are frequent. Bush establishment is very irregular and shows local variations in density, so that a heterogeneous patchwork of scrub arises over the invaded area. In time the bush-crowns expand and coalesce, and the complete bush covering is known as 'fen carr'. The early stages of carr development are marked by the killing out of the ground flora. If the bushes are sufficiently dense, there follows a phase in which much bare peat is visible, and a new shade-enduring flora comes to occupy the ground. *Cladium* is usually completely killed out under dense carr. This shade flora includes characteristically the Marsh Fern, *Thelypteris palustris*, and also shade-tolerant species such as *Lysimachia vulgaris, Symphytum officinale, Iris pseudoacorus, Phragmites, Calystegia sepium, Rubus caesius* and *Urtica dioica*. The rarities remain to be mentioned. *Myrica gale* and *Carex appropinquata* are now found nowhere else in the county, and a large tussock of the latter by the main drove has probably been looked at by more botanists than has any other Cambridgeshire plant. The Fen Violet, *Viola stagnina*, and the Fen Orchid, *Liparis loeselii*, used to occur abundantly at Wicken and elsewhere in the Wicken and Burwell area, but both are now extinct in the county. There is good evidence that their former abundance was associated with the peat-digging industry, which provided special habitats of bare wet peat and mossy hollows in the old peat-cuttings.

There is a good deal of published literature on Wicken. References to the important papers can be conveniently found in the short *Guide to Wicken Fen* published by the National Trust (ed. 3, 1959), which

20

contains an excellent abbreviated account of the vegetation, as well as much other information.

(ii) *Chippenham Fen*

Chippenham Fen is situated four or five miles due north of Newmarket and occupies a depression which forms a marginal extension of the Cambridgeshire fens. It lies just within the area of the Breck sands, and typical Breckland species occur right to the edge of the peat fen.

A large part of the Fen is covered with the deep parallel trenches of former peat-cutting, but apart from this there seems to have been little interference in later years. In the trenched area *Cladium* is dominant, and on the ridges large tussocks of *Molinia* are abundant, together with *Schoenus nigricans* and *Juncus subnodulosus*. There is a generous sprinkling of such species as *Angelica sylvestris*, *Eupatorium cannabinum*, *Lythrum salicaria*, *Urtica dioica*, *Valeriana officinalis*, *Serratula tinctoria* and *Scrophularia nodosa*. The common moss is *Acrocladium cuspidatum*, and *Phragmites* grows abundantly throughout.

The whole assemblage resembles the mixed sedge at Wicken, and like that is subject to bush and tree invasion. The woody species are *Fraxinus excelsior*, *Alnus glutinosa*, *Viburnum opulus* and *Betula* spp. The birches in some places form a fringe invading the Fen, but the bush colonization in general is far sparser than at Wicken and *Frangula* is by no means common.

Much of the Fen is covered by woodland, originally planted at the end of the eighteenth century, and containing non-native trees such as larch and spruce. The undergrowth shrubs include *Ligustrum vulgare*, *Viburnum opulus*, *Corylus avellana* and *Sambucus nigra*. Tree species which are regenerating naturally include *Alnus glutinosa*, *Fraxinus excelsior*, *Taxus baccata* and *Prunus padus*. There is little ground vegetation, but the common Stinging Nettle is abundant, and climbers or scramblers, *Lonicera peryclymenum*, *Bryonia dioica* and *Solanum dulcamara*, are abundant. It is a good locality for fungi especially in the autumn.

On the edges of the Fen are wet meadows with a peaty soil, and a vegetation clearly derived from fen vegetation. Species present are *Molinia* (abundant locally), *Thalictrum flavum*, *Potentilla erecta*, *P. anserina*, *Silaum silaus*, *Angelica sylvestris*, *Valeriana dioica*, *Succisa pratensis*, *Cirsium dissectum*, *C. palustre*, *Symphytum officinale*, *Mentha aquatica*, *Prunella vulgaris*, *Ajuga reptans*, *Carex panicea*, *Briza media* and *Ophioglossum vulgatum*. This vegetation is sparsely invaded by bushes and is dissected by drains containing relict *Cladium* and *Schoenus*.

INTRODUCTION

The general resemblance to the litter and droves of Wicken Fen is striking, and similar types may be recognized in other Cambridgeshire fens and in the Norfolk Broads. Chippenham Fen has a special interest in that *Pinguicula vulgaris*, *Aquilegia vulgaris*, *Selinum carvifolia*, *Parnassia palustris* and *Menyanthes trifoliata* all grow there, but are rare or absent from the rest of the county.

Particularly round the fen margins and near local sources of water supply, such as springs and streams, sites exist where fens have locally developed, such as Dernford Fen, Teversham, Fulbourn and Wilbraham Fens, Quy Fen, Thriplow Peat Holes and Fowlmere water-cress beds. Here remain more or less representative collections of fen species, but the communities are greatly modified or reduced.

There is a good deal of evidence that local acid bog development formerly took place in many parts of the Cambridgeshire peat fens, as at Hinton, Sawston and Teversham Moors, Wicken, Burwell and Chippenham Fens and at Sutton Meadlands. The following are some of the species that were characteristic of these acid bogs: *Drosera rotundifolia*, *D. anglica*, *D. intermedia*, *Eriophorum angustifolium*, *Carex dioica*, *Sphagnum palustre* and *Splachnum ampullaceum*. But acid bog communities of this type have now been completely destroyed by human activity, and the merest trace remains of this earlier phase in, for example, the persistence of small patches of Bog Myrtle, *Myrica gale*, at Wicken. (See also note on p. xvi.)

The Washes between the Old and the New Bedford Rivers still afford habitats for a number of fen and marsh species formerly more widespread. Thus the only recent records for *Bidens cernua* and *Polygonum minus* come from the Welney Washes just within the county boundary.

Most of the Fenland is arable and we find there a number of characteristic weeds which are rare or absent from the uplands. This list includes *Brassica nigra*, *Chenopodium ficifolium*, *C. polyspermum*, *Fumaria muralis* ssp. *boraei*, and *Veronica agrestis*. These species also occur on the silts, but two species seem to be rather restricted to the peat lands, viz. *Galeopsis speciosa* and *Stachys arvensis*.

The nearer one approaches the shore of the Wash the more recent becomes the landscape. This effect seems to be reflected in the flora, for the farther north one proceeds in Cambridgeshire the poorer that flora becomes, and it may well be that the lands bordering the Wash, having been reclaimed from the sea for agriculture, have never had a natural vegetation, so that even roadside verges and lode banks show a dull monotony.

22

INTRODUCTION

With the exception of the maritime and submaritime species there are no species peculiar to the siltlands, but there are thirty-four species which are frequent throughout the county except for the siltlands. Some of these are arable weeds, e.g. *Lycopsis arvensis*, *Anthemis cotula*, *Matricaria recutita*; others are water plants including *Baldellia ranunculoides*, *Berula erecta*, *Rorippa microphylla* and *Groenlandia densa*. But by far the largest group is of marsh plants. Twenty-three of the absentees are plants of wet habitats including *Filipendula ulmaria*, *Lotus uliginosus*, *Lysimachia vulgaris*, *Myosoton aquaticum*, *Senecio aquaticus*, *Succisa pratensis* and *Valeriana officinalis*. Such marshy communities as are present in the area are strikingly poor in species.

(2) *Chalk*

The botany of the southern and eastern parts of Cambridgeshire is dominated by the great area of chalk, which enters the county at Odsey, west of Royston, and sweeps north-eastwards for thirty miles through Newmarket to Kennett. In addition, there is a ridge extending westwards from Harston which reaches Arrington and Wimpole. Another outcrop occurs between Comberton and Madingley.

The chalk consists of three distinct strata: the Upper, Middle and Lower Chalk. The Upper Chalk lies nearest to the county boundary and is largely overlain by chalky boulder clay. The Middle Chalk resembles the Upper in having a low clay content, but differs from it in having no extensive bands of flint. It is the most important of the three botanically, being the stratum which underlies the Roman Road, Fleam Dyke, Devil's Dyke and Newmarket Heath, where the most extensive examples of our chalk flora are to be found. In many areas, however, it is covered with sands and gravels, outwash materials from the higher boulder-clay lands to the south and east. Such areas occur at Chrishall Grange, Thriplow, Six Mile Bottom, Dullingham and Kennett. The soils of the Middle Chalks are, when derived from the parent material, shallow rendzinas often less than 8 in. in depth with a high percentage of $CaCO_3$ even to the surface, and a pH between 7·2 and 7·5. The majority of soils, however, are derived at least in part from sandy or loamy calcareous drifts, which give rise to brown calcareous soils very similar in appearance to brown earths, but differing from them in being more calcareous. True rendzinas are very rare. The more clayey Lower Chalk or 'Chalk Marl' forms the skirt lands of the Fens and nearly all has been under cultivation for a very long period. There is, however, a hard band of rock, the Burwell Rock, which is the source

INTRODUCTION

of the material of which the local clunch is made. It is one of the few local building stones of any importance and was used in several medieval church interiors. Thus many of these clunch pits are of great age and were first excavated when the surrounding land was chalk grassland. They have now a characteristic chalk flora, whereas more recent pits, dug after the grassland had disappeared under the plough, contain few species of interest.

Until about 1800, when the last wave of Enclosure Acts changed the landscape of the county, most of the chalk district was open sheep-walk, covered with chalk grassland rich in species which are rare or local today. The best impression of the past can now be gained by standing in the middle of Newmarket Heath. Here, on the parts of this immense plain which are little used by racehorses, as for example on the Beacon Course, the aboriginal turf can still be found. It contains *Thesium humifusum, Asperula cynanchica, Veronica spicata, Hippocrepis comosa, Scabiosa columbaria,* and *Astragalus danicus,* as well as many other characteristic chalk-loving species. Many of our rarest chalk grassland species are almost confined to the unploughable artificial habitats like the Fleam and Devil's Dykes and the Roman Road. The list is a long one and includes *Pulsatilla vulgaris, Senecio integrifolius, Himantoglossum hircinum, Hypochaeris maculata, Carex ericetorum, Linum anglicum, Potentilla tabernaemontani, Geranium sanguineum* and *Juniperus communis.* The last two are not now known elsewhere in East Anglia.

Bryophytes show a comparable situation: the rarest chalk species, such as *Tortella tortuosa, T. inclinata, Trichostomum crispulum, Weissia sterilis, Thuidium philibertii* and *Entodon concinnus,* are confined to these artificial habitats, or to old lawns made from chalk grassland that formerly existed nearby.

Another remarkable group of rarities is found in and around the chalk pits at Cherry Hinton. Here in grassy patches, in shrubby places or on artificial chalky banks, can be found *Seseli libanotis, Bunium bulbocastanum, Lonicera caprifolium, Falcaria vulgaris* (a relatively recent arrival now firmly established), *Muscari atlanticum* and the moss *Tortula vahliana.* The *Muscari* is perhaps a relic from the Saffron fields which stretched from Cherry Hinton to Saffron Walden as late as the latter half of the eighteenth century. Elsewhere in the county Candytuft, *Iberis amara,* occurs, and is almost entirely confined to railway banks; such species as *Inula conyza, Cerastium arvense, Minuartia hybrida* and *Lathyrus nissolia* are usually associated with past disturbance.

The character of our chalk grassland is now changing rapidly. The change began with the Enclosure Acts after which the Dykes which were once part of the open fields ceased to be grazed by cattle and sheep. However, the rabbit continued to hold the bushes and dominant grasses in check. Since the almost complete disappearance of the rabbit about 1953, scrub development has been rapid and it is becoming a considerable threat to our rarer species. The dominant grass is mainly *Bromus erectus*. The grass *Brachypodium pinnatum* also occurs but is local on the chalk and it develops much more vigorously on the damper clayey soils in the west of the county.

Compared with other parts of England the main feature of the chalk grassland is its dryness. This is partly topographical, for here we have no deep sheltered combes with damp north-facing slopes, and partly climatic. Orchids do not occur in great abundance or variety: *Anacamptis pyramidalis* and *Gymnadenia conopsea* are the two species most frequently found, though *Ophrys apifera*, the Bee Orchid, usually appears in a number of places each year. The Man Orchid, *Aceras anthropophorum*, is known in two localities and the Lizard Orchid persists. However, though these five species still remain, at least another five have disappeared from the chalk: *Orchis ustulata, Ophrys sphegodes, Herminium monorchis, Coeloglossum viride* and *Spiranthes spiralis*. The first three are almost certainly extinct in the county.

Some of the most exciting species in the county are to be found in and around the large square post-enclosure fields of the chalk. Four species of *Fumaria* are known to occur: *F. officinalis, F. micrantha, F. parviflora* and *F. vaillantii*. Poppies too are much in evidence and all five 'native' species occur: *Papaver rhoeas, P. argemone, P. hybridum, P. lecoqii* and *P. dubium*. *Legousia hybrida* occurs in most fields, as do *Silene noctiflora*, and *Kickxia elatine* and *K. spuria*. Rarer species include *Valerianella dentata, Galeopsis angustifolia, Torilis arvensis*, the Cornflower, *Centaurea cyanus*, and the Lesser Broomrape, *Orobanche minor*, which is sometimes abundant in clover leys. The other broomrape *O. elatior* is found in field margins and old chalk pits, where it is parasitic on *Centaurea scabiosa*. Other species of field margins include *Nepeta cataria, Calamintha nepeta* and *C. ascendens, Salvia horminoides, Astragalus glycyphyllos*, and the very rare Ground Pine, *Ajuga chamaepitys*. The rare Tuberous Thistle, *Cirsium tuberosum*, although a species of open grassland elsewhere in Britain, is with us only known from one field margin on the chalk marl.

The bryophytes of arable fields on the chalk include a number of annual species. Most characteristic is *Phascum floerkeanum*, which is sometimes abundant in stubble fields between August and November, particularly after wet summers.

25

INTRODUCTION

There is little evidence about the natural woodland vegetation of the chalk. The few beech woods which exist are generally of an even age and were probably planted in the late eighteenth or early nineteenth centuries. Beech was unknown to Ray in Cambridgeshire, and although there are quite high percentages of Beech pollen in the upper peats by the fen margin, it cannot be regarded as a native tree at the present day. Despite the unnatural origin of these woods, a characteristic species of the ground flora of such woods, *Cephalanthera damasonium*, is to be found there.

Where grassland is neglected it soon reverts to a scrub which includes the calcicolous shrubs such as *Cornus sanguinea* (Dogwood), *Euonymus europaeus* (Spindle-tree), *Rhamnus catharticus* (Buckthorn), and the Roses, *Rosa canina* and *R. rubiginosa*, and *Viburnum lantana* (Wayfaring Tree). The Field Maple, *Acer campestre*, develops into a fine tree in our climate when allowed to grow: it and Ash, *Fraxinus excelsior*, are the dominant trees of old woodland in Lime Kiln Close, Cherry Hinton, and it seems likely that ash woodland would often develop on the chalk if disturbance ceased.

(3) *Boulder Clay*

There are two large areas of boulder clay in the county: the western one on the low plateau ground between Coton, Papworth and Tadlow, and the other on the eastern border of the county from Linton to Cheveley and Ashley.

Since the Cambridgeshire woodlands are almost confined to the boulder clay, and since little other natural or semi-natural vegetation remains upon it, this account is mainly confined to the boulder-clay woodlands.

These woodlands have been systematically exploited by man, and are all of the type known as 'coppice with standards'. A limited number of standard trees, mostly the Oak, *Quercus robur*, are retained, and they grow large open crowns with abundant sharply bent main branches, formerly in great demand for the building of wooden ships. The shrub-layer consists largely of Hazel, *Corylus avellana*, probably largely planted, which used to be cut at intervals of ten or twenty years, for use as small wood, firing, kindling, hedge stakes, pea and bean sticks, etc. *Fraxinus*, *Acer campestre* and other natural components of the woodlands all suffered this treatment to some degree, but none resisted the coppicing so well as *Corylus* and *Fraxinus*.

At the present day much less coppicing is done. In many places the woods are used as pheasant preserves, and the suppression of enemies

26

of game birds allows an increase of ground vermin, such as mice, which greatly lessens the chances of natural regeneration of the woodland trees.

All these woods on the chalky boulder clay are derived from oak–ash woods, but have been converted to the oak–hazel type by prolonged coppicing. This derivation is suggested by their richness in ash, and by calcicolous herbs and shrubs which are absent from the pure oak woods. Flora and soil alike occupy a middle position between the oak woods of loam soils and the ash woods on limestone. *Ulmus* species occurring frequently in these woods are probably all planted.

The spring ground flora of these woods is remarkable. Its beginning is sombre in colouring with the small shrub *Daphne laureola*, two rarities *Helleborus foetidus* and *H. viridis*, and large areas of *Mercurialis perennis*; all these species have inconspicuous greenish flowers.

This is followed by a blaze of colour with great sheets of *Endymion non-scriptus*, *Primula elatior* and its hybrid with *P. vulgaris*, sprinkled through with *Ranunculus ficaria*, *Arum maculatum*, *Anemone nemorosa*, *Viola reichenbachiana*, *V. riviniana*, *Ranunculus auricomus*, *Luzula pilosa*, *Carex sylvatica*, *Paris quadrifolia*, *Galeobdolon luteum*, *Sanicula europaea*, *Fragaria vesca* and *Potentilla sterilis*, and the orchids *Orchis mascula*, *Dactylorchis fuchsii*, *Platanthera chlorantha* and *Listera ovata*.

Other species characteristic of these woods are *Rumex sanguineus*, *Moehringia trinervia*, *Campanula trachelium*, *Hypericum hirsutum*, *Scrophularia nodosa*, *Iris foetidissima*, *Epipactis helleborine*, *Bromus ramosus*, *Festuca gigantea* and numerous bryophytes. Shrubs which occur frequently include *Rosa micrantha*, *Viburnum opulus*, *Rhamnus catharticus*, *Salix cinerea*, *S. caprea*, *Crataegus monogyna* and *C. oxyacanthoides* and their hybrid. On the other hand, some of our woodland species are surprisingly rare or local. These include *Lathyrus sylvestris*, *Allium ursinum*, *Galium odoratum*, *Euphorbia amygdaloides*, *Hypericum maculatum* ssp. *obtusiusculum*, *Athyrium filix-femina*, *Milium effusum*, *Melica uniflora*, *Neottia nidus-avis* and *Rosa tomentosa*. A small group of species is almost confined to the eastern boulder clay and includes *Lysimachia nemorum*, *Veronica montana*, *Dryopteris carthusiana*, *Melampyrum pratense* and *Silene dioica*. Finally there are some species which have either never occurred, or are now extinct, which, nevertheless, are frequent in woods elsewhere in Britain and are probably excluded from Cambridgeshire for climatic reasons, e.g. *Agrimonia odorata*, *Carex strigosa*, *Chrysosplenium oppositifolium*, *Dryopteris borreri* and *Peplis portula*.

The margins of these boulder-clay woods and the wide rides cut

INTRODUCTION

through them, provide a special habitat for one or two species, e.g. *Melampyrum cristatum*, *Carex pallescens* and *C. pendula*.

Water-logging permits the occurrence of such species as *Juncus conglomeratus* and *J. effusus*, and in some places *Iris pseudacorus*, *Lychnis flos-cuculi*, *Valeriana officinalis* and *Callitriche* species.

Good examples of the western boulder-clay woods are Hardwick Wood, Hayley Wood near Longstowe, and Buff Wood at East Hatley, and those of the east, Ditton Park Wood and Borley Wood. Gamlingay Wood is also mostly boulder clay and is the only known locality for *Sorbus torminalis* in Cambs.

In arable fields the mosses *Phascum cuspidatum*, *Pottia davalliana* and various *Bryum* species are abundant, and *Euphorbia platyphyllos* (rare), *Lithospermum officinale*, *L. arvense*, *Galium tricornutum* and *Ranunculus arvensis* are characteristic weeds.

Roadsides and tracks produce *Trifolium ochroleucon*, *Cichorium intybus*, *Lathyrus aphaca*, *Hordeum secalinum*, *Cirsium eriophorum*, *Bromus commutatus*, *Vicia tenuissima*, *Cruciata chersonensis*, *Allium vineale*, *Betonica officinalis*, *Pimpinella major*, *Sanguisorba officinalis* and *Ononis spinosa*.

Besides the two large areas described at the beginning of this section, the boulder clay is to be found forming small islands in the Fens, as around Ely, Sutton and March. It has, however, little effect on the flora, although a few of the characteristic species are present. The best of the boulder-clay woods in the Fens was at Doddington, but it was cut down during the last war.

(4) *Breckland*

The 'Breckland' is a tract of heath-covered country about 300 square miles in area, embracing the towns of Brandon, Thetford and Mildenhall, and stretching just far enough to the south-west to cross the River Kennett and cover a few square miles of Cambridgeshire. The area is situated in the driest part of England and the surface layers of soil are porous sands; thus they are very dry. Most of the area in Cambridgeshire is under cultivation and the only places where some of the original vegetation remains are by the edges of sand pits around Chippenham and Kennett. Characteristic species of such places are *Teesdalia nudicaulis*, *Cerastium semidecandrum*, *Trifolium arvense*, *Sedum acre*, *Galium saxatile*, *Hypochaeris glabra*, *Myosotis ramosissima*, *Vicia lathyroides*, *Rumex tenuifolius* and *Ornithopus perpusillus*, while *Silene conica*, *gallica* and *otites*, *Medicago minima*, *Trifolium subterraneum*, *striatum*

INTRODUCTION

and *scabrum*, *Potentilla argentea*, *Crassula tillaea* (*Tillaea muscosa*), *Herniaria glabra* and *Carex arenaria* are recorded. The characteristic weeds of arable and waste land are *Arabidopsis thaliana*, *Descurainia sophia*, *Spergula arvensis* and *Scleranthus annuus*, and *Montia* (*Claytonia*) *perfoliata* is frequent in hedgerows. Few of the bryophytes characteristic of the Breckland have been recorded from the Cambridgeshire Breck fringe, but *Polytrichum juniperinum*, *Pleuridium acuminatum* and *Tortula ruraliformis* occur there, and *Rhytidium rugosum* has been recorded, although not seen recently. For an account of the interesting ecology of the Breckland the papers of A. S. Watt (1936, 1940) should be consulted.

At Hildersham, the 'Furze Hills' provide a small, isolated, botanically famous locality for some of the species characteristic of the Breckland; and in spite of quarrying operations (now fortunately finished) many of these species can still be found there, including *Dianthus deltoides*, *Phleum phleoides*, *Potentilla argentea* and, of course, the Gorse or Furze itself, *Ulex europaeus*. Geologically the Furze Hills are of some interest; they are made of 'caps' of glacial gravels, containing pebbles of flint and other rocks and much sand, clay and ground-up chalk.

(5) *Greensand*

The small area of Lower Greensand within the county is found around the village of Gamlingay. The most obvious difference in landscape from the rest of the county is the presence of quantities of Bracken, *Pteridium aquilinum*, Furze, *Ulex europaeus* and Broom, *Sarothamnus scoparius*. The most interesting localities are the Great Heath Wood and its surrounding area, White Wood and The Cinques. Gamlingay Wood is partly on the clay and partly on the sand, and an interesting ecological account of it is given by Adamson (1911). Species that are almost or completely confined to the greensand are *Myosurus minimus*, *Turritis glabra*, *Hypericum humifusum* and *pulchrum*, *Senecio sylvaticus*, *Hieracium perpropinquum*, *Digitalis purpurea*, *Plantago coronopus*, *Nardus stricta*, and *Carex demissa*, *nigra*, *pilulifera* and *echinata*, and the bryophytes *Polytrichum formosum*, *Campylopus flexuosus* and *C. pyriformis*, *Pleurozium schreberi*, *Hypnum cupressiforme* var. *ericetorum*, *Calypogeia fissa*, *Cephalozia bicuspidata* and *Cephaloziella* species. Many calcifuge species such as *Dicranella heteromalla*, *Pohlia nutans*, *Mnium hornum*, and *Plagiothecium denticulatum* are more abundant here than elsewhere in the county. *Quercus cerris* and *Castanea sativa* are frequently planted, while *Quercus petraea* and *Tilia cordata* are

29

INTRODUCTION

native. The weeds of arable and waste land are similar to those of the Breckland sands, *Scleranthus annuus*, *Spergula arvensis* and *Spergularia rubra* being common, while in heathland *Cerastium semidecandrum*, *Galium saxatile*, *Calluna vulgaris* and *Ornithopus perpusillus* occur. In former times there was a large well-developed heath with quaking acidic bogs, but the latter were drained in the eighteen-fifties. In the bogs and on the heaths occurred such species as *Lycopodium inundatum* and *clavatum*, *Teesdalia nudicaulis*, *Moenchia erecta*, *Hypericum elodes*, *Genista anglica*, *Carex dioica*, *Equisetum hyemale*, *Potentilla argentea*, *Peplis portula*, *Solidago virgaurea*, *Arnoseris minima*, *Vaccinium oxycoccus*, *Erica cinerea* and *tetralix*, *Anagallis minima*, *Littorella uniflora*, *Hammarbya paludosa*, *Narthecium ossifragum* and *Juncus squarrosus*, and the bryophytes *Sphagnum* spp., *Polytrichum commune*, *Rhacomitrium canescens*, *Splachnum ampullaceum*, *Philonotis fontana*, *Aulacomnium palustre*, *Lophozia ventricosa* and *Odontoschisma sphagni*. Most of these disappeared after the drainage but a few lingered on for many years in isolated habitats.

(6) *Salt-marsh*

By the banks of the tidal River Nene and its tributaries north of Wisbech occur communities of halophytes not found elsewhere in the county. The locally dominant grasses which form more or less distinct zones on the river bank are *Puccinellia distans* nearest the water, followed by a belt of *Parapholis strigosa* and, farther out, *Agropyron pungens*. Other maritime species occurring here are: *Cochlearia anglica*, *Spergularia media* and *marina*, *Aster tripolium*, *Limonium vulgare*, *Armeria maritima*, *Glaux maritima*, *Plantago maritima*, *Atriplex littoralis*, *Halimione portulacoides*, *Salicornia dolichostachya* and *ramosissima*, *Juncus gerardii* and *Triglochin maritima*.

30

EXPLANATION OF TEXT

With few exceptions the system and nomenclature in the following account follows that of J. E. Dandy (1958) for the Vascular Plants, P. W. Richards and E. C. Wallace (1950) for the Mosses, and E. W. Jones (1958) for the Liverworts. Following the scientific name, a well-known vernacular one may be given. If there is a well-known name for a number of species all in one genus (e.g. Willow-herb for *Epilobium* species) it is placed in parentheses after the generic name. Following the specific vernacular name is the first known record for the plant in the county. This may be based on either a published record, a field record or a herbarium specimen. On the next line follows a list of synonyms that are used by Babington (1860), Evans (1939) or Clapham, Tutin & Warburg (1952). Old binomials and pre-Linnaean names by the earlier authors are given in Babington (1860). An account of the habitat of the species is then given with any notes on rare, difficult or interesting species.

The distribution is indicated by listing the Ordnance Survey Grid Reference numbers given on the Map. Note that most of the 10 km. squares are in 100 km. square 52; but a few squares in the north of the county belong to 100 km. square 53, and these are given at the end of the main (52) series. Numbers in square brackets are given in a separate series. If the numbers are in Roman type, 25, the most recent record is since 1950, if in italics, *25*, it is between 1930 and 1950, if in square brackets, [25], it is before 1930. There are ten squares which are represented within the county boundary by a small portion only. Their Grid Reference numbers are 15, 23, 33, 43, 63, 69, 75, 76, 31, 50. These squares are not taken into account when a species is said to be in 'all squares' or 'all squares except...'. Following the phrase 'all squares except...' a number in Roman type, 25, means that there is no record at all for that square, a number in italics, *25*, means that the most recent record was between 1930 and 1950, and a number in square brackets, [25], means that the most recent record was pre-1930. 'All squares except...' is only used when the species is absent from not more than seven squares. All records available up to the end of 1962 have been included.

Two other special points need clarification. All the Cambridgeshire part of square 43 (Great and Little Chishill), although now within the *administrative* county of Cambridgeshire, is in v.c. 19 (north Essex). Vascular plants recorded for Wicken Fen are given for 56 only, although the Fen extends into 57.

31

EXPLANATION OF TEXT

For many local plants, an abbreviation for a well-known locality is given after the distribution data; these localities are indicated on the Map and a list of them is given opposite the Map with their precise grid references. If a dagger (†) is given after its number the species is considered to be extinct within the county. Finally, some idea of the distribution of the species in the British Isles is given. If the species occurs throughout the greater part of the British Isles the word 'general' is used. If the abbreviation CGE occurs in the text it refers to the Cambridge University Herbarium. The approximate acreage for crop plants mentioned is for the year 1959; in such notes 'S. Cambs' is used to indicate the administrative County of Cambridgeshire excluding the Isle of Ely.

Keys, designed to be useful in the field, have been supplied in the following cases known to present difficulty to the student: the larger Ferns, *Ranunculus, Papaver, Fumaria, Viola, Cerastium, Stellaria* (part), *Chenopodium, Trifolium, Vicia* (part), *Aphanes, Rosa, Epilobium, Callitriche, Ulmus, Myosotis, Calystegia, Mentha, Potamogeton, Juncus, Dactylorchis, Eleocharis* (and part of *Scirpus*), *Carex*. In addition to these keys, there are many notes on identification under individual species or genera.

1 (*a*)   The Wind-pump, Wicken Fen, September 1956. *W. M. Lane*

1 (*b*)   Spring scene in Buff Wood, April 1963. *P. D. Sell*

2   *Asplenium ruta-muraria* on the steps of the Senate House, Cambridge, 1952.   *F. H. Kendon*

3 *Trifolium ochroleucon*, Hardwick, 1952. *M. C. F. Proctor*

4 *Hottonia palustris* flowering in a peaty ditch near Upware, 1950.
M. C. F. Proctor

5   *Lathraea clandestina* by the river, Cambridge, 1951.   *M. C. F. Proctor*

6   *Aceras anthropophorum*, Abington, 1958.   *J. C. Faulkner*

7   *Cephalanthera damasonium* in Beech woodland, Gog Magog Hills, 1950.
M. C. F. Proctor

8   *Acorus calamus*, Earith, 1951.   *M. C. F. Proctor*

PTERIDOPHYTA

LYCOPODIACEAE

LYCOPODIUM L. (Clubmoss)

1†. Lycopodium inundatum L. Relhan, 1785
Formerly in bogs of the Great Heath, Gamlingay, where it became extinct after their drainage between 1850 and 1880. It was last recorded by Babington (1860). [25.] Scattered throughout lowland Britain, rare in Ireland.

2†. Lycopodium clavatum L. T. Martyn, 1763
Formerly on the Great Heath, Gamlingay; last recorded by Babington (1860). [25.] General, but rare and decreasing in lowland areas.

EQUISETACEAE

EQUISETUM L. (Horsetail)

1†. Equisetum hyemale L. Ray, 1660
Recorded by Relhan from Gamlingay and Stretham Ferry; no specimens exist. [25, 57.] Scattered over the British Isles but very rare in S. England and Wales.

2. Equisetum fluviatile L. Water Horsetail T. Martyn, 1763
E. limosum L.
Fairly common by ditches, edges of ponds and in marshes. The var. **aphyllum** (Roth) Alston & Ballard with simple stems is frequent. *25, 29, 34,* 36, 37, 39, 44, 45, *46,* 47, 48, 56, 57, 59, 68, 20, [35, 38, 49, 54, 55, 58, 67]. General.

3. Equisetum palustre L. Marsh Horsetail T. Martyn, 1763
Fairly common in marshes, fens and damp meadows. 25, 29, 34–7, 44–8, 55, 56, 59, 64, 66, 67, 69, 50, [49, 54, 57, 65]. General.

4†. Equisetum sylvaticum L. Wood Horsetail
J. Martyn in T. Martyn, 1763
Three very old records from Madingley (J. Martyn), Chesterton (Relhan), and Kingston Wood (Mr Vernon in Babington notes),

EQUISETACEAE

undated and unsubstantiated by herbarium specimens. [35, 36, 46.] General, but very local in S. England.

5. Equisetum arvense L.   Common Horsetail   Ray, 1660
Frequent in arable land, hedgerows and waste places but not an abundant, troublesome weed as in many parts of Britain. Often occurs in wet habitats where sterile material can be distinguished from *E. palustre* by the appearance of the stem in cross-section and the rhizome. In *E. arvense* the central hollow of the stem is larger than the outer ones and the rhizome is hairy; whilst in *E. palustre* the central hollow is equal to or smaller than the outer ones and the rhizome is glabrous. All squares. General.

6. Equisetum telmateia Ehrh.   Great Horsetail   Ray, 1660
Recorded by W. W. Newbould at Eversden, by Henslow at Ely, and by Shrubbs at Milton in 1888. Only the last record is represented by a specimen in CGE. [35, 46, 58.] General, but local or rare in Scotland.

KEY TO THE LARGER FERNS STILL TO BE FOUND IN CAMBS

1 Sori continuous, covered by the inrolled indusium-like margin.
                                    Pteridium aquilinum
   Sori discrete, not covered by the margin of leaf.   2

2 Leaves solitary from a long creeping underground rhizome; petiole and rhizome with few or no scales; on fen peat.
                                    Thelypteris palustris
   Leaves in a crown at the apex of the short, erect rhizome; petiole and rhizome with numerous scales.   3

3 Sori oblong-reniform.   **Athyrium filix-femina**
   Sori nearly circular.   4

4 Leaves deeply 2-pinnatifid, or 2-pinnate with toothed pinnules; pinnae decreasing markedly in length towards the base; pinnules (except sometimes the lowest pair of each pinna) not at all narrowed at the base.   **Dryopteris filix-mas**
   Leaves 3-pinnate or 2-pinnate with pinnules lobed to nearly halfway; basal pinnae not or a little shorter than the longest pair; pinnules markedly narrowed at the base or stalked.   5

5 Margin of indusium glandular and usually toothed; scales of the petioles usually darker in the middle than near the margin; pinnules flat or convex.   **D. dilatata**
   Margin of indusium entire or sinuate, rarely with a few glands; scales of the petioles uniformly coloured; pinnules flat or concave.   **D. carthusiana**

34

OSMUNDACEAE

OSMUNDA L.

1†. Osmunda regalis L.   Royal Fern   P. Dent in Ray, 1685
Recorded by Ray (1685) from the corner of Gamlingay Park. No other botanist has seen it since. It is suggested by both Babington and Evans that the identification was an error, but it is difficult to understand how this could happen with so characteristic a species. [25.] General, but local or rare except in Ireland and parts of W. Britain where it is still frequent.

DENNSTAEDTIACEAE

PTERIDIUM Scop.

1. Pteridium aquilinum (L.) Kuhn   Bracken   Ray, 1660
Pteris aquilina L.

Widespread but very local as our soils are mainly unsuitable for it; locally abundant only on the greensand at Gamlingay, on the light soils of the 'Breckland fringe' of the county, and in several woods in the south-east where the boulder clay has a higher sand content. Small plants are often to be seen on town walls and churches. 25, 29, 34, 35, *36*, 37, 43–5, 49, 54–8, 64–6, 76, 20, 40, 41, [46]. **G.** General.

ADIANTACEAE

ADIANTUM L.

1. Adiantum capillus-veneris L.   Maidenhair Fern

In the crevices of brickwork at Old North Road Station. It was first seen about 1920 according to Evans, and was last seen in 1953. 35. Native on limestone near the sea in W. England, Wales and Ireland; its scattered occurrences elsewhere are attributable to chance dispersal of spores to a suitable damp shady habitat on an old mortared wall.

BLECHNACEAE

BLECHNUM L.

1†. Blechnum spicant (L.) Roth   Hard Fern   Ray, 1663
B. boreale Sw.

Recorded from Gamlingay by Ray but not seen since. [25.] General, calcifuge.

ASPLENIACEAE

ASPLENIACEAE

Phyllitis Hill

1. Phyllitis scolopendrium (L.) Newm.   Hart's-tongue   Ray, 1663
Scolopendrium vulgare Sm.
Widespread but local on old walls especially in damp, shady positions.
.The fronds of Cambridgeshire plants are generally less than 6 in. long.
23, 25, 26, 35, *36*, 39 44–7, 54–6, 20, 40, [34, 37, 38, 48, 58]. **Cam.**
General, but rare over large areas of Scotland.

Asplenium L.

1. Asplenium adiantum-nigrum L. ssp. **adiantum-nigrum**   Black
Spleenwort   Ray, 1685
Several old records on walls; only a few recent ones. 29, 34, 37, 45, *46*,
[36, 38, 48, 54, 65, 40, 41]. **Cam.** General.

2. Asplenium trichomanes L.   Maidenhair Spleenwort   Ray, 1660
Several old records on walls; only two recent ones. 45, *55*, 40, [34,
36–8, 58, 65, 41]. **Cam.** General.

3. Asplenium ruta-muraria L.   Wall-Rue   Ray, 1660
Local on old walls, especially those of churchyards, and on bridges and
railway platforms. The locality 'on the steps of the Senate House,
Cambridge' was first recorded by Babington a hundred years ago. It
may still be seen there (plate 2). 44, 45, 48, 54, 55, 65, 20, 30, 40, [34,
36–8, 56]. **Cam.** General.

ATHYRIACEAE

Athyrium Roth

1. Athyrium filix-femina (L.) Roth   Lady Fern   Ray, 1685
Rare in wet woods. In Cambridgeshire the rarest of the three generally
common British woodland ferns (cf. *Dryopteris filix-mas* and *D.
dilatata*). 36, 43, 54, *66*, [25]. **C.** General.

ASPIDIACEAE

ASPIDIACEAE

DRYOPTERIS Adans.

1. Dryopteris filix-mas (L.) Schott   Male Fern   Ray, 1660
Lastrea filix-mas (L.) C. Presl
Throughout the county in woods, copses and on walls, though nowhere abundant. All squares except 29, 37, *46*, 48, 56, 59, 68. General.

2. Dryopteris carthusiana (Vill.) H. P. Fuchs   Narrow Buckler-fern
S. W. Wanton, 1850
D. spinulosa Watt; *D. lanceolatocristata* (Hoffm.) Alston; *Lastrea spinulosa* C. Presl
Rare, found in only two localities, 65, [55]. **DPW**. Throughout the British Isles but generally local, particularly so in N. Scotland, and with a preference for peaty soils.

3. Dryopteris dilatata (Hoffm.) A. Gray   Broad Buckler-fern
Ray, 1677
D. aristata Druce, non Kuntze; *D. austriaca* auct.; *Lastrea dilatata* (Hoffm.) C. Presl
Known to Evans in only two places but now recorded in a number of localities in woodland. This seems to be a real increase, possibly associated with the abandonment of the traditional 'coppice with standards' woodland management. 25, 26, *35*, 36, 44, 45, 54, *64*, 65, 66. **T**. General.

POLYSTICHUM Roth

1†. Polystichum aculeatum (L.) Roth   Hard Shield-fern
T. Martyn, 1763
P. lobatum (Huds.) Chevall.
There are very old records from Gamlingay by T. Martyn and Relhan unsupported by herbarium sheets. It has not been seen since. [25.] General.

THELYPTERIDACEAE

THELYPTERIS Schmidel

1†. Thelypteris limbosperma (All.) H. P. Fuchs   P. Dent in Ray, 1685
Lastrea oreopteris (Ehrh.) Bory; *Dryopteris oreopteris* (Ehrh.) Maxon; *T. oreopteris* (Ehrh.) Slosson

37

THELYPTERIDACEAE

Recorded by the eighteenth-century botanists from Gamlingay but not known to Babington. [25.] General, but very rare or absent in parts of E. Britain and some parts of Ireland.

2. Thelypteris palustris Schott   Marsh Fern   Relhan, 1802

Lastrea thelypteris (L.) Bory; *Dryopteris thelypteris* (L.) A. Gray

Relhan's record of *Polypodium thelypteris* (1802) from near Fulbourn, and on Fulbourn and Teversham Moors, was presumably this species. In his ed. 3 (1820), Relhan refers this plant to *Aspidium oreopteris*, but Babington (in notes) states that this is incorrect, and that a specimen from Fulbourn in the Relhan herbarium is *T. palustris*. It was found at West Fen, Ely, by W. Marshall in the first half of the nineteenth century, by S. W. Wanton at Wicken Fen in 1850, and later at the Wimblington Fire Lots by Fryer. It is now known only at Dernford Fen where it is very local and Wicken Fen where it is abundant. This fern also occurs abundantly with several other fern species in the 'Fen Valley Wood' at Cavenham, on the Breckland in W. Suffolk. 45, 56, [49, 55, 58]. W. Locally abundant on peat in fens, or carr developed over fen or bog, in Britain south of Cumberland and Yorks, and also in Perth and Angus. In Ireland from N. Kerry, Clare and Kildare northwards.

POLYPODIACEAE

POLYPODIUM L.

1. Polypodium vulgare L. sens. lat.   Polypody   Ray, 1660

Very rare on old walls, formerly perhaps more widespread. A specimen in CGE collected at Wood Ditton in 1837 has the morphological characters of the hexaploid ssp. **prionodes** (Aschers.) Rothm. (*P. interjectum* Shivas), as has a specimen from the roof of Milton church collected in 1961, while another from Shudy Camps collected in 1828 shows both tetraploid, ssp. **vulgare**, and hexaploid on the same sheet. Ray's locality (Garret Hostel Lane, Cambridge) survived until 1949, when the old wall was demolished during rebuilding. Unfortunately no specimen of this plant was preserved. 25, 39, *45*, 46, *54*, *41*, [23, 36, 48, 58, 64, 65, 40]. General.

AZOLLACEAE

AZOLLA Lam.

1. Azolla filiculoides Lam.   G. Goode, 1913

A free-floating aquatic fern first found in ditches between Jesus College and Jesus Green, Cambridge, and subsequently recorded

OPHIOGLOSSACEAE

near Waterbeach, at Stretham Pits, and its only extant locality, Clay-hythe. *45, 46*, 56, 57. Naturalized in many places mainly in S. England; native of western N. America.

OPHIOGLOSSACEAE

BOTRYCHIUM Sw.

1†. Botrychium lunaria (L.) Sw.   Moonwort   Ray, 1685
Formerly in a few localities in the south-east of the county; last recorded by Babington (1860). [54, 66.] General, though local or rare everywhere.

OPHIOGLOSSUM L.

1. Ophioglossum vulgatum L. ssp. **vulgatum**   Adder's Tongue
                                     Ray, 1660
Scattered over the county in damp grassland, fens and scrub. 25, 34, *35*, 36, 44–6, *55*, 56, *57*, 64, 65, *66*, [37, 54, 58, 40]. W. General, but rare in Scotland.

SPERMATOPHYTA

GYMNOSPERMAE

PINACEAE

PINUS L.

1. Pinus sylvestris L.   Scots Pine   Shrubbs, 1890
(This is the oldest record, but it must have been known at least as a planted tree much earlier.)
Occurs throughout the county but probably always planted. 23, 25, 26, 34–6, 38, 43–5, 49, 54, 55, 57, 64–6, 68, 20, 40, 41. The dominant tree of many parts of N. Scotland (ssp. *scotica* (Schott) E. F. Warburg); in England usually supposed to be introduced but possibly persisting in small quantity since Boreal times. Planted and naturalized in many places.

(Other species of *Pinus* are planted, particularly **P. nigra** Arnold ('Corsican' and 'Austrian' Pines of the foresters). Other coniferous trees commonly planted in Cambs include **Larix decidua** Mill. (European Larch) and **Picea abies** (L.) Karst. (Norway Spruce). Large coniferous plantations are not found in the county because of the unsuitability of the soils. The sandy Breckland soils of Norfolk and Suffolk, however, carry very large planted coniferous forests.)

CUPRESSACEAE

JUNIPERUS L.

1. Juniperus communis L. ssp. **communis**   Juniper   Ray, 1660
Formerly in a number of localities on the chalk, now only found on the Fleam Dyke (cf. Walters, 1961). 55, [45, 54]. F. Throughout most of Great Britain but local; absent from a number of counties. In Ireland in the west and north, absent from the south-east. The Fleam Dyke locality is the only one now known in East Anglia.

TAXACEAE

Taxus L.

1. Taxus baccata L.   Yew   Ray, 1660

In woods and copses on the chalk, perhaps introduced; also widely planted, especially in churchyards. 25, 34, 35, 39, 44, 45, 49, 56, 64–6, 30, [36, 55]. Native in Britain from Perth and Argyll southwards but rather local and mainly on limestone. In Ireland mainly in the west, rare in the east. It is frequently planted.

ANGIOSPERMAE

DICOTYLEDONES

RANUNCULACEAE

Caltha L.

1. Caltha palustris L. ssp. **palustris**   Marsh Marigold   Ray, 1660

Scattered in marshes, ditches, and other wet places. Two variants occur, one with the perianth segments contiguous and carpels erect when ripe, the other with non-contiguous segments and spreading carpels. 34, 35, 44–6, 48, 54–8, 65–7, [25, 47, 40]. General.

Helleborus L.

1. Helleborus foetidus L.   Stinking Hellebore   Ray, 1663

A rare plant of woodland, probably originally planted. *34*, 35, 44, 45, 54, 55. GG. England and Wales, local; ? naturalized in Scotland.

2. Helleborus viridis L. ssp. **occidentalis** (Reut.) Schiffn.
    Green Hellebore   S. Corbyn, 1656

A rare plant of boulder-clay woods. 25, *34*, 35, *36*, [45, 46]. **B.** England and Wales, local; ? naturalized in Scotland.

Eranthis Salisb.

1. Eranthis hyemalis (L.) Salisb.   Winter Aconite   Shrubbs, 1879

Naturalized in plantations and copses, abundant in a few places. 34, 44, 45, 54, 56, 65, [38]. GG. Naturalized in many parts of Great Britain; native of S. Europe.

41

RANUNCULACEAE

DELPHINIUM L.

1. Delphinium ambiguum L.   Larkspur   J. Sherard, *c*. 1724
D. ajacis auct.; *D. gayanum* Wilmott; *D. consolida* sensu Bab.
Formerly an occasional cornfield weed, now only found as a casual.
44–6, *54*, 64, [34, 35, 38, 56, 66]. Widespread but rare casual; native of the
Mediterranean region.

ACONITUM L.

1. Aconitum napellus L. sens. lat.   Monkshood   Shrubbs, 1895
According to Shrubbs (in notes) this species was cultivated in fields
near Harston and Shepreth from about 1860 to 1890. He records it
from Hauxton (1895) and Shepreth (1910). There are undated specimens in CGE from the back of Hauxton church (coll. unknown) and a
meadow near Foxton (E. M. Walker), presumably an escape from
cultivation. [34, 44, 45.] Native in S.W. Britain and throughout much
of continental Europe.

ANEMONE L.

1. Anemone nemorosa L.   Wood Anemone   S. Corbyn, 1656
Common in boulder-clay woods. 25, 26, 34–6, 43, *44*, 45, 54, 55, 64, 65, [66].
General, but rare in S. Ireland.

PULSATILLA Mill.

1. Pulsatilla vulgaris Mill.   Pasque Flower   J. Gerarde, 1597
Anemone pulsatilla L.
Very local in old chalk grassland; one of the classical rarities of the
Cambridge area, now sadly reduced by ploughing and by thoughtless
destruction. 45, 54–6, 66, [44]. A local plant chiefly of east and central
England.

CLEMATIS L.

1. Clematis vitalba L.   Old Man's Beard   Ray, 1660
Hedgerows, thickets and wood-margins; common on the chalk. 24, 29,
34–6, 38, 43–6, 54–6, 64–7, 76, [57]. Native in lowland England north
to Yorks; naturalized in Ireland.

RANUNCULUS L. (Buttercup, Crowfoot)

1 Plants terrestial or in marshes; flowers yellow; achenes not transversely wrinkled.   2
Plants aquatic or subaquatic; flowers white; achenes strongly
wrinkled transversely.   *13*

2 Sepals 3; petals 7–12.   *3*
   Sepals 5; petals usually 5.   *4*

3 Leaves with no axillary bulbils after flowering.
      17. R. ficaria ssp. **ficaria**
   Leaves with axillary bulbils after flowering.
      17. R. ficaria ssp. **bulbifer**

4 Leaves palmately lobed or divided.   *5*
   Leaves simple, entire or toothed.   *12*

5 Receptacle elongate-cylindrical.   **10. R. sceleratus**
   Receptacle short and convex.   *6*

6 Sepals strongly reflexed during flowering.   *7*
   Sepals not strongly reflexed during flowering.   *9*

7 Stem swollen just below ground surface; achenes smooth.
      3. R. bulbosus
   Stem not swollen; achenes tuberculate.   *8*

8 Flowers 12–25 mm. diam.; achenes with a ring of tubercles near
    the conspicuous green border.   **5. R. sardous**
   Flowers 3–6 mm. diam.; achenes covered with hooked tubercles.
      6. R. parviflorus

9 Plants with long runners; flower-stalk furrowed.   **2. R. repens**
   Plants without runners; flower-stalk not furrowed.   *10*

10 Flowers 18–25 mm. in diam.   **1. R. acris**
   Flowers less than 14 mm. in diam.   *11*

11 Achenes glabrous but spiny; an annual arable weed. **4. R. arvensis**
   Achenes shortly hairy but not spiny; a perennial herb of woodland
    and hedgerow.   **7. R. auricomus**

12 Flowers more than 20 mm. in diam.   **8. R. lingua**
   Flowers 7–18 mm. in diam.   **9. R. flammula** ssp. **flammula**

13 Plant with no finely dissected submerged leaves. **11. R. hederaceus**
   Plant with finely dissected submerged leaves.   *14*

14 Submerged leaves all small, circular in outline, their segments lying
    in one plane.   **13. R. circinatus**
   Submerged leaves with segments not lying in one plane.   *15*

15 Leaves exceeding 80 mm.   **12. R. fluitans**
   Leaves less than 35 mm.   *16*

16 Petals rarely exceeding 5 mm., not contiguous; no floating leaves.
      14. R. trichophyllus
   Petals usually exceeding 5 mm., contiguous; floating leaves usually
    present.   *17*

43

RANUNCULACEAE

17 Achenes glabrous, slightly winged. (**R. baudotii**)
Achenes (especially when immature) with at least some hairs, not
winged. *18*

18 Peduncle in fruit usually less than 50 mm.; petals rarely more than
10 mm.; nectaries circular. **15. R. aquatilis**
Peduncle in fruit usually more than 50 mm.; petals usually more than
10 mm.; nectaries elongated, ± pyriform. **16. R. pseudofluitans**

1. Ranunculus acris L. Meadow Buttercup Ray, 1663
Meadows and roadsides throughout the county, but not as abundant as
R. bulbosus and *R. repens*. All squares. General.

2. Ranunculus repens L. Creeping Buttercup Ray, 1660
An abundant weed of cultivated land, and in waste places, particularly
common in wet grazed or trampled ground. All squares. General.

3. Ranunculus bulbosus L. Bulbous Buttercup Ray, 1660
Very abundant in the south of the county where it is the most common
yellow Buttercup of meadows and pastures; widespread but less abundant in the Fens. All squares except 39, 30. General, but rare in the
north and west.

4. Ranunculus arvensis L. Corn Crowfoot Ray, 1660
A local cornfield weed on the chalk and boulder clay; frequent in the
Fens. 25, 26, 34–7, 45–9, 57, 59, 65, 30, *41*, [38, 58, 40]. General, but
very rare in Scotland and Ireland.

5. Ranunculus sardous Crantz Ray, 1663
R. hirsutus Curt.
Weed of arable land and waste places; no recent records. [35, 45, 46,
55.] A very local and often casual weed, scattered throughout Great
Britain, particularly near the sea.

6. Ranunculus parviflorus L. Ray, 1663
On roadsides, trackways and in waste places; no recent records. [25,
35, 36, 45, 47.] Throughout England and Wales; rare in Ireland; local.

7. Ranunculus auricomus L. Goldilocks Ray, 1660
Woods, copses and hedgebanks, widespread on the boulder clay, rare
elsewhere. An aggregate of apomictic microspecies. The common
plant in Cambridgeshire is almost apetalous, but a petaloid form also
occurs. 25, 26, 34–6, 45, 46, *54*, 64, 65, 76, [38, 47–9, 55–7, 40]. General.

44

8. Ranunculus lingua L.   Greater Spearwort   S. Corbyn, 1656
Marshes and edges of fen lodes; formerly rather widespread in the
Fens; now rare. *55*, 56, 57, [25, 29, 44, 45, 47–9, 58, 67]. **W.** General, but
always very local.

9. Ranunculus flammula L. ssp. **flammula**   Lesser Spearwort
Ray, 1660
Marshes and wet places; said by Babington and Evans to be plentiful
in suitable habitats, now very local. 25, 44, *45*, *49*, 55–7, 66, 67, [38,
47, 48]. **W.** General.

10. Ranunculus sceleratus L.   Celery-leaved Crowfoot   Ray, 1660
By streams, ditches, and at the edges of shallow ponds and marshes,
particularly where disturbed or trampled; rather common. All squares
except 26, 64, [55, 66]. General, but rare in the north and west.

11†. Ranunculus hederaceus L.   Ivy-leaved Water Crowfoot
Ray, 1660
On mud at the edges of ponds and streams, not seen this century.
[25, 35, 44, 45, 55.] General.

The next five species, belonging to the subgen. *Batrachium*, are very difficult
taxonomically. Although some of the data, unsupported by herbarium
material, may be inaccurate, a fair idea of the distribution is given. We are
indebted to Mr C. D. K. Cook for the determination of our herbarium
material of this group.

12. Ranunculus fluitans Lam.   Ray, 1685
Scattered throughout the county in flowing water. 24, 29, 45, 57, 66, 76,
[34, 38, 44, 46–8, 54, 55, 58]. Throughout Great Britain south of the
Clyde; very rare in Ireland.

13. Ranunculus circinatus Sibth.   T. Martyn, 1763
Common in deep fen lodes, but rare in the south of the county. 25, 29,
37–9, *45*, *46*, 47, 48, *49*, 55–7, *65*, *67*, 68, 20, 30, 31, 50, [36, 38, 54, 58].
Scattered throughout the British Isles except the north.

14. Ranunculus trichophyllus Chaix   Ray, 1660
R. drouetii F. W. Schultz
Frequent in ditches and ponds throughout the county. All squares
except 24, 26, *57*, 66, 68, [25]. General.

RANUNCULACEAE

15. Ranunculus aquatilis L.   Ray, 1660

R. heterophyllus Weber; *R. radians* Revel; *R. baudotii* sensu Evans; *R. floribundus* sensu Bab.

Ditches, dykes and ponds throughout the county, but generally less common than *R. trichophyllus*. 25, 29, 35–8, 43, *45*, 46, 48, 49, *56*, 58, 40, *41*, [47, 55, 65, 67]. General.

16. Ranunculus pseudofluitans (Syme) Newbould ex Bak. & Foggitt

                                             Babington, 1855

Chalk streams, local. 34, 44, [54]. Scattered throughout Great Britain south of Edinburgh, and Ireland.

(**R. baudotii** Godr. The plant collected by Babington in a brick-pit by the Ely road to the north of Chesterton was correctly named, but it does not seem to have persisted and may have been accidentally introduced. The plant from Upwell collected by Gilmour and Tutin and recorded as var. *marinus* by Evans is *R. aquatilis*. *R. baudotii* should be looked for around Wisbech.)

17. Ranunculus ficaria L.   Lesser Celandine   Ray, 1660

Common in woodlands, on hedgebanks and by the sides of ditches throughout most of the county, but there are very few records north of Ely. 25, 26, 34–9, 43–8, 54–8, 64–6, 75, 76, 41, [49]. General.

The distribution of the following two subspecies is imperfectly known.

ssp. **ficaria**

R. ficaria var. *fertilis* Clapham nom. nud.

25, 34, 35, *36*, 44, 45, *54*, 56, *65*.

ssp. **bulbifer** Lawalrée

R. ficaria var. *ficaria* sensu Clapham

26, 35, 36, 44–6, 65.

(**R. ficaria** ssp. **ficariaeformis** Rouy & Fouc. occurs in Cambridge gardens where it was presumably planted.)

ADONIS L.

1. Adonis annua L.   Pheasant's Eye   Skrimshire, *c.* 1820

Casual in cornfields and waste places. 26, 35, 45, 65, [23, 40]. Naturalized or a casual in scattered localities in England and rarely Ireland; native of S. Europe and S.W. Asia.

MYOSURUS L.

1. Myosurus minimus L.   Mouse-tail   Ray, 1660
Arable fields, especially on boulder clay; rare. 25, 37, [35, 38, 44–6, 48, 54, 58]. G. Throughout E. and C. England, local.

AQUILEGIA L.

1. Aquilegia vulgaris L.   Columbine   Ray, 1660
Woods and wet places, now rather rare but formerly more widespread. Probably a relic of cultivation or garden escape in all localities except perhaps at Chippenham Fen. 24, 26, 45, 56, 66, 50, [25, 35, 44, 48, 55, 65]. General, but questionably native in some areas.

THALICTRUM L.

1. Thalictrum flavum L.   Meadow Rue   Ray, 1660
Sparingly distributed in relict areas of fen and marsh in the south, but frequent along ditch sides in the Fenland. 29, 38, 39, 45–8, *49*, 55–8, 66–9, 20, 30, 40, 41, 50, [35, 44]. W. Throughout Great Britain north to Inverness; Ireland.

2. Thalictrum minus L. ssp. **minus**   Lesser Meadow Rue   Ray, 1660
T. minus ssp. *montanum* (Wallr.) Clapham; *T. saxatile* sensu Bab.;
    T. babingtonii Butcher
Local in disturbed chalk grassland and on chalky banks. 23, 44, 45, 55, 66, [54]. GG. General, but always local.

BERBERIDACEAE

BERBERIS L.

1. Berberis vulgaris L.   Barberry   Ray, 1660
Occasionally found in hedgerows. 45, 46, *65*, [34, 44, 54–6]. GG. Local and probably native in hedges and scrub in some parts of Great Britain, introduced in Ireland.

MAHONIA Nutt.

1. Mahonia aquifolium (Pursh.) Nutt.   Oregon Grape
                                    J. S. L. Gilmour, 1932
Much planted in dry woodland margins and plantations for pheasant

47

BERBERIDACEAE

cover. 23, *24*, 25, 34, 35, 44, 45, 54–6, 64, *65*, 66, 41. F. Commonly planted and naturalized in a number of places in the British Isles; native of western N. America.

NYMPHAEACEAE

NYMPHAEA L.

1. Nymphaea alba L.   White Water-lily   Ray, 1660

Abundant in some fen lodes, rare elsewhere. 35, 46, 47, 56–9, 40, [25, 34, 44, 45, 48, 55, 66]. W. General.

NUPHAR Sm.

1. Nuphar lutea (L.) Sm.   Yellow Water-lily   Ray, 1660

Abundant in fen lodes and drains, scattered elsewhere. *Nuphar* has large, broadly elliptical floating leaves and translucent, submerged leaves, while *Nymphaea* has large nearly circular floating leaves only. 29, 34, 35, 37–9, 44–9, 56–9, 68, 69, 20, 30, 40, [25, 36]. W. General, but rare in Scotland.

CERATOPHYLLACEAE

CERATOPHYLLUM L.

1. Ceratophyllum demersum L.   Horn-wort   Ray, 1660

Rather local in rivers and lodes, chiefly in the Fens. 29, 37–9, 45, 46, 48, 49, 56–9, 20, 30, 31, 40, 50, [34, 36, 44, 47, 55, 66, 41]. W. Scattered throughout England and Ireland, rare in Wales and Scotland.

(Fryer's record of *C. submersum* L. was an error.)

PAPAVERACEAE

PAPAVER L. (Poppy)

1	Leaves toothed or slightly lobed.	2
	Leaves once or twice pinnately lobed or divided.	3
2	Leaves broad and clasping at base.	**6. P. somniferum**
	Leaves narrow and not clasping at base.	**7. P. atlanticum**
3	Peduncles with spreading hairs.	4
	Peduncles with appressed hairs.	5

4 Hairs of peduncles colourless or yellowish.

             1. P. rhoeas forma **rhoeas**

Hairs of peduncles reddish.   **1. P. rhoeas** forma **pryorii**

5 Petals with black spot at base; capsule covered with stiff hairs or
 bristles.   6
 Petals without black spot at base; capsules glabrous.   7

6 Capsule almost globose with numerous bristles.   **4. P. hybridum**
 Capsule narrowly obovoid-oblong with few bristles. **5. P. argemone**

7 Leaves light green, stiffly hairy, the segments broad with blunt
 lobes; latex white.   **2. P. dubium**
 Leaves dark green, pubescent, the segments narrow with acute
 lobes; latex yellow.   **3. P. lecoqii**

1. Papaver rhoeas L.   Field Poppy   Ray, 1660
Formerly an abundant weed of arable land and in waste places, now
much reduced by weed-killers. The forma **pryorii** (Druce) C. E. Salmon
occurs. All squares. General, but rare in the north and west.

2. Papaver dubium L.   Babington, 1859
Given by Babington only from Chippenham but now recorded from
many places, though not as frequent as *P. rhoeas.* 24–6, 34–6, 38,
44–8, *54,* 56, *57,* 59, 65–7, 40, [41]. General.

3. Papaver lecoqii Lamotte   Babington, 1836
An arable weed especially on the chalk. The distribution of this species
and *P. dubium* has not been satisfactorily worked out. 35, 44, 45, *46,* 47,
56, 65, 68, [34, 36, 38, 39, 48, 54, 57, 66]. GG. Mainly on calcareous
soils in south and east England and C. Ireland.

4. Papaver hybridum L.   S. Corbyn, 1656
A local weed of arable land, mainly on the chalk. 34, 44, 45, *46, 54,*
55, 56, 66, [35, 47, 48, 57]. GG. Rather rare in England, S. Scotland
and Ireland.

5. Papaver argemone L.   S. Corbyn, 1656
A widespread weed of arable land. 25, 29, 33, 34, 44–6, 48, 49, *54,*
55–8, 65, 66, 30, 40, 41, [35, 38, 47]. GG. General, but rare in the north.

6. Papaver somniferum L.   Opium Poppy   Ray, 1660
A widespread casual of waste ground and rubbish heaps, formerly
cultivated in the Fens. All our specimens seem to be referable to ssp.

PAPAVERACEAE

hortense Hussenot. 35, 36, 39, 45, *46*, 55, 56, 58, 59, 41, 50, [38, 66].
Casual or garden escape throughout the British Isles; native of Europe
and Asia.

7. Papaver atlanticum (Ball) Coss.
Garden escape. 45.

ROEMERIA Medic.

1†. Roemeria hybrida (L.) DC.　　Violet Horned Poppy　Ray, 1660
Found by the nineteenth-century naturalists in a few fields between
Swaffham Prior and Burwell. [56, 66.] Formerly a rare arable weed in
E. Anglia, now apparently only a very rare casual; native of C. and
S. Europe.

(Henslow (1829) records **Glaucium flavum** Crantz (*G. luteum* Crantz) as a
native of the county, but gives no locality. The only likely place for it would
have been by the banks of the River Nene north of Wisbech.)

CHELIDONIUM L.

1. Chelidonium majus L.　　Greater Celandine　　　Ray, 1660
Frequent in hedgerows or on banks and walls, chiefly near habitations;
presumably always introduced. All squares except 24, 26, 37, *54*, 59, 68.
Probably native, but also widely planted and found as a garden escape;
in Great Britain southwards from Sutherland, and in Ireland.

ESCHSCHOLTZIA Cham.

1. Eschscholtzia californica Cham.
Garden escape. *45*.

FUMARIACEAE

CORYDALIS Medic.

1. Corydalis solida (L.) Sw.
Garden escape. *45*.

2. Corydalis lutea (L.) DC.　　Yellow Fumitory　Skrimshire, 1818
Garden escape. 25, 35, 36, 39, 44–6, *47*, 55, 58, 67, [40]. Scattered
throughout the British Isles; native of S. Europe.

FUMARIACEAE

FUMARIA L. (Fumitory)

1 Flowers more than 9 mm. in length.　　　　**1. F. muralis** ssp. **boraei**

　　Flowers not over 9 mm. in length.　　　　　　　　　　　　　　　　*2*

2 Sepals more than 2 × 1 mm.; flowers at least 6 mm.　　　　　　　*3*

　　Sepals not over 1·5 × 0·75 mm.; flowers 5–6 mm.　　　　　　　　*5*

3 Bracts longer than fruit-stalks; fruit rounded at apex; leaf-segments
　　channelled; sepals almost as broad as long.　　　**2. F. micrantha**

　　Bracts shorter than fruit-stalks; fruit truncate or emarginate at
　　apex; leaf segments flat; sepals about ¼ as broad as long.　　　*4*

4 Plant robust; racemes many-flowered; sepals 2–3·5 × 1–1·5 mm.;
　　fruit distinctly broader than long.

　　　　　　　　　　　　　　3. F. officinalis ssp. **officinalis**

　　Plant more slender; racemes fewer-flowered; sepals 2 × 1 mm.; fruit
　　about as broad as long.　　　　**3. F. officinalis** ssp. **wirtgenii**

5 Bracts shorter than fruit-stalks; leaf-segments flat; flowers pink;
　　racemes lax, short-stalked.　　　　　　　　　　　　　　　　　*6*

　　Bracts about equalling fruit-stalks; leaf-segments channelled, very
　　narrow; flowers white or pink-flushed; racemes dense, c. 20-
　　flowered, subsessile.　　　　　　　　　　　　　　　　　　　*7*

6 Leaves 2-pinnatisect, with distant leaflets; racemes 6–12-flowered;
　　pedicels short, suberect or erect-spreading.

　　　　　　　　　　　　　　4. F. vaillantii var. **vaillantii**

　　Leaves 2–3-pinnatisect, with closer leaflets; racemes 10–16-flowered;
　　pedicels longer and more erect.　　**4. F. vaillantii** var. **chavinii**

7 Leaflets linear-acute; sepals 1 × 0·6 mm.; fruit suborbicular,
　　rounded-obtuse, the keel drawn into a short but persistent
　　apiculus.　　　　　　　　　　**5. F. parviflora** var. **parviflora**

　　Leaflets short, thick and sometimes divaricate; sepals 1·5 × 0·75
　　mm.; fruit suborbicular, obtuse with the well-marked keel drawn
　　into an extremely short, blunt and notched beak.

　　　　　　　　　　　　　　5. F. parviflora var. **symei**

(**F. capreolata** L. All the old records are probably erroneous, but the
species occurred as a weed in the Botanic Garden, Cambridge, in 1931. The
plant from Elm referred to this species by Babington is **F. muralis** ssp. **boraei**
(Jord.) Pugsl.

F. bastardii Bor. The plant recorded from Chatteris by Fryer is also
F. muralis ssp. **boraei** (Jord.) Pugsl.)

1. Fumaria muralis Sond. ex Koch ssp. **boraei** (Jord.) Pugsl.

　　　　　　　　　　　　　　　　　　　　A. Peckover, 1860

F. capreolata sensu Bab. pro parte

FUMARIACEAE

A rare weed of arable land in the Fens. 68, [38, 40]. General, but commonest in the west.

2. Fumaria micrantha Lag.   Henslow, 1831

A local weed of arable land. 23, *24*, 33, 34, 44, 45, 55, 56, 68, [48, 54, 67]. Scattered throughout Great Britain and the northern half of Ireland, more common in the east than the west.

3. Fumaria officinalis L.   Ray, 1660

A common weed of arable land and waste places; the ssp. **wirtgenii** (Koch) Arcang. is the commonest subspecies. All squares except 26, 29, 37, 48, 59, [38, 54]. General.

4. Fumaria vaillantii Lois.   Henslow, 1831

A local weed of arable land entirely restricted to the chalk. The var. **chavinii** (Reut.) Rouy & Fouc. has been recorded. 34, 35, 44–6, 55, 66, 67, [24, 54, 56]. GG. Very local in the southern half of England.

5. Fumaria parviflora Lam.   Henslow, 1826

A rare arable weed now almost entirely confined to the chalk. The var. **symei** Pugsl. occurs, and is said by H. W. Pugsley to be especially characteristic of the Gog Magog Hills. 44, 45, 54–6, [24, 34, 54, 58, 66, 76]. GG. A rare plant of England and S.E. Scotland.

CRUCIFERAE

BRASSICA L.

1. Brassica oleracea L.   Ray, 1660

A relic of cultivation almost certainly occurring more frequently than has been recorded. The var. **oleracea** (Cabbage), var. **acrocephala** DC. (Kale), var. **gemmifera** Zenker (Brussels Sprouts), var. **capitata** L. (Savoy Cabbage), var. **botrytis** L. (Cauliflower), var. **italica** Plenck (Sprouting Broccoli) and var. **gongylodes** L. (Kohlrabi) are all grown as crops. Over 2500 acres of Brussels Sprouts are grown, mainly in the south of the county, while 1850 acres of Kale are fairly widely distributed. The rest are much less cultivated as large crops, but are frequent in gardens. 34, 45, 55. Native of maritime cliffs in south and south-west England, and Wales; elsewhere as a relic of, or escape from, cultivation.

52

CRUCIFERAE

2. Brassica napus L.   Ray, 1660

Cultivated as 'Rape', var. **napus** (1700 acres), and 'Swedish Turnip' or 'Swede', var. **napobrassica** (L.) Rchb. There are several old records for this species and Evans (1939) gives it as a mere relic of cultivation. It is now less extensively grown as a crop but absence of modern records is probably due to lack of attention by botanists. [45, 46, 55.] Cultivated in many parts of the world, where it can be found as an escape or naturalized; origin unknown.

3. Brassica rapa L.   Ray, 1660

Var. **rapa** is the cultivated turnip found as a crop throughout the county, while wild plants are var. **campestris** (L.) Koch with a non-tuberous tap-root. The scarcity of records must be due to lack of attention. 34, 39, 45. Found throughout Europe, N. Africa, W. Asia and China; possibly originated from N. and C. Europe.

4. Brassica elongata Ehrh.

Casual. [66.]

5. Brassica juncea (L.) Coss. & Czern.

Casual, occasionally cultivated as a mustard crop. 67.

6. Brassica nigra (L.) Koch   Black Mustard   Ray, 1660

Given by Babington (1860) for a few localities; said by Evans to be 'fairly common on peat and clay field-sides in 1875, but now (1939) a rare casual'. Now widespread in the Fens in waste places and by field-sides; rare elsewhere. 29, 34, 46–8, 56, 57, 64, 68, 69, 20, 31, 41, 50, [35, 38, 45, 49, 55, 58, 40]. Probably native on cliffs by the sea and on stream banks. Throughout England and Wales, but common, probably as an escape, in waysides and waste places. Only in the S. of Scotland and S. and E. Ireland. Long cultivated for its seeds which yield the black mustard of commerce and also an oil used in medicine and soap-making.

ERUCASTRUM C. Presl

1. Erucastrum gallicum (Willd.) O. E. Schulz   W. Marshall, 1885

Brassica erucastrum sensu Evans; *E. pollichii* Schimper & Spenner

Naturalized on the Devil's Dyke since 1885; casual elsewhere. *55, 56, 58*, 66. D. A frequent casual; native of W. and C. Europe.

53

CRUCIFERAE

SINAPIS L.

1. Sinapis arvensis L.　　Charlock　　　　　　　　Ray, 1660
Brassica arvensis (L.) Rabenh., non L.
Common weed of arable land and waste ground, although in recent years much reduced by selective weed-killers. All squares except 30, 41. General.

2. Sinapis alba L.　　White Mustard　　　　　　　Ray, 1660
Brassica alba (L.) Rabenh.
A common weed of arable land and waste places. It is often a relic of cultivation as it is grown for both green fodder and green manure, as well as mustard derived from its seeds. 23, 24, 34–6, 38, 43–6, 48, 54–8, 65, 68, 31, 40, [47, 49, 66]. General, introduced; probably native of Mediterranean region.

DIPLOTAXIS DC.

1. Diplotaxis muralis (L.) DC.　　Wall Rocket　　　Babington, 1846
A weed of waste and cultivated ground and on walls, widespread but rarely abundant. Biennial or perennating plants are recorded with numerous leaves on their branching stems; they are referable to var. **caulescens** Kitt. (var. *babingtonii* (Syme) Marquand). Care should be taken not to confuse non-fruiting specimens of this variety with *D. tenuifolia*. 25, 29, 33–9, 44–6, 48, 49, 55, 56, *58*, 66–8, 76, 40, 50, [54, 57, 65]. Naturalized, especially in S. England; by railways in Ireland; native of S. and C. Europe.

2. Diplotaxis tenuifolia (L.) DC.
　　　　　　　　　　　　　　P. W. Richards & J. Rishbeth, 1946
In a few places on the chalk. Presumably recently introduced, as all its localities have long been famous botanically and it would surely have been noticed earlier if present. 35, 44, 45, 55, 66. **D.** Doubtfully native in S. England, casual farther north and in Scotland and Ireland; native in S. and C. Europe.

ERUCA Mill.

1. Eruca sativa Mill.
Casual. [45.]

RAPHANUS L.

1. Raphanus raphanistrum L.　　Wild Radish　　　　Ray, 1660
Scattered throughout the county as a weed of arable land and waste places, but never abundant. Both white- and yellow-flowered varieties occur. 24, 26, 29, 34–9, 44–7, 49, 54, 56, 65–7, 30, 31, 40, 50, [25, 48]. General, doubtfully native.

RAPISTRUM Crantz

1. Rapistrum perenne (L.) All.
Casual. 45, [54].

2. Rapistrum rugosum (L.) All.
Casual. 24, 45, [44, 56, 67].

CONRINGIA Adans.

1. Conringia orientalis (L.) Dumort
Erysimum orientale (L.) Crantz, non Mill.
Casual. [44–6, 48.]

LEPIDIUM L.

1. Lepidium sativum L.　　Garden Cress
Casual or garden escape. 45, 56.

2. Lepidium campestre (L.) R.Br.　　Pepperwort　　Henslow, 1829
In a few localities in arable land and waste places. 29, 38, 47, 54, 55, [45, 48, 49, 56, 57, 65]. Throughout most of the British Isles though absent from N. Scotland and rare in S. Scotland and Ireland.

(**L. smithii** Hook. The plant recorded by A. Bennett and Fryer from Chatteris (1884) was later, according to Evans (1939), stated by them to be *L. campestre*.)

3†. Lepidium ruderale L.　　　　　　　　　T. Martyn, 1763
Known to the eighteenth- and early nineteenth-century botanists around Wisbech where it was last recorded by Babington (1860). In 1914, Moss recorded it from Chesterton where it was probably casual. [46, 40, 41.] Native in waste places and by waysides, particularly near the sea throughout England, but especially in East Anglia and around the Thames estuary. Rare and doubtfully native in Scotland; casual in Ireland.

CRUCIFERAE

4. Lepidium neglectum Thell.

Casual. *45.*

5. Lepidium latifolium L.   Dittander   P. Dent in Ray, 1685

P. Dent recorded this species from the Maids' Causeway, where it was probably a garden escape. Skrimshire's record from near Wisbech in 1795 is possibly in a more natural habitat. In 1944 H. Gilbert-Carter recorded the plant growing by the railway on a grassy bank at the edge of Homerton College playing fields, Cambridge; it still survives there. 45, [41]. **Cam.** Native in salt-marshes and wet sand in N. E. England and from Norfolk and Wales southwards, and S. Ireland. Formerly cultivated as a condiment.

6. Lepidium perfoliatum L.

Casual. [36.]

7. Lepidium graminifolium L.

Casual. [45.]

CORONOPUS Zinn.

1. Coronopus squamatus (Forsk.) Aschers.   Swine-cress

S. Corbyn, 1656

Senebiera coronopus (L.) Poir.; *Coronopus procumbens* Gilib.

A common weed of waste ground, especially trampled places such as tracks, gateways, and farmyards. All squares except 54. Common in S. England but infrequent in the north although reaching Inverness and Ross; Ireland.

2. Coronopus didymus (L.) Sm.   Lesser Swine-cress   Shrubbs, 1888

A rare weed of waste and cultivated ground recorded from the Botanic Garden, Cambridge, Shelford, and Thriplow. 44, 45. An introduced weed widespread in S. England and extending north to Ross; also in Ireland. Probably native only in S. America.

CARDARIA Desv.

1. Cardaria draba (L.) Desv.   Hoary Cress

W. W. Newbould, 1857

Since its introduction in Cambs, just over a hundred years ago, this plant has spread rapidly and is now a common weed of roadsides, waste places and field margins. Specimens collected at Cherry Hinton in 1955 with rounded or truncate, not cordate, fruits have been determined as ssp. **chalepensis** (L.) Thuill. (*C. chalepensis* (L.) Hand.-Mazz.). All squares except 26, 66–8, [64]. Spreading rapidly as a weed of arable land and now throughout England and Wales; more rarely in Scotland and Ireland.

ISATIS L.

1. Isatis tinctoria L.   Woad   Ray, 1669
Formerly grown in the Fens for the blue pigments (woad) formed when the partially dried leaves are crushed into a paste and exposed to the air. Ray tells us it was grown at Littleport and Relhan at Ely, while Evans reports that Woad Mills were once a feature of the Fen country at Parson's Drove near Wisbech, being in full swing in 1908, but out of use by the time of the Great War in 1914. [58, 30, 40.] Naturalized on cliffs in the Severn Valley and in a chalk-pit at Guildford; native of C. and S. Europe.

IBERIS L.

1. Iberis amara L.   Candytuft   W. H. Coleman, 1837
Formerly in a number of localities on the chalk, now known on the edge of Morden Grange Plantation, and on banks of railway cuttings in the neighbourhood of Royston, where it is locally abundant. 33, 34, [23, 24, 35, 45, 57]. Local on chalk and limestone in S. England, casual elsewhere.

2. Iberis umbellata L.
Garden escape. 67, [45].

THLASPI L.

1. Thlaspi arvense L.   Field Penny-cress   Relhan, 1785
A widespread but not abundant weed of arable land and waste places. 25, 29, 34, 38, 39, 44–8, 56, 57, 58, 64, 65, 68, 20, [54, 55, 66, 40]. General, but rare in the north and west.

TEESDALIA R.Br.

1. Teesdalia nudicaulis (L.) R.Br.   Shepherd's Cress   Ray, 1660
Small annual herb of sandy or gravelly soil germinating in autumn and overwintering as a rosette. Reported by Relhan from near Litlington

CRUCIFERAE

and known to Babington and Evans at Gamlingay. It is now known only from a few localities in the Chippenham and Kennett areas. 66, 76, [25, 34]. **K.** Throughout Great Britain northwards to Ross, but rare in Scotland; rare and local in N. Ireland.

CAPSELLA Medic.

1. Capsella bursa-pastoris (L.) Medic.   Shepherd's Purse
Ray, 1660
An abundant weed of arable land, waste places and waysides. Very variable with a strong tendency for distinctive local populations to arise because of self-pollination. All squares. General.

COCHLEARIA L.

1. Cochlearia danica L.   E. G. Jefferys, 1946
Found as a casual on railway tracks by Hayley Wood in 1946. *25.* Native on sandy and rocky shores, or walls and banks by the sea all round the British Isles.

2. Cochlearia anglica L.   Skrimshire in Relhan, 1802
Found only by the Nene at Foul Anchor where it is abundant. 41. On muddy shores and estuaries all round the British Isles.

NESLIA Desv.

1. Neslia paniculata (L.) Desv.
Casual. [44, 58.]

BUNIAS L.

1. Bunias erucago L.
Casual. *58.*

2. Bunias orientalis L.   Henslow, 1825
Rare alien persisting at Cambridge and by the Devil's Dyke. 44, 45, 66.

LUNARIA L.

1. Lunaria annua L.   Honesty
Frequent garden escape. 34, *35*, 36, 45.

ALYSSUM L.

1. Alyssum alyssoides (L.) L.   Babington, 1855
A. calycinum L.

There are a few records in the nineteenth century from sandy fields in the east of the county where it may have been native. Elsewhere a casual or garden escape. [45, 54–6, 67.] Rare in scattered localities throughout S. and E. England and E. Scotland northwards to Moray, generally only a casual. Native throughout Europe (although said to be introduced in the N.W.) and W. Asia.

LOBULARIA Desv.

1. Lobularia maritima (L.) Desv.
Alyssum maritimum L.
Garden escape. 45, [46].

BERTEROA DC.

1. Berteroa incana (L.) DC.
Casual. *45.*

EROPHILA DC.

1. Erophila verna (L.) Chevall. sens. lat.   Whitlow Grass
                                         Ray, 1660
Draba verna L., incl. *E. boerhavii* et *praecox* sensu Evans
Common as a weed of arable land and waste places on the dry sandy soils at Gamlingay and around Chippenham; elsewhere mainly on walls, railway tracks and dry, disturbed ground. An extremely variable species, the variants coming more or less true to seed owing to prevalent self-pollination. Of the three species described in Clapham, Tutin & Warburg (1952) specimens have been seen from Cambs referable to E. **verna** (L.) Chevall. sens. strict. and to **E. spathulata** Lang, but not to **E. praecox** (Stev.) DC. 23, 25, 33–9, 44–7, *49,* 54–6, 64–7, 76, *40,* 50, [48, 57]. General.

ARMORACIA Gilib.

1. Armoracia rusticana Gaertn., Mey. & Scherb.   Horse-radish
                                         Ray, 1660
Cochlearia armoracia L.
Naturalized in fields, waste places and roadsides. All squares except

CRUCIFERAE

66, [54]. General, but rare in the north and west; introduced. Cultivated for the condiment prepared from the roots.

CARDAMINE L.

1. Cardamine pratensis L.   Cuckoo-flower; Lady's Smock   Ray, 1660
In wet pasture and by the edges of pits, rivers and lodes. A plant occurs with larger leaflets and large white flowers (var. *dentata* sensu Bab., et Evans). 25, 29, 34–7, 39, 44–6, *47*, 48, 49, 56–9, 65, 66, 68, [38, 40]. General.

2†. Cardamine amara L.   W. Sargent, 1861
No recent records; formerly known in a few localities in wet places. Last seen by Fryer at Sutton in 1879. It is still locally common at St Ives in Huntingdonshire and also just over the Suffolk border in the Icklingham area. [45–8.] Throughout England except the S.W., very rare in Wales; in Scotland absent from the N.W.; only in N.E. Ireland.

3. Cardamine flexuosa With.
                        Card index of the County High School for Boys, 1905
Walls and river banks in a few localities, rare. *45*, 46, 54. General.

4. Cardamine hirsuta L.   Skrimshire in Relhan, 1802
A local weed of cultivated land and waste places, or on walls; almost entirely confined to the Fens and sands. 38, 44, 45, *49*, *54*, 55, 56, 65, 66, 40, 50, [25, 35, 48, 58, 76, 41]. General.

BARBAREA R.Br.

1. Barbarea vulgaris R.Br.   Yellow Rocket   Ray, 1660
Frequent on stream banks, in hedgerows, and in waste places, but unrecorded for large areas of the Fens, and probably absent from the silts. 24, 25, 29, 34–7, 39, 43–6, 48, 54, *55*, 56–9, 64–6, 75, 40, [47, 49, 50]. General in lowlands.

2. Barbarea stricta Andrz.
Casual. 46, [?38, 45].

3. Barbarea intermedia Bor.
A. O. Chater, F. H. Perring and S. M. Walters, 1955
Well established at Long Drove near Over in 1955. 37. A local species
occurring throughout Great Britain except the extreme north, and in
Ireland.

4. Barbarea verna (Mill.) Aschers.
Garden escape. [44.]

ARABIS L.

1. Arabis turrita L.   J. Andrews and S. Dale, 1722
Naturalized on old walls in St John's and Trinity College grounds and
still to be found in the 'Wilderness' of the former. First recorded by
S. Dale in 1722 to whom it was shown by J. Andrews. 45. Formerly
also naturalized on old walls at Oxford and Cleish Castle, Kinross;
now apparently extinct except in Cambridge. Native in C. and
S. Europe.

2. Arabis hirsuta (L.) Scop.   Hairy Rock-cress   T. Martyn, 1763
Uncommon, mainly on disturbed peaty or chalky soils, particularly
by droves in the Fens; also on walls. *44*, 45, 54–7, 65–7, *40*, 41, [35, 39,
48, 68]. General.

TURRITIS L.

1. Turritis glabra L.   Tower Mustard   Evans, 1912
Arabis glabra (L.) Bernh.
A patch was found by Evans in 1912 on the side of a lane at Gamlingay
where it was still persisting in 1952. 25. A very local plant of England
and Scotland north to Perth; absent from Wales and Ireland.

RORIPPA Scop.

1. Rorippa nasturtium-aquaticum (L.) Hayek sens. lat.   Watercress
                                           Ray, 1660
Nasturtium officinale sensu Bab., et Evans
Common throughout the county in running water. All squares except
26, 41. General.

61

CRUCIFERAE

The following segregates are recorded.

R. nasturtium-aquaticum (L.) Hayek sens. strict.   A. Hosking, 1903
Nasturtium officinale R.Br.; et sensu Bab., et Evans pro parte
Common in streams or other running water. Cultivated as 'Green or
Summer-cress' and propagated by cuttings even though it produces
good seed. 29, 34, *35*, 38, 44–6, 48, 49, 54–8, 65–7, 69, 20, 30. Through-
out Great Britain and N. Ireland.

R. microphylla (Boenn.) Hyland.   Shrubbs, 1885
Nasturtium officinale sensu Bab., et Evans pro parte; *N. microphyllum*
(Boenn.) Rchb.
Apparently an allotetraploid hybrid of *R. nasturtium-aquaticum* and
some other species, usually flowering about 2 weeks later than the
former. Widespread in streams but absent from much of the north of
the county and more local than *R. nasturtium-aquaticum*. 24, 25, 34,
37, 38, 43–7, 55–9. General.

× **nasturtium-aquaticum**   H. W. Howard, 1946
Nasturtium microphyllum × *officinale*; *R.* × *sterilis* Airy Shaw
The triploid hybrid which is cultivated as 'Brown- or Winter-cress'. Its
fruit is much deformed and dwarfed and has an average of less than one
good seed per fruit. The plant spreads vegetatively and in cultivation is
propagated by cuttings. Both species and this hybrid grow together
in ditches on Coldham's Common, Cambridge. 45, 56. General,
especially in the north and west.

2. Rorippa sylvestris (L.) Bess.   S. Corbyn, 1656
Nasturtium sylvestre (L.) R.Br.
Frequent in the Fens on moist ground, rare as a garden weed; absent
from the dry areas of the south. 29, 37, 38, 43, 45–8, 58, 30, 40, [36, 49,
56, 65, 67]. Throughout Great Britain north to Argyll and Angus; in
most of Ireland but rare in the west.

3. Rorippa islandica (Oeder) Borbás   Relhan, 1786
Nasturtium palustre (L.) DC., non Crantz
Absent from most of the south of the county; occasional on open peaty
ground in the southern half of the Fens, rarer in the north. It occurs on
the stonework of the river locks at Baitsbite, north of Cambridge.
37, 39, 44–6, 48, *49*, 54, 56, 58, 67, 40, [47, 57, ?30]. General, but rare in
the north.

62

4. Rorippa amphibia (L.) Bess.   S. Corbyn, 1656

Nasturtium amphibium (L.) R.Br.; *Armoracia amphibia* (L.) G. F. W. Mey.

Absent from the south of the county; local by ponds, streams and rivers in the southern half of the Fens, rare in the north. 36, 37, 44, 45, *46*, 47, 48, 54, *57*, 58, 59, 30, [29, 56, 40]. Local from Somerset and Kent north to Westmorland and Berwick, rare farther north; Ireland.

HESPERIS L.

1. Hesperis matronalis L.   Dame's Violet

Frequent garden escape. 35, 44, 45, 55, 64, 66.

ERYSIMUM L.

1. Erysimum cheiranthoides L.   Treacle Mustard   Ray, 1660

A widespread weed of arable land and waste places, common in the Fens. 29, 34, 36–9, 44–9, 56–9, 65, 66, 68, [25]. Locally common in S. England, rarer in north; reaching Caithness but absent from much of Scotland; rare in Wales and W. Ireland. Probably introduced.

CHEIRANTHUS L.

1. Cheiranthus cheiri L.   Wallflower   Ray, 1660

Escape from cultivation sometimes naturalized on walls. 45, 56, 58, [34, 38, 46, 55, 40]. Introduced and well established on walls throughout lowland Britain, widespread in Ireland. Probably native of the E. Mediterranean region.

ALLIARIA Scop.

1. Alliaria petiolata (Bieb.) Cavara & Grande   Jack-by-the-Hedge
Ray, 1660

Sisymbrium alliaria (L.) Scop.; *Alliaria officinalis* Andrz. ex Bieb.

Common in hedgerows, waste places and wood margins. All squares except 57, 59, 30, [40]. Great Britain northwards to Ross, and in Ireland.

SISYMBRIUM L.

1. Sisymbrium officinale (L.) Scop.   Hedge Mustard   Ray, 1660

Common in hedgebanks, by roadsides, in waste places and as a weed of arable land. All squares except 26. General in the lowlands.

63

CRUCIFERAE

2. Sisymbrium irio L.   London Rocket   Skrimshire, 1797

Recorded by Skrimshire from Wisbech in 1797 and by J. Watson from Barnwell in 1818. There are no specimens to support these records. [46, 40.] Roadsides, walls and waste places in various parts of Great Britain and around Dublin in Ireland; doubtfully native. Probably native in the Mediterranean region.

3. Sisymbrium orientale L.   1928 (spec. in CGE, no collector given)

Frequent casual in waste places. 29, 35, 45, *46*, *58*, 20, 40, 41. Well-established alien especially in S. England; native of S. and S.E. Europe.

4. Sisymbrium altissimum L.   C. E. Moss, 1914

Frequent casual. 29, 44, 45, *46*, 67, 76. Well-established alien, especially in England; native of E. Europe and the Near East.

ARABIDOPSIS (DC.) Heynh.

1. Arabidopsis thaliana (L.) Heynh.   Thale Cress   Ray, 1663

Sisymbrium thalianum (L.) Gay

An abundant weed of arable land and waste places on the sands around Chippenham; elsewhere local. 25, 33, 38, 44, 45, 54–6, 65–7, 76, 20, 41, 50, [46–9, 58]. General, especially on dry soils.

CAMELINA Crantz

1. Camelina sativa (L.) Crantz   Gold of Pleasure   Skrimshire, 1796

Mentioned by Babington as occasionally found in crops and by Evans as occurring fairly often about horse-pickets during the Great War (1914–18). It is now a very rare casual. *45*, *46*, 54, [48, 58, 66, 40]. An introduced weed throughout Britain northwards to Stirling and Fife, and in Ireland. Probably native of E. Europe and W. Asia.

DESCURAINIA Webb & Berth.

1. Descurainia sophia (L.) Webb ex Prantl.   Flixweed   Ray, 1660

Sisymbrium sophia L.

An abundant weed in arable land and waste places on the sands around Chippenham, frequent in the Fens, elsewhere widespread but never abundant and usually a casual of waste places. 23, 29, 38, 39, 44, 45, *46*, 48, 49, 55, 56, *57*, 59, 65–9, 40, 50, [34, 35, 47, 58]. Widespread in the east of the British Isles but rare in Wales, and absent from W. Scotland and from all but the eastern Irish counties.

RESEDACEAE

RESEDA L.

1. Reseda luteola L.   Dyer's Rocket; Weld   Ray, 1660
In waste places, disturbed chalk grassland and by chalk-pits, almost as
widespread as *R. lutea* but less abundant and rarely found as an arable
weed. 25, 26, 29, 34, 35, 44–9, 55–7, 64, 66, 67, 76, 40, [41]. General,
though rare in the north.

2. Reseda lutea L.   Wild Mignonette   S. Corbyn, 1656
A common weed of arable land and waste places especially on the
chalk. 23–5, 29, 33–5, 38, 43–6, 48, 54–8, 65–7, 30, 40, [47, 49]. Great
Britain northwards to S. Scotland and probably adventive farther
north; also in Ireland.

VIOLACEAE

VIOLA L. (Violet, Pansy)

1 Style hooked or obliquely truncate; stipules entire to fimbriate, not
   lobed; lateral petals spreading ± horizontally ('Violets').   *2*
   Style expanded above into a globose head; stipules pinnatifid or
   palmatifid; lateral petals directed towards the top of the flower
   ('Pansies').   *11*

2 Sepals obtuse; plant acaulous (i.e. leaves and flowers all radical);
   petioles and capsules pubescent.   *3*
   Sepals acute; plant normally caulescent though in small forms
   appearing acaulous; petioles and capsules glabrous.   *7*

3 Plant with long stolons; flowers sweet-scented; hairs on petioles
   deflexed.   *4*
   Plant without stolons; flowers odourless; hairs on petioles
   spreading.   *6*

4 Flowers violet.   **1. V. odorata** var. **odorata**
   Flowers white with coloured spur.   *5*

5 Inner side of petals bearded at base. **1. V. odorata** var. **dumetorum**
   Inner side of petals not bearded at base.
                                      1. V. odorata var. **imberbis**

6 Corolla 10–15 mm., not cross-like, spur much longer than appendages; usually flowering in April.   **2. V. hirta** ssp. **hirta**
   Corolla not over 10 mm., spur shorter or scarcely longer than
   appendages; usually flowering in late April–May.
                                      2. V. hirta ssp. **calcarea**

VIOLACEAE

7 Main axis ending in a rosette of leaves, not growing out into a
flower stem; leaves ovate-rotund; teeth of stipules usually filiform, flexuous, spreading.   8
Main axis without basal rosette, growing out into a flower stem;
leaves ovate to lanceolate; teeth of stipules triangular-subulate,
±straight, ascending.   9

8 Corolla blue-violet, spur paler, stout, usually furrowed or
notched; appendages of sepals generally rather long and ac-
crescent in fruit (more than 1 mm.).
                                      3. V. riviniana ssp. **riviniana**
Corolla lilac, spur darker, slender, not furrowed or notched; appendages of sepals generally short and remaining very small in
fruit.   **4. V. reichenbachiana**

9 Spur short, ±conical, not or scarcely longer than the appendages
of calyx; plant with slender underground rhizome; leaves thin in
texture, ovate-lanceolate, rarely subcordate.   **6. V. stagnina**
Spur about twice as long as the appendages of calyx; rhizome
relatively short, thick; leaves thick, usually cordate at base.   10

10 Plant not over 10 cm.; corolla 7–18 mm., deep or bright blue;
margin of leaf concave towards the apex; heath and dry grassland.   **5. V. canina** ssp. **canina**
Plant often more than 15 cm.; corolla 15–22 mm., pale blue; margin of leaf straight or convex; fens.   **5. V. canina** ssp. **montana**

11 Petals longer than sepals, often purplish.   **7. V. tricolor**
Petals shorter than or equalling sepals, yellowish white.
                                            8. V. arvensis

1. Viola odorata L.   Sweet Violet   Ray, 1660
Common in woods, copses, plantations and on banks by hedges and
ditches, on the chalk and boulder clay; local in the Fens and on the
sands; undoubtedly often planted. Var. **odorata** and var. **dumetorum**
(Jord.) Rouy & Fouc. are probably equally common, while var.
imberbis (Leighton) Henslow is rare. 23, 24, 25, 26, 34–6, 39, 43–7, 49,
54–8, 64, 65, 75, 76, 20, 40, 41, [38, 48, 66]. In Britain from Dunbarton and Banff southwards, and all over Ireland.

2. Viola hirta L.   Hairy Violet   Ray, 1688
ssp. **hirta**
Frequent in hedgerows, plantations and on open grassland on the
chalk, less so on the boulder clay and nearly absent from the Fens.
23–6, 33–6, 43–6, 54–6, 64–6, 67, [47]. In Britain from Kirkcudbright
and Kincardine southwards on suitable soils; rare in Ireland.

ssp. **calcarea** (Bab.) E. F. Warb.   Henslow, 1827

V. hirta var. *calcarea* Bab.; *V. calcarea* (Bab.) Greg.

Usually flowering 1 or 2 weeks later than ssp. *hirta*; on the chalk, rare. 45, *55*, [66]. GG. Calcareous pastures from Cornwall and Kent to Glamorgan, N. Lancs, and Yorks, local.

Viola hirta × **odorata** (*V.* × *permixta* Jord.). Fertility variable, usually partially sterile. *24*, 45, 55, [25, 35].

3. Viola riviniana Reichb. ssp. **riviniana**   Henslow, 1826

Common in boulder-clay woods. 25, 26, 34–6, 43, 44, *49*, 54, 55, 64–6, *67*, 75, 40. Throughout the British Isles in woods and hedgebanks; replaced in more open habitats by ssp. *minor* (Murbeck) Valentine.

4. Viola reichenbachiana Jord. ex Bor.   Babington, 1848

V. sylvatica sensu Bab. pro maxima parte

Abundant in woods and copses on the boulder clay, rare elsewhere. Flowers 2–3 weeks earlier than *V. riviniana*, a closely allied species with which it was long confused. 25, 33–6, 43–5, 54, 55, 64, 65, 75, 76, [46, 49]. Rather common in S., C. and E. England, local in Scotland, Wales and Ireland.

× **riviniana** (*V. intermedia* Reichb., non Krock). Probably frequent where the parents occur together, and not entirely sterile. *25*, *26*, *35*, *64*.

5. Viola canina L.   Dog Violet   L. Jenyns, 1824

V. flavicornis auct.

The plants long known at Newmarket Heath, Quy Fen, and Gamlingay (both the latter now probably extinct) are referable to ssp. **canina** (var. *ericetorum* sensu Evans). According to material in CGE, the plants formerly known in a number of localities in the Fens are probably referable to ssp. **montana** (L.) Hartm. (var. *crassifolia* sensu Evans); they are not, however, as extreme as those from Woodwalton Fen, Hunts. P. M. Hall, when determining the Cambs specimens as var. *crassifolia*, made the following note: 'From the Grunty Fen and Gamlingay specimens it appears that *canina* occurs as var. *ericetorum* [ssp. *canina*] on the margins of the fens, and as var. *crassifolia* [? ssp. *montana*] in the actual fens.' 25, 56, 66, [38, 47–9, 58]. General, but local.

× **riviniana**. Recorded from Gamlingay, 25.

× **stagnina** (*V. ritschliana* W. Becker; *V. lactea* sensu Fryer). There are two specimens of this hybrid in CGE, one collected by Babington at

VIOLACEAE

Bottisham Fen in 1851 and the other by Fryer at Chatteris in 1885. The Fryer specimen was originally named *V. lactea*, but was determined as *V. canina* × *stagnina* by P. M. Hall. [38, 56.]

6†. Viola stagnina Kit.   Fen Violet   Babington, 1829
Known to Babington (1860) in a number of localities in the Fens, and listed by him as 'abundant' at Wicken Fen. Last seen at Sutton Dole Fen by Fryer in 1895. Still grows at Woodwalton Fen, Hunts, and at one place in W. Suffolk. [46, 47, 56, 58, ?66.] Fens in a few counties in E. and C. England and in damp grassy hollows on limestone in a few counties in Ireland, mainly in the valley of the River Shannon.

7. Viola tricolor L. ssp. **tricolor**   Wild Pansy   Mrs Casborne, 1831
V. curtisii var. *pesneaui* sensu Evans; *V. variata* sensu Evans; ?*V. lloydii* sensu Evans
A rare weed in arable land. 44–6, 66. General.

8. Viola arvensis Murr.   Field Pansy   Ray, 1660
V. tricolor sensu Bab.; *V. agrestis* sensu Evans; *V. agrestis* var. *segetalis* sensu Evans; *V. deseglisei* sensu Evans; *V. ruralis* sensu Evans; *V. anglica* Drabble; *V. deseglisei* var. *subtilis* sensu Evans pro parte
An abundant weed of arable land. All squares except 26. General.

× **tricolor** (*V. deseglisei* var. *subtilis* sensu Evans pro parte; *V. obtusifolia* sensu Evans; ?*V. contempta* sensu Evans)
Only recorded from 45, 46, 65, 66, but probably elsewhere. Hybrids between these two species are fertile and are often found without one or other of the parents.

POLYGALACEAE

POLYGALA L.

1. Polygala vulgaris L.   Milkwort   S. Corbyn, 1656
Local in remnants of chalk grassland, by chalk-pits, and in undisturbed grassland in the Fens. 23, 24, 33–5, 44, 45, 55, 56, 66, 67, 41, [25, 46, 47, 49, 54, 57, 40]. **GG**; **W**. General.

2. Polygala serpyllifolia Hose   W. A. Leighton, 1831
The only certain records for this species (which is usually found on lime-free soils) are Shelford Common in 1831, Gog Magogs in 1833, and Gamlingay where it still occurs. 25, [45]. **G**. General.

68

POLYGALACEAE

(**P. calcarea** F. W. Schultz. Recorded from Chippenham Fen by Fryer (1899), and Gog Magogs by A. Hosking (1903). As no specimens have been traced these records must be regarded as doubtful.)

GUTTIFERAE

HYPERICUM L. (St John's Wort)

1. Hypericum androsaemum L.   Tutsan   W. H. Mills, 1947
Recorded only from a wet place in Longstowe Wood, presumably introduced. *35*. Scattered throughout the British Isles; rarer or absent in the north and east.

2. Hypericum hircinum L.
Garden escape. *46*.

3. Hypericum calycinum L.
Naturalized garden escape. 45, *66*.

4. Hypericum perforatum L.   Common St John's Wort   Ray, 1660
Common in open woods, hedgebanks and grasslands on the chalk and boulder clay;   apparently absent from the open Fenland in the extreme north of the county. 23–5, 34–6, 38, 44, 45, *46*, 49, 54–6, 58, 64–6, 75, [48]. Throughout most of the British Isles, but rare in the west and north.

5. Hypericum maculatum Crantz   R. A. Pryor, 1873
Ssp. **obtusiusculum** (Tourlet) Hayek (*H. dubium* Leers; *H. quadrangulum* sensu Evans) occurs in a few woods on the boulder clay, while ssp. **maculatum** is known only from the orchard of Christ's College, Cambridge, where it was found by C. E. Raven in 1948. *35*, 45, 65, [36]. DPW. Scattered throughout the British Isles.

× **perforatum** (*H.* × *desetangsii* Lamotte). Found at Swansley Wood around 1930 by W. H. Mills and West Wickham Wood in 1931 by T. G. Tutin. *36*, *65*.

6. Hypericum tetrapterum Fr.   Ray, 1660
H. quadrangulum sensu Bab.
Not uncommon by ditches, rivers, ponds and other damp places. 25, 34–6, 39, 43–8, *49*, 54–7, 64–6, 68, 76, 20, [38, 58, ?41]. General, except in the extreme north.

69

GUTTIFERAE

7. Hypericum humifusum L.   Ray, 1685
A very rare species of gravelly or sandy soils. It was refound in Ray's original locality on the first Furze Hill at Hildersham in 1956. 25, 54, 66, [44, 45]. G. General on suitable soils.

8. Hypericum pulchrum L.   Ray, 1660
A plant of acid heathland formerly recorded at Hinton and still to be found at Gamlingay. 25, [45]. G. General on suitable soils.

9. Hypericum hirsutum L.   Hairy St John's Wort   Ray, 1663
A typical plant of our boulder-clay woods, rare elsewhere; apparently absent from the extreme north of the county. 25, 26, 34–6, 43–5, 54, 55, 64, 65, [23, 38, 47–9, 58, 66]. Scattered throughout Great Britain, mostly on basic soils, though absent from N.W. Scotland; very rare in Ireland.

10†. Hypericum elodes L.   Marsh St John's Wort   Ray, 1660
Known only from Great Heath Wood, Gamlingay, where it was last seen by W. B. Gourlay in 1930. *25*. Scattered through England, Wales and western Scotland; in Ireland except for the central plain.

CISTACEAE

HELIANTHEMUM Mill.

1. Helianthemum chamaecistus Mill.   Rockrose   S. Corbyn, 1656
H. vulgare Gaertn.
Locally common on remnants of chalk grassland; occurs also on the sands. *24*, 34, 35, 44, 45, 54–6, 65, 66, [23, 25, 47, 76]. GG. Throughout most of Great Britain mainly on basic soils; in Ireland known only in one locality, in Donegal.

FRANKENIACEAE

FRANKENIA L.

1†. Frankenia laevis L.   Sea Heath
                              J. Martyn, in T. Martyn, 1763
Recorded by J. Martyn and Relhan at Tydd Gote near Wisbech. [41.] On margins of salt-marshes on the coasts of S. and E. England, local.

CARYOPHYLLACEAE

Silene L.

1. Silene vulgaris (Moench) Garcke   Bladder Campion   Ray, 1660
S. inflata Sm.; *S. cucubalus* Wibel
Common plant of field margins and waste places throughout the county.
The common form is that with nearly glabrous leaves but plants with
hairy leaves are frequent, especially on the chalk. All squares except 25,
37, 39, 47, 20, 41, [58]. General but rare in the north.

(**S. maritima** With. Recorded by Skrimshire from the river bank at East-
field, Wisbech, in 1818. There is no specimen to support the record. [41.])

2. Silene conica L.   Evans and Moss, 1909
Only known in a small area on the sands between Chippenham and
Freckenham. It is frequent on the Breckland just over the border in
Suffolk. 67. Scattered throughout Britain north to Moray, mainly in
the east. Often occurs as a casual.

3. Silene dichotoma Ehrh.
Casual. 44, [45].

4. Silene gallica L.   P. Dent in Ray, 1670
An arable weed recorded from time to time especially on the sands
around Chippenham and Newmarket. Last recorded near Chippenham in 1951. The Cambridgeshire material is referable to var. **anglica**
(L.) Mert. & Koch (*S. anglica* L.). 66 [47, 65, 41]. A variable arable
weed, widespread in S. and C. Europe and introduced elsewhere.

5. Silene pendula L.
Casual or garden escape. [45.]

6. Silene otites (L.) Wibel   Spanish Catchfly   Mr Sare, 1650
Known to the seventeenth- and eighteenth-century botanists at Newmarket, and still to be found in this county near Freckenham, on a
sandy roadside bank. It occurs in several places in the Breckland just
over the border in Suffolk. 67. Native in the Breckland of W. Norfolk,
W. Suffolk and Cambs.

7. Silene italica (L.) Pers.
Casual. [45.]

CARYOPHYLLACEAE

8. Silene noctiflora L.   Relhan, 1786

Melandrium noctiflorum (L.) Fr.

A common weed of arable land, except in the Fens where it is very rare. 34–6, 38, 44–9, 54–7, 59, 64–8, [25, 39, 58, 40, 41]. Throughout Great Britain northwards to Aberdeen and in Ireland.

9. Silene dioica (L.) Clairv.   J. Clarke, 1860

Lychnis dioica L.; *Melandrium rubrum* (Weigel) Garcke

This species, which is common in most English counties, is rare in Cambs, and the neighbouring areas of Hunts and S. Lincs. Permanent populations are found only at Hildersham and in a few woods near the Suffolk border; elsewhere there are occasional records of odd plants only. 35, 44, 54, 55, 57, 65, 66. Throughout the British Isles generally but local or rare in the west of Ireland (see map, page 12).

10. Silene alba (Mill.) E. H. L. Krause   White Campion   Ray, 1660

Lychnis vespertina Sibth.; *Lychnis alba* Mill.; *Melandrium album* (Mill.) Garcke

A common weed of waysides, arable land and waste places throughout the county. All squares except 41. Throughout most of the British Isles but absent from the Highlands and parts of the far West.

× **dioica** (*Melandrium album* × *rubrum*). Pink-flowered plants otherwise closely resembling *S. alba* are not infrequently seen in Cambridgeshire, particularly in the few areas where *S. dioica* occurs. Occasional pink-flowered plants in large *S. alba* populations are assumed to originate from long-distance pollination with *S. dioica* pollen. The complete interfertility of the two species means that plants with any combination of intermediate characters can occur. 44, 45, 47, 54, 55, 65.

LYCHNIS L.

1. Lychnis flos-cuculi L.   Ragged Robin   Ray, 1660

Scattered throughout the county in marshy places. 25, 34–6, 39, 43, 44, 45, 46, 49, 55–7, 64, 65, 67, 30, 40, [38, 48, 66, 41]. General.

AGROSTEMMA L.

1. Agrostemma githago L.   Corn Cockle   Ray, 1660

Lychnis githago (L.) Scop.

72

Rare cornfield weed, formerly more frequent. *25*, 34, 35, *36*, *44*, *45*, *46*, *54*, *66*, [38, 48, 49, 56, 65, 40]. Throughout most of the British Isles but now rare, and absent from some parts of Wales and Scotland.

DIANTHUS L.

1†. Dianthus caryophyllus L.   Clove Pink   Relhan, 1802
Formerly on old walls at Chippenham Park and at Leverington, where it was recorded by Relhan. [66, 41.] Occasionally naturalized on old walls; native of S. Europe and N. Africa.

2. Dianthus deltoides L.   Maiden Pink   Ray, 1685
Formerly found in a few localities in dry places. Now only at Ray's locality at Hildersham Furze Hills where var. **deltoides** with rose-coloured flowers and only slightly glaucous leaves, and var. **glaucus** L. with white flowers and very glaucous leaves, grow side by side. 54, [25, 45, 58]. **FH.** A local lowland plant of grassy fields and banks, and hilly pastures, from Kent and Somerset northwards to Inverness.

VACCARIA Medic.

1. Vaccaria pyramidata Medic.
Saponaria vaccaria L.
Casual or garden escape. 45, [46, 55].

SAPONARIA L.

1. Saponaria officinalis L.   Soapwort   Ray, 1660 ˙
Occurs in hedgerows, roadsides, and waste places, probably always a garden escape. The double-flowered form occurs at Fordham. 34, 45, 55, *67*, *40*, [25, 35, 36, 44, 46, 56, 66, 41]. Perhaps native in S.W. England and Wales; as a garden escape northwards to Ross; native in Europe and Asia.

KOHLRAUSCHIA Kunth

1. Kohlrauschia prolifera (L.) Kunth
Casual. *58*.

GYPSOPHILA L.

1. Gypsophila muralis L.
Casual. [46.]

CARYOPHYLLACEAE

CERASTIUM L. (Mouse-ear Chickweed)

1   Petals about twice as long as sepals; flowers more than 12 mm. in
diam.; perennials with prostrate non-flowering shoots.   *2*
Petals not over 1·5 times as long as sepals; flowers not over 12 mm.
in diam.; annuals or biennials except *C. holosteoides.*   *3*

2   Stem and leaves densely white-felted.   **2. C. tomentosum**
Stem and leaves shortly hairy or nearly glabrous.   **1. C. arvense**

3   Perennial herb with prostrate non-flowering shoots, usually eglandular (occasionally a few glands above); sepals with glabrous tips,
eglandular; capsule 9–12 mm. long.   **3. C. holosteoides**
Annual or biennial herbs lacking non-flowering shoots, usually
glandular; sepals glabrous or hairy at tip; capsule not exceeding
10 mm., commonly much shorter.   *4*

4   Flowers in compact clusters which remain compact in fruit, so that
the fruit-stalk does not exceed the sepals; stamens 10.
      4. C. glomeratum
Inflorescence lax at least in fruit; fruit-stalk exceeding sepals;
stamens 4–5.   *5*

5   Bracts entirely herbaceous; fruit-stalks usually erect; parts of flowers
usually in 4's, sometimes in 5's.   **5. C. atrovirens**
At least the upper bracts with broad scarious margins and tips;
fruit-stalks at first recurved or sharply deflexed; parts of flowers
in 5's.   **6. C. semidecandrum**

1. Cerastium arvense L.   Ray, 1660
Scattered over the county in dry sandy, gravelly and chalky places.
23–6, 33, 34, 44, *45*, 54–6, *58*, 65–7, 76, [48, 57, 40]. Throughout most of
Britain, but much rarer in the west and almost absent from Wales and
W. Scotland; rare in Ireland.

2. Cerastium tomentosum L.   Collector unknown, 1941
Garden escape occasionally naturalized. 23, 34, 45, 46, 68. Widespread
in the British Isles as a garden escape; native of S.E. Europe and the
Caucasus.

3. Cerastium holosteoides Fr.   Ray, 1660
C. triviale sensu Bab.; *C. vulgatum* sensu Evans
A common, often abundant, weed of arable land and waste places, on
walls, and in grassland, throughout the county. Our plants, which are

hairy on the stem, leaves and sepals, are referable to var. **vulgare** (Hartm.) Hyl. All squares. General.

4. Cerastium glomeratum Thuill.   Ray, 1663

C. viscosum sensu Evans

Scattered throughout the county as a weed of arable land and waste places, in grassland and on walls, but never abundant like *C. holosteoides*. 23, 25, 33, 35–9, 43–5, 47, 48, 54–6, 64–6, 40, 41, 50, [46, 58]. General.

5. Cerastium atrovirens Bab.   P. Duval, 1947

C. tetrandum Curt.

In the last ten years this plant has been found on railway tracks in a number of localities in the south of the county. 23, 25, 33, 34. In sandy and stony places by the sea all round the British Isles; rarely inland.

6. Cerastium semidecandrum L.   I. Lyons, 1763

A fairly common weed of waste places on the sands at Gamlingay and in the east of the county; elsewhere a rare plant of walls and dry places. Some of the records, especially those from the Fens, should perhaps, in the absence of supporting herbarium sheets, be regarded as doubtful. 23, 25, 44, 45, *49*, 54, 56, *66*, 67, 76, *41*, [38, 46–8, 58, 65]. G. General in dry habitats, confined to the coast in the north and west.

MYOSOTON Moench

1. Myosoton aquaticum (L.) Moench   Ray, 1660

Malachium aquaticum (L.) Fr.; *Stellaria aquatica* (L.) Scop.

Scattered through much of the county by ditches and rivers and in wet places, though absent from the chalk uplands and apparently rare in the extreme north of the county. 35, 37, 38, 44, 45, *46*, 48, 56–9, 66, 67, 20, [34, 36, 47, 49, 55, 40]. W. Lowlands of England, Wales and S. Scotland, rare in the west.

STELLARIA L. (Stitchwort)

The first three species are closely allied but can be distinguished as follows:

Sepals usually 3–5 mm.; stamens (0-)3–5(–10); seeds usually 0·8–1·3 mm. diam.   **1. S. media**

Sepals usually not over 3 mm.; stamens 1–3(–5); seeds usually not over 0·8 mm. diam.   **2. S. pallida**

CARYOPHYLLACEAE

Sepals usually more than 5 mm.; stamens usually 10; seeds usually more than 1·1 mm. diam. **3. S. neglecta**

1. Stellaria media (L.) Vill.   Chickweed   Ray, 1660

A very variable and abundant weed of cultivated ground and waste places. All squares. General.

2. Stellaria pallida (Dumort.) Pire   Babington, 1879

S. apetala auct.; *S. boraeana* Jord.

Rare weed of dry places and on walls. It can usually only be recognized in March and April and may well be under-recorded. 25, 44, 45, 54, 40. FH. On sandy ground, chiefly in S. and E. England but north to Sutherland, and in Ireland.

3. Stellaria neglecta Weihe   Babington & J. Downes, 1831

Rare, recorded from Cambridge in 1831 and 1900, from Barton in 1831 and from Hinxton in 1915. Specimens in CGE support all these records. [44, 45.] Throughout Great Britain but commonest in the west; also in S.W. Ireland.

4. Stellaria holostea L.   Ray, 1660

Occasional in woods and hedgerows on the boulder clay and sands, elsewhere very local. Cambridgeshire is one of the very few English counties in which this species is not a conspicuous hedgerow flower in May. 25, 35, 36, 43–5, *46*, 48, 49, 54–6, *64–6*, *67*, 75, 76, [38]. General.

5. Stellaria palustris Retz.   T. Martyn, 1763

S. glauca With.; *S. dilleniana* Moench, non Leers

Formerly rather widespread in marshy places, especially in the Fens; last seen near Sutton by A. J. Crosfield in 1909. *S. graminea* is often mistakenly recorded as this species, especially when growing in damp places. The ciliate bracts of *S. graminea* are, however, an entirely reliable character; those of *S. palustris* are quite glabrous. [25, 29, 34, 38, 44–9, 56–8, 40.]. A local plant of marshes and base-rich fens in Great Britain from Perth southwards; in Ireland chiefly in the central districts and absent from the south-west. Becoming scarcer everywhere.

6. Stellaria graminea L.   Ray, 1660

Frequent on the sandy soils at Gamlingay and in the east of the county; scattered in the Fens in a few localities; elsewhere very rare. 25, 35, 36, 39, 46, 47, *49*, 66, 76, [29, 34, 38, 45, 48, 58, 40]. G. General.

76

STELLARIA

7. **Stellaria alsine** Grimm   Ray, 1660
S. uliginosa Murr.
A rare plant of wet places. 25, 45, 48, 54, [56, 58]. C. General.

MOENCHIA Ehrh.

1†. **Moenchia erecta** (L.) Gaertn., Mey. & Scherb.   Ray, 1685
Formerly occurred at Gamlingay. Recorded by Babington (1860),
'By the road from White Wood to Gamlingay', and by Evans (1939),
'at the Cinques Gamlingay'. Evans's statement suggests that he saw the
plant but there is no certain evidence of this, and there are no recent
records. The only Cambs herbarium material was collected by Babington
and Henslow at Gamlingay in 1830. [25.] Local species of England
and Wales.

SAGINA L. (Pearlwort)
1. **Sagina apetala** Ard.   Ray, 1660
Widely distributed, but very rarely abundant, in dry places, often on
walls and in crevices in stonework. 23, 25, 26, 29, 36, 38, 43, 45, 47, 56,
58, 65, 66, 20, 30, 40, 41, 50, [35, 44, 46, 48, 49, 76]. Throughout most
of the British Isles but rare in the north.

2. **Sagina ciliata** Fr.   Babington, 1849
Dry sandy and gravelly places; now known only from the sandy areas
in the east of the county, formerly at Gamlingay. 44, 54, 56, 65, 66,
[25, 76]. FH. Scattered throughout the British Isles north to Inverness.

3†. **Sagina maritima** Don   Babington, 1853
The plant is said by Babington (1860) to have been found at Wisbech
by Henslow; and by Evans (1939) to be found from Foul Anchor to
Wisbech. There is one specimen in CGE collected by Babington at
Tydd Marsh in 1853. [41.] All round the coasts of the British Isles but
local; also very rare on Scottish mountains up to 4000 ft.

4. **Sagina procumbens** L.   Ray, 1660
Widespread as a weed of waste ground and gardens, and especially in
trampled places such as paths and lawns; also in the crevices of pave-
ments and on walls. 23, 25, 26, 29, 34–6, 38, 39, 44–6, *49*, 54, 58, 65, 66,
20, 30, 40, 41, 50, [47, 48, 76]. General.

77

CARYOPHYLLACEAE

5. Sagina nodosa (L.) Fenzl   Ray, 1660

Formerly in the east of the county, in a number of localities in the Fens and at Gamlingay; now known only in the Newmarket area. 66, 67, [25, 38, 44, 45, 48, 49, 56, 57, 76, 40, 41]. D. General.

MINUARTIA L.

1. Minuartia hybrida (Vill.) Schischk.   Ray, 1660

Alsine tenuifolia (L.) Crantz; *Arenaria tenuifolia* L.; *Minuartia tenui-folia* (L.) Hiern, non Nees ex Mart.; incl. *Arenaria tenuifolia* var. *laxa* sensu Evans

In a few scattered localities mainly on disturbed chalky ground. 23, 33, 34, 54–6, 65, 66, 76, [35, 44, 45, 57, 58, 67]. F. In Great Britain southwards from York, Derby and Caernarvon but especially in E. England; also as an introduction in S. and C. Ireland where it is naturalized along railway lines.

MOEHRINGIA L.

1. Moehringia trinervia (L.) Clairv.   Ray, 1663

Arenaria trinervia L.

A local plant of woods on the boulder clay and sands. Resembles *Stellaria media* in general appearance, but is readily distinguished by the leaf venation; it has three subparallel veins, whereas *Stellaria* has a more normal pattern of mid-vein and diverging laterals. 25, 36, 43–5, 48, 54, 55, 58, *64*, 65, 66, [35, 38, 46, 49]. Throughout most of the British Isles.

ARENARIA L.

1. Arenaria serpyllifolia L. sens. lat.   Ray, 1660

incl. *A. leptoclados* (Reichb.) Guss.

Common weed of waste places and arable land especially where the soil is dry and practically bare. It also occurs on walls and crevices in stonework. Two species are sometimes recognized within this aggregate. **A. serpyllifolia** L. sens. strict. has the flowers 5–8 mm. in diam., sepals ovate-lanceolate, capsules flask-shaped with curving sides, and seeds 0·5–0·7 mm., while **A. leptoclados** (Reichb.) Guss. has the flowers 3–5 mm. in diam., sepals lanceolate, capsule conical and straight-sided and the seeds 0·3–0·5 mm. Both are recorded for Cambs and it is probable that the former is the more frequent. 23–6, 29, 33–8, 43–9, 54–9, 64–8, 75, 40, 41, [39, 76]. General.

SPERGULA L.

1. Spergula arvensis L.   Corn Spurrey   Ray, 1660

A common weed of arable land and waste places on the sands around Gamlingay, Hildersham and Chippenham, frequent in the Fens and local elsewhere. The var. **vulgaris** (Boenn.) Mert. & Koch is recorded. It has brownish-black seeds covered with pale club-shaped deciduous papillae, and is grass-green in colour, and only slightly viscid. 25, 36, 38, 39, 47–9, 54, 55, 58, 65–7, 20, 30, 31, 40, [29, 35, 45]. G. A locally abundant and troublesome calcifuge weed of arable land throughout the British Isles.

SPERGULARIA (Pers.) J. & C. Presl

1. Spergularia rubra (L.) J. & C. Presl   Sand Spurrey   Ray, 1660

Lepigonum rubrum (L.) Wahl.

A plant of sandy or gravelly habitats and thus rare in the county. Native on the sands in the south-west and east where it is found in arable land or waste places; elsewhere in sand- or gravel-pits where it is possibly only casual. 25, 35, 45, 66, [23, 29, 54, 58]. G. A locally common plant of sandy soils throughout Britain; rare in Ireland.

2. Spergularia media (L.) C. Presl   ?Ray, 1660

First certain record J. S. L. Gilmour and T. G. Tutin, 1930 *S. marginata* Kittel; *Lepigonum medium* (L.) Fries

Known only by the banks of the tidal River Nene north of Wisbech. 41. FA. All round the coasts of the British Isles.

3. Spergularia marina (L.) Griseb.   ?Ray, 1660

First certain record, Henslow, 1831 *S. salina* J. & C. Presl; *Lepigonum marinum* (L.) Wahl.

Now occurs only on the banks of the tidal River Nene and its tributaries near Wisbech. Also, in Babington (1860), from Cambridge (Relhan) and Ely (L. Jenyns). Evans (1939) refers to these records under *S. media* (*S. marginata*) as no doubt due to error, but *S. marina* is sometimes found as a casual inland and the records could be correct. In the absence of herbarium sheets they should perhaps however be regarded as doubtful, though a specimen (in CGE), collected by W. W. Newbould at Stuntney near Ely in 1859, is *S. marina*. 31, 41, [? 45, 57]. FA. All round the coasts of the British Isles, occasionally in salty areas inland, or as a casual.

79

CARYOPHYLLACEAE

POLYCARPON L.

1. Polycarpon tetraphyllum (L.) L.
Casual. *45*, [40].

ILLECEBRACEAE

HERNIARIA L.

1. Herniaria glabra L.   Rupture-wort   W. W. Newbould, 1855
Known only from W. W. Newbould's first record at Six Mile Bottom until found in a sand pit at Chippenham by P. H. Oswald in 1952 where it still occurs. It is found in some quantity in a few localities just over the border in Suffolk. 66, [55]. A rare and local plant in dry sandy places in East Anglia, S. Lincs and Cumberland.

SCLERANTHUS L.

1. Scleranthus annuus L.   Ray, 1660
Fairly common on arable and waste land on the sandy soils around Gamlingay, Chippenham, Hildersham and Kennett; rare elsewhere. 25, *47*, 54, 65–7, 76, [44–6, 48]. G. Throughout Great Britain north to Ross, rare in Ireland.

PORTULACACEAE

MONTIA L.

1. Montia fontana L.   Blinks   Ray, 1663
Formerly in a few localities on seasonally damp acid soils; last seen at Sutton Meadlands in 1945. All specimens examined are referable to ssp. **chondrosperma** (Fenzl) Walters, which has a mainly southern and lowland distribution in Britain. *25*, *47*, [29, 38, 45, 48, 58]. The species occurs throughout the British Isles.

2. Montia perfoliata (Willd.) Howell   Evans, *c.* 1900
Claytonia perfoliata Donn ex Willd.
First reported near the Suffolk boundary at Kennett where it was said by Evans to have been possibly introduced with pheasant food. It is known only from a few localities, mainly on sandy soils. In the Breckland of Suffolk it is locally common. 25, 45, 66, 67, 76. K. Introduced. Scattered throughout Great Britain, very rare in Ireland; native of N. America.

AMARANTHACEAE

AMARANTHACEAE

AMARANTHUS L.

1. Amaranthus retroflexus L.
A frequent casual. A specimen from Hardwick has been determined as var. **delilei** (Richter & Loret) Thell. by J. P. M. Brenan. 35, 45, 55, 56, 59.

2. Amaranthus albus L. (? *A. blitum* sensu Bab.)
Rare casual. The *Blitum rubrum minus* of Ray and the *A. blitum* of later authors are probably referable to this species. *66*, [?45].

3. Amaranthus blitoides S. Watson
Casual. *66*.

4. Amaranthus hybridus L. (*A. chlorostachys* Willd.)
A specimen collected in a carrot field at Chippenham in 1949 has been referred to var. **pseudoretroflexus** subvar. **aristulatus** Thell. *66*.

5. Amaranthus quitensis H., B. & K.
Casual. *66*.

6. Amaranthus graecizans L. ssp. **sylvestris** (Vill.) Brenan
Casual. 45.

CHENOPODIACEAE

CHENOPODIUM L.

Mature plants with ripe seed are needed to identify most species of this genus satisfactorily.

1 At least some leaves cordate or truncate at base. *2*
Leaves never cordate, ±cuneate at base. *3*

2 Perennial; stigmas long; seeds vertical except in the terminal flowers. **1. C. bonus-henricus**
Annual; stigmas short; seeds all horizontal. **9. C. hybridum**

3 Inflorescence axis and perianth glabrous (rarely ± mealy in *C. urbicum* and then perianth not completely enclosing fruit). *4*
Inflorescence axis and perianth mealy, at least when young; fruit usually entirely or almost entirely enclosed by perianth. *9*

CHENOPODIACEAE

4 Leaves entire or at most with a single obscure tooth on each side; stems 4-angled; seeds black. **2. C. polyspermum**
   Leaves, except the uppermost, not entire (very rarely entire or nearly so in *C. rubrum* and *C. botryodes* but then seeds red-brown); stems ridged but not 4-angled.   5

5 All flowers with 5 perianth segments and 5 stamens. **8. C. urbicum**
   All flowers except the terminal with 2–4 perianth segments and 2–3 stamens.   6

6 Leaves mealy-glabrous beneath, green above. **11. C. glaucum**
   Leaves green on both sides.   7

7 Fruiting perianth fleshy, turning scarlet; flowers in sessile heads forming a spike, leafless at top. **12. C. capitatum**
   Fruiting perianth not fleshy, not turning scarlet; inflorescence leafy.   8

8 Flowers arranged in sessile, globose, axillary cymes.
   13. C. foliosum
   Flowers arranged in elongated axillary spikes. **10. C. rubrum**

9 Leaves entire or nearly so, smelling strongly of bad fish.
   3. C. vulvaria
   Leaves toothed or lobed, not smelling strongly of bad fish.   10

10 Inflorescence branches short, numerous and divaricate; testa densely covered with minute pits. **7. C. murale**
   Inflorescence branches usually long and themselves little branched; testa with furrows or elongate pits but not with dense, minute pits.   11

11 Leaves with a long oblong middle lobe with short lateral lobes.
   6. C. ficifolium
   Leaves ovate-lanceolate or triangular- or rhomboid-ovate, middle lobe (if present) short.   12

12 Leaves typically small, rhomboid-ovate, about as long as broad, glaucous-mealy beneath; pericarp adherent to the testa.
   5. C. opulifolium
   Leaves typically larger and longer than broad, not glaucous-mealy; pericarp easily detachable from ripe seed. **4. C. album**

1. Chenopodium bonus-henricus L.   Good King Henry   Ray, 1660

In waste places, particularly in and around old cottage gardens, in a number of scattered localities mainly in the Fens. *34, 35, 37, 45, 47, 48, 55, 56, 58, 65, 30, 31, [23–5, 36, 44, 46, 54, 57, 66, 40]*. Long-established and well naturalized throughout England and Wales, northwards to southern and eastern Scotland; rare in N. and W. Scotland and in Ireland. Origin not certain but occurs in Europe, W. Asia and N. America. Cultivated as a pot herb.

2. Chenopodium polyspermum L. Ray, 1660
Scattered over the county as a weed of arable land and waste places, but
most frequent in the Fens. Two varieties are recorded, var. **polyspermum**
(*C. polyspermum* var. *cymosum* Chevall.) with obtuse leaves and a
usually prostrate or decumbent habit, and var. **acutifolium** (Sm.)
Gaudin (*C. polyspermum* sensu Evans) with acute leaves and usually an
erect habit. 25, 29, 35, 37–9, 46–9, 55, *57*, 58, 59, 68, 69, 20, 40, [45,
47]. South Cornwall and Kent to Cheshire and Lincoln, N. Yorks and
Berwick, and Mid Lothian, probably introduced in the north and in
Ireland.

3. Chenopodium vulvaria L. Stinking Orache Ray, 1660
C. olidum Curt.
Recorded several times in the Cambridge area where it was first known
near Peterhouse Tennis Court by Ray. Last seen at Mount Pleasant,
a locality known to Babington, by R. S. Winteringham in 1958. 45. A
native of S. and E. England, particularly near the coast, northwards to
Lincs. A rare casual in N. England, Scotland and Ireland.

4. Chenopodium album L. sens. lat. Fat Hen Ray, 1660
Very variable annual herb; a common weed of arable land and waste
places throughout the county. Two segregates, **C. reticulatum** Aellen
and **C. suecicum** J. Murr. (*C. album* var. *viride* sensu Bab.), have been
recorded. All squares. General.

5. Chenopodium opulifolium Schrad. ex Koch & Ziz
Casual. [25, 46, 58.]

6. Chenopodium ficifolium Sm. I. Lyons, 1763
Fairly common as a weed of arable land and waste places throughout
most of the county. 29, 34, 35, 37–9, 44, 45, *46*, 47–9, 55–9, 67, 68, 20,
30, 31, 40, 41, 50, [25]. In England from Sussex and Kent to Somerset,
Leicester and Lincoln; rare and probably casual in S.W. England,
Wales and northern England. Probably more abundant and persistent
in the Cambridgeshire Fenlands than anywhere else in the British Isles.

7. Chenopodium murale L. Ray, 1660
Ray and Relhan suggest that this species was common, as they mention
no localities for it. There are a few other records, the only recent ones
being as weeds in Cambridge gardens. 45, *76*, [38, 46, 47, 56]. Scattered

CHENOPODIACEAE

throughout lowland England and Wales, very rare in Scotland and Ireland. Probably only native on the coast from Somerset to Norfolk, and certainly not native from Lancashire and Yorkshire northwards or in Ireland.

8†. Chenopodium urbicum L.   I. Lyons, 1763

Recorded by I. Lyons from Hinton and by Relhan from Barnwell, Coton and Cottenham. [45, 46.] In lowland districts from Cornwall and Kent to Lancs and Yorks; rare and casual from Yorkshire northwards.

9. Chenopodium hybridum L.   J. Sherard in Ray, 1724

In a few scattered localities in the Fens and around Cambridge. The var. **paeskii** Aschers. & Graeb. with a more compact inflorescence and the whole plant reddish has been found recently at Waterbeach and Cambridge. 44–6, 56, 58, *66*, 41, [38, 47, 67]. **Cam.** Questionably native from Devon and Kent to Shropshire and Lincs; casual elsewhere.

10. Chenopodium rubrum L.   Ray, 1660

Very variable herb; scattered through the county as a weed of arable land and waste places. The var. **pseudobotryoides** Wats. ex Bab. is recorded; it is a small plant with shorter inflorescences and usually grows by the edges of ponds. 29, 34–6, 38, 39, 44–8, 49, 55–9, 66, 68, 20, 30, 31, 40, 41, 50. Not uncommon in England, rare and sometimes casual in Wales, Scotland and Ireland.

11. Chenopodium glaucum L.

Casual. [46.]

12. Chenopodium capitatum (L.) Aschers.

Casual. 44.

13. Chenopodium foliosum (Moench) Aschers.

Casual. [45.]

BETA L.

1. Beta vulgaris L. ssp. maritima (L.) Thell.   Beet
                                           Skrimshire in Relhan, 1802

B. maritima L.

The only record of this plant is from the salt-marshes by the Nene below Wisbech, by Skrimshire. The ssp. **vulgaris** ('sugar-beet') is grown as a crop throughout the county (*c.* 26,000 acres in Isle of Ely and 15,900 in S. Cambs). [41.] Throughout maritime areas of the British Isles except the north of Scotland.

2. Beta trigyna Waldst. & Kit.
Recorded as an established alien since 1942 by the Cambridge to Babraham road. 45.

ATRIPLEX L. (Orache)
1. Atriplex littoralis L.   Relhan, 1785
Found only by the banks of the tidal River Nene near Wisbech. 41.
Around the coasts of the British Isles but local, and now confined to the east coasts in Scotland and Ireland.

2. Atriplex patula L.   Ray, 1660
A. erecta Huds.; *A. angustifolia* Sm.
Very variable; found throughout the county as a weed of cultivated and waste places. All squares except *24*, 39, 64, 67, [65]. General.

3. Atriplex hastata L.   Ray, 1660
A. deltoidea Bab.
Scattered over the county as a weed of waste and cultivated land. 29, 34–8, 44–6, *47*, 48, *56*, 57–9, 67, 30, 31, 40, 41, 50, [49]. General.

4. Atriplex glabriuscula Edmondst.   J. Martyn in T. Martyn, 1763
A. babingtonii Woods
Recorded only from the banks of the River Nene both above and below Wisbech. Babington writes as though it was still there in 1860, as does Evans in 1939. No specimens have been seen. [40, 41.] All round the coasts of the British Isles.

5. Atriplex hortensis L.
Casual. 45, *49*, 57.

AXYRIS L.
1. Axyris amarantoides L.
Casual. *35*.

HALIMIONE Aellen

1. Halimione portulacoides (L.) Aellen   J. Martyn in I. Lyons, 1763
Obione portulacoides (L.) Moq.; *Atriplex portulacoides* L.
Found only by the banks of the River Nene near Wisbech. 41. **FA.**

CHENOPODIACEAE

Coasts of England, Wales and S.W. Scotland; in Ireland on southern coasts from Galway Bay to Co. Down.

2†. **Halimione pedunculata** (L.) Aellen   Skrimshire in Relhan, 1802
Obione pedunculata (L.) Moq.; *Atriplex pedunculata* L.
Found by Skrimshire by the riverside near Wisbech where he last recorded it in 1826. [41.] A very rare plant of salt-marshes recorded from Kent, Suffolk, Norfolk, Cambs, and Lincs. Now almost extinct; only seen in one locality in East Anglia in the last 20 years.

SUAEDA Forsk. ex Scop.

1. **Suaeda maritima** (L.) Dumort.   Seablite
J. Martyn in I. Lyons, 1763
Found only by the River Nene near Wisbech. 41. FA. Around the coasts of the British Isles.

SALSOLA L.

1. **Salsola pestifer** A. Nels.
Casual. *35, 46, 66.*

SALICORNIA L. (Glasswort; Marsh Samphire)

Salicornia was first recorded from near Wisbech by Relhan in 1788. It was also recorded from this locality by Babington (1860) as *S. herbacea* L., and by Evans (1939) as *S. stricta* Dum. Both these names must be taken to be in the aggregate sense and herbarium sheets in CGE cannot be determined. The following two segregates were determined by P. W. Ball and T. G. Tutin from live material collected by Miss E. J. Gibbons in 1959. Spirit-preserved specimens are kept in Herb. Leicester.

1. **Salicornia dolichostachya** Moss   E. J. Gibbons, 1959
Banks of the River Nene below Wisbech. 41. FA. Distribution not fully known but on coasts of much of Britain and in S. and E. Ireland.

2. **Salicornia ramosissima** Woods   E. J. Gibbons, 1959
Banks of the River Nene below Wisbech. 41. FA. Widespread and common on the east and south coasts of England and S. Wales. Similar plants occur in S. and E. Ireland.

TILIACEAE

Tilia L. (Lime)

1. Tilia platyphyllos Scop.   A. Ley, 1891
There are a few records of this tree, all of which are almost certainly of planted origin. It is commonly planted in the Cambridge area. 35, 45, 46, [37, 64]. Native in S. Wales and a few other places in C. England north to Derbyshire.

2. Tilia cordata Mill.   T. Martyn, 1763
Still occurs in its original locality at White Wood, Gamlingay, where it is probably native. Elsewhere in a very few places in hedgerows where its status is doubtful. Occasionally planted. 25, *35*, 36, 45, [46]. G. Scattered throughout England and Wales north to the Lake District and Yorks, and as a planted tree northwards to Perth.

× **platyphyllos**   Common Lime   T. Martyn, 1763
T. europaea L.; *T. vulgaris* Hayne
Commonly planted over much of the county. 23, 25, 26, 29, 34–6, 39, 43–6, 49, 55, 58, 65, 69, 20, 30, 40, 41, 50, [56]. Throughout the British Isles except the far north; usually planted.

MALVACEAE

Malva L.

1. Malva moschata L.   Musk Mallow   Ray, 1663
In a few localities mainly in the south of the county, on roadsides, in hedgerows and in waste places. 34, 35, 44, 46, 54, [23, 25, 29, 36, 48, 65]. General except for north.

2. Malva sylvestris L.   Common Mallow   Ray, 1660
Common throughout the county on roadsides and in waste places. All squares. General, but local in the extreme north.

3. Malva neglecta Wallr.   Ray, 1660
M. rotundifolia sensu Bab., et Evans
Throughout the county on roadsides and in waste and cultivated ground. 25, 26, 29, 35, 36–8, 39, 44–6, *47*, 48, 49, 55–9, *66*, 67, 68, 20, 30, 40, 41, 50. General, but absent from the north and west.

MALVACEAE

ALTHAEA L.

1. Althaea officinalis L.   Marsh Mallow   Ray, 1660

Formerly in a number of localities in the Wisbech area and refound in one of them in 1959. Recorded by Relhan from 'Cow Fen' and 'Trumpington Meadow'. 41, [45, 30, 40]. Native at or near the coast of southern England and Wales and in S.W. Ireland. Occasional as an introduction elsewhere.

2. Althaea rosea (L.) Cav.   Hollyhock

Garden escape. 34, 45, 58, 67, 50.

MALOPE L.

1. Malope trifida Cav.

Casual. 45.

LINACEAE

LINUM L. (Flax)

1. Linum bienne Mill. (*L. angustifolium* Huds.)

Casual. [35, 45.]

2. Linum usitatissimum L.   Flax   Ray, 1660

Recorded from a number of localities in or by fields where it was a relic of cultivation. Formerly grown for the production of linen or linseed. 34, 45, 54, *40*, [46]. A cultivated plant of unknown origin.

3. Linum anglicum Mill.   S. Corbyn, 1656

L. perenne sensu Bab.

Locally frequent in a few localities in disturbed chalk grassland. A nationally famous plant which is fortunately still common quite near to Cambridge. 45, *54*, 55, [66]. GG. E. England from N. Essex to Durham extending west to Leicester, Westmorland and Kirkcudbright; very local.

4. Linum catharticum L.   S. Corbyn, 1656

Scattered over the county on most soils but particularly in chalk grassland and in fens. 23–6, 33–5, *36*, 43–6, 54–7, 59, 64–7, 76, 41, [38, 48, 49]. General.

LINACEAE

5. Linum austriacum L.
Garden escape. 44.

(**Radiola linoides** Roth (*R. millegrana* Sm.). Herb of seasonally wet, acid sandy places; the only record is by J. Watson from near Newmarket in 1821. It may or may not have been in the county. [66.] General, but very local.)

GERANIACEAE

GERANIUM L. (Cranesbill)

1. Geranium pratense L.   Meadow Cranesbill   Ray, 1660
Locally frequent on roadsides and in grassland in the S.W. of the county, mainly on the boulder clay. 25, 26, 29, 34, 35, 45, 46, 56, [36, 44, 58, 65, 41]. Widespread but local in Britain showing a northern and eastern tendency, being absent as a native from the south-west, much of Wales and W. Scotland; very rare in Ireland, native only in Antrim.

2. Geranium endressii Gay
Garden escape. *45.*

3. Geranium phaeum L.   Relhan, 1788
A garden escape occasionally naturalized. 45, 54, [55, 58]. A garden escape; throughout the British Isles, except the extreme north and west.

4. Geranium sanguineum L.   Ray, 1660
Recorded from Church Meadow, Balsham, by R. B. Smart, from Wood Ditton by L. Jenyns and from the Devil's Dyke where it was first recorded by Ray and is still to be found. 66, [55, 56, 65]. D. A mainly coastal species of N. and W. Britain, the east and west coasts of Ireland, sometimes inland on limestone. The Cambridgeshire locality is the only one known to be extant in England south and east of a line joining the mouth of the Humber to the Severn estuary.

5. Geranium pyrenaicum Burm. f.   W. Pulling in Relhan, 1820
Frequent on roadsides, field margins and in waste places on the chalk, boulder clay and sands, but almost absent from the Fens. The forma **pallidum** Gilmour & Stearn with white flowers occurs on roadsides on the Gogs. 25, 34, 35, 37, 44–6, 48, 49, 54, 56, 58, 65, 66, 40, [55]. Common in S. and E. England, rare westwards and northwards to Inverness and Clyde Islands; scattered over Ireland.

89

GERANIACEAE

6. Geranium columbinum L.   W. W. Newbould, 1857

Recorded for a few localities but found recently only at Hardwick and Chippenham. 35, *55*, 66, [34]. In Britain north to Moray; Ireland, but local especially in the north.

7. Geranium dissectum L.   Ray, 1660

Throughout the county as a weed of cultivated and waste land, and in grassland. All squares. General.

8. Geranium rotundifolium L.   I. Lyons, 1763

Recorded from a number of localities. Most of these records are as casuals but those from Gamlingay and Newmarket may well be native. *45, 55, 58,* [25, 38, 44, 57, 66]. From Cornwall and Kent to Carmarthen, Northampton and Suffolk, local; casual farther north; in Ireland in Cork, Clare, Wexford and Kilkenny.

9. Geranium molle L.   Ray, 1660

Common weed of arable land, waste places and grassland throughout the county. All squares except *24*, 26, 57. General.

10. Geranium pusillum L.   Ray, 1660

Throughout the county in arable or waste land. It can conveniently be distinguished from the closely allied *G. molle* by the hairs of the pedicels all being short, and not mixed long and short as in *G. molle*. 23, *24*, 25, 26, 35, 36, 39, 44–7, 54–6, 58, 65, *66*, 67, 76, 30, *40*, *41*, 50, [34, 38, 48, 49, 64]. Widespread and common in England; local in Wales and Scotland; in a few scattered localities in Ireland, mainly near the coast.

11. Geranium lucidum L.   Ray, 1660

Recorded from a number of scattered localities; plentiful on a roadside bank at Hildersham. 24, 45, 54, [34, 46, 56, 40]. General, but rare in the north.

12. Geranium robertianum L. ssp. robertianum Herb Robert Ray, 1660

Throughout the çounty except the open Fens, in shady woods, copses and hedgebanks, but not really abundant as in most other English counties. All squares except 37, 59, *67*, 68, 30, [38, 48]. General.

90

ERODIUM L'Hérit. (Storksbill)

1†. **Erodium moschatum (L.) L'Hérit.**
C. Miller and T. Martyn, in I. Lyons, 1763
Recorded from a number of localities where it was presumably either
an escape from cultivation or a casual. Last seen by Evans in 1909.
[25, 47, 48, 56, 57, 30, 41.] Native in waste places mainly near the sea,
from Cornwall to Kent, S. Lancs, and the Isle of Man, local; also in
many inland counties from Yorks southwards, but scarcely native;
round most of the Irish coast.

2. **Erodium cicutarium (L.) L'Hérit. ssp. cicutarium**   S. Corbyn, 1656
Scattered over the county in dry places, but abundant only on the
Breckland soils around Newmarket. Ssp. **dunense** Andreas occurs at
Chippenham. 23, 25, 29, 33, 39, 44–8, 54–6, 65–7, 76, 40, 41, [34, 38,
49, 58, 64]. K. All round the coasts of the British Isles; inland wide-
spread but local, particularly in Ireland.

OXALIDACEAE

OXALIS L. (Sorrel)

1. **Oxalis acetosella L.**   Wood Sorrel   Ray, 1660
Recorded from a few woods on the boulder clay, mainly in the east of
the county, but surprisingly local in Cambs. 35, 64, 65, [55, 66].
General.

2. **Oxalis corniculata L.**
Rare casual or garden weed. The var. **microphylla** Hook. f. is recorded
as a weed in a Cambridge garden. 37, 39, 45, [?38, 46, 40].

3. **Oxalis europaea** Jord. (*O. stricta* auct.)
Casual or garden weed. 47, *57*, 58, 50, [?38, 45, 46, 48].

4. **Oxalis articulata** Savigny
Casual. 45.

BALSAMINACEAE

IMPATIENS L.

(**Impatiens capensis** Meerb. (*I. biflora* Walt.). Recorded from The Backs,
Cambridge, in 1946, but almost certainly in error for *I. parviflora* DC. Well-
established on the Great Ouse in Beds and Hunts and likely to be found in the
county near Swavesey and Over. ?45. Widely naturalized in lowland England; native of eastern N. America.)

91

BALSAMINACEAE

1. Impatiens parviflora DC.   Babington, 1856
Locally well naturalized in shady gardens and damp woods, as in the
grounds of Quy Hall. Babington's record from Sawston in 1856 is
only the third British record. 34, 44, 45, *47*, 56, [54]. Scattered throughout Great Britain; native of C. Asia.

2. Impatiens glandulifera Royle   R. S. Adamson, 1910
I. roylei Walp.

Not well established in Cambs; an escape from cottage gardens in
which it is still cultivated. Recorded from the left bank of the River
Granta just below Great Chesterford station. 45, 54, [58]. Widespread;
native of the Himalayas.

ACERACEAE

Acer L.

1. Acer pseudoplatanus L.   Sycamore   Ray, 1660
Commonly planted and naturalized throughout the county. All
squares except 37. Planted throughout the British Isles and naturalized
in many places; native of the mountains of C. and S. Europe.

2. Acer platanoides L.
Planted tree. 45.

3. Acer laetum C. A. Mey.
Planted tree. [57].

4. Acer campestre L.   Common or Field Maple   Ray, 1660
Usually a hedgerow bush or small tree, but occasionally well-grown
specimens make trees more than 30 ft. high. In woods, copses and
hedgerows throughout the county. All squares except 29, 37, *38*, 59,
68, 30, [48, 40]. Common in S., E. and C. England, rare and often
introduced in the rest of the British Isles.

HIPPOCASTANACEAE

Aesculus L.

1. Aesculus hippocastanum L.   Horse-chestnut
                                                G. W. Chapman, 1932
Planted in hedgerows, around meadows and fields and in parks
throughout the county; often also self-sown. All squares except 29,

HIPPOCASTANACEAE

37, 59. Commonly planted over much of the British Isles; native of Albania and Greece.

(The red-flowered Horse-chestnut **Aesculus carnea** Hayne is also frequently planted.)

AQUIFOLIACEAE

ILEX L.

1. Ilex aquifolium L.   Holly   Ray, 1660

Probably always planted, and fairly frequent in gardens and hedgerows, but perhaps native in Great Heath Wood, Gamlingay. 25, 34, 36, 39, 44, 45, 49, 56, 64, 65, 76, 20, 40, [46]. Throughout most of the British Isles, but often planted.

CELASTRACEAE

EUONYMUS L.

1. Euonymus europaeus L.   Spindle-tree   S. Corbyn, 1656

Rather local in woods, spinneys and hedgerows, absent from the open Fenlands. 25, 26, 34–6, 43–5, 54, 55, 57, 64–6, [48, 56, 58]. In Britain north to the Clyde and Forth, and in Ireland.

BUXACEAE

BUXUS L.

1. Buxus sempervirens L.   Box   Henslow, early nineteenth century

Planted in many parts of the county and much grown as a hedge round gardens. Box foliage was discovered in a Romano-British burial ground at Great Chesterford and a specimen of the material is preserved in CGE. This is, however, insufficient evidence for the native status of *Buxus* in the county. 23, *25*, 34, *35*, 36, 45, 54, 55, 64–6. Native in beechwoods and scrub on chalk and oolitic limestone in Kent, Surrey, Bucks and Gloucester; elsewhere commonly planted and sometimes naturalized.

RHAMNACEAE

RHAMNUS L.

1. Rhamnus catharticus L.   Buckthorn   Ray, 1660

Throughout the county in hedges, copses, woods and scrub. Rare in the open Fenland but abundant at Wicken where it is one of the later

RHAMNACEAE

dominants of fen carr. 24, 26, 34–9, 43–6, 48, 54–8, 65–7, [47, 49, 64, 40]. Over much of England on calcareous soils, but frequent only in midland, southern and eastern counties; rare in Wales, doubtfully native in Scotland; in Ireland mainly in the central plain.

FRANGULA Mill.

1. Frangula alnus Mill.   Alder Buckthorn   Ray, 1685
Rhamnus frangula L.

Formerly recorded from Odsey, Gamlingay, Hildersham, Fulbourn and Wilbraham; now known only at Chippenham and Wicken Fens, by a pit just north of Wicken, and a single tree (probably introduced) near March. In Wicken Fen it is the dominant shrub of young 'carr'. The extraordinary abundance of this plant at Wicken seems to be a rather peculiar phenomenon. In 1860 *Frangula* was not mentioned at all by Babington in a list of plants from Wicken Fen. The spread is therefore relatively recent, and is associated with the cessation of regular cutting of *Cladium* and a generally lower water-table. At Chippenham *Frangula* is still rather rare, though increasing. 49, 56, 57, 66, [23, 25, 44, 54, 55]. W. General in England, Wales and S.W. Scotland, rare in Ireland. Not normally a species of base-rich soils.

LEGUMINOSAE

LABURNUM Medic.

1. Laburnum anagyroides Medic.   Collector unknown, 1916
A small tree frequently planted, and sometimes self-sown. *24, 25*, 34, 35, 44–6. Native of C. and S. Europe.

GENISTA L.

1. Genista tinctoria L.   Dyer's Greenweed   Ray, 1660
Rare in permanent grassland, mainly on the boulder clay and sands. 25, 35, 54, [36, 45, 57, 65, 66]. FH. Throughout England and Wales, rare in S. Scotland, absent in the north and from Ireland.

2. Genista anglica L.   Petty Whin   Ray, 1660
Recorded from Hildersham Furze Hills where it was last seen in 1930, and Gamlingay where it was last seen in 1932. *25, 54*. Scattered throughout Great Britain north to E. Ross. Absent from the W. of Scotland, the Isle of Man and Ireland.

94

LEGUMINOSAE

ULEX L.

1. Ulex europaeus L.   Furze; Gorse   Ray, 1660
Rather rare; mainly on the sands at Gamlingay and in the east of the county. 25, 44, 45, 49, 54–6, 66, 67, 76, [35, 36, 65, 41]. **G.** Throughout the British Isles though rare in some areas, and probably only an introduction in the far north.

2†. Ulex minor Roth   Relhan, 1785
U. nanus T. F. Forst.
There are old records for Linton, Thriplow Heath, Newmarket Heath, Ickleton Grange and Wisbech. Only the Ickleton Grange record is supported by a herbarium sheet. [44, 54, 66, 40.] S. and E. England north to Cumberland; N. Ireland.

SAROTHAMNUS Wimm.

1. Sarothamnus scoparius (L.) Wimm. ex Koch ssp. **scoparius** Broom
Ray, 1660
Cytisus scoparius (L.) Link
Rare, occurring mainly on the sands at Gamlingay and in the east of the county. 25, 44, 55, 65–7, [58, 76]. **G.** General.

ONONIS L. (Restharrow)

1. Ononis repens L. ssp. **repens**   Ray, 1660
O. arvensis L.
Frequent on roadsides, tracks and in waste places on the chalk, boulder clay and sands, rare in the Fens. 23–6, 33–7, 43–5, *46*, *47*, 54–8, 64–8, 75, 76, [40, 41]. General, but rare in the north and west.

2. Ononis spinosa L.   Ray, 1660
O. campestris Koch
On roadsides and in waste places, especially on the boulder clay. 25, 26, 34–6, 44–7, 54–8, 64, 66, 67, 41, [24, 38, 48, 40]. Scattered throughout England except the south-west, rare in Wales and S. Scotland.

(Intermediate plants, apparently hybrids, occur where both species grow together, as at Chippenham and Thriplow.)

95

LEGUMINOSAE

MEDICAGO L.

1. Medicago falcata L.   Ray, 1660

On roadsides or in grassy places in a few localities, mainly on the sands in the east of the county. *45, 54, 55, 66*, 67, [56]. In grassy places on the Breckland of Norfolk, Suffolk and Cambs; introduced elsewhere.

× **sativa**

M. sylvestris Fr.; *M. varia* Martyn

This fertile hybrid has been frequently recorded on the sands in the east of the county around Chippenham, Fordham and Six Mile Bottom. It has also occurred in the Cambridge area. *45, 46, 55*, 67, [*54, 66*].

2. Medicago sativa L.   Lucerne   Henslow, 1829

Widely grown as a fodder crop and now naturalized on roadsides, trackways and waste places throughout the county. All squares except 37, 48, 49, 58, 59, 30, 41. Cultivated and naturalized in much of the British Isles; probably native of the Mediterranean region and W. Asia.

3. Medicago lupulina L.   Black Medick   Ray, 1660

An abundant weed of grassy and waste places throughout the county. Similar in appearance vegetatively to *Trifolium dubium*, from which it can be distinguished by the leaflets having a mucro in the centre of the emarginate apices. All squares except 39. General, but rare in the north and west.

4. Medicago minima (L.) Bartal.   Ray, 1660

Recorded from several localities on the sands around Chippenham and Newmarket, and from a gravel pit at Wilbraham by Ray. Only seen recently at Isleham gravel pit. 67, [55, 66]. In sandy fields and heaths from S. Kent to Norfolk and in the Channel Islands; elsewhere as a casual.

5. Medicago polymorpha L.

M. hispida Gaertn.; *M. denticulata* Willd.

Casual. 39, 45, [49].

6. Medicago arabica (L.) Huds.   Ray, 1660

M. maculata Sibth.

A weed of arable land and waste places in a number of scattered localities. 25, 34, 36, 37, 44–6, 55, 65, 76, [35, 36, 48, 54, 67]. **Cam.** Widespread in England and Wales, particularly in the south, naturalized in a few localities in Scotland and Ireland.

7. Medicago tribuloides Desr.
Casual. 45.

MELILOTUS Mill. (Melilot)

1. Melilotus altissima Thuill.   Ray, 1660
M. officinalis sensu Bab.
Frequent on roadsides, by the edges of fields, and in waste places
throughout most of the county. 24, 26, 34–6, 43–7, 49, 54–8, 64, 66, 67,
[23, 25, 29, 38, 48, 65]. Probably introduced. In England generally
distributed, but rare in Scotland and Wales; in Ireland naturalized in
a few places. Native of continental Europe and Asia.

2. Melilotus officinalis (L.) Pall.   Babington, 1848
M. arvensis Wallr.
Through much of the county on roadsides, field margins, and in waste
places. Similar to *M. altissima*, from which it can be distinguished by
its glabrous, not hairy pod, and the keel being shorter than the other
petals, not with wings, standard and keel equal as in the other species.
24, 29, 34–6, 38, 44–6, 54–6, 58, 65, 31, 50, [25, 57, 64, 66]. Introduced.
Naturalized in much of Britain north to Moray, and in S. and E.
Ireland; native of continental Europe and Asia.

3. Melilotus alba Medic.   A. Fryer, 1872
A weed of waste places in scattered localities over much of the county.
29, 34, *35*, 36, 44–7, 49, 55, 66, 67, 31, [38]. Naturalized in much of
England and Wales but mainly in the south-east; rare casual in Scotland
and Ireland. Native of most of Europe and Asia, east to Tibet.

4. Melilotus indica (L.) All.   Shrubbs, 1888
Recorded from a few localities in fields and waste places. 45, *46*, 54,
76, [57]. Naturalized in England, Wales and E. Scotland; rare casual
in Ireland. Native of the Mediterranean region and W. Asia.

TRIFOLIUM L. (Clover; Trefoil)

Two species not recorded for Cambs, *T. glomeratum* L. and *T. suffocatum* L.,
are included in this key as they occur in the Breckland just over the Suffolk
border, and may well be found in the Chippenham area.

1	Flowers bright yellow, not over 8 mm.	2
	Flowers white, cream, pink or purplish-red; if yellowish more than 10 mm.	5

LEGUMINOSAE

2 Standard not folded lengthwise; flowers 5–7 mm.; heads *c*. 40-flowered. *3*

  Standard folded lengthwise; flowers 2–3 mm.; heads 2–20-flowered. *4*

3 Stipules half-ovate; style much shorter than the pod.
               13. T. campestre

  Stipules linear-oblong; style barely shorter than the pod.
               14. T. aureum

4 Pedicels rather stout, shorter than the calyx-tube; heads usually 5–20-flowered. **15. T. dubium**

  Pedicels very slender, equalling the calyx-tube; heads usually 2–6-flowered. **16. T. micranthum**

5 Fertile flowers 2–6, creamy white, 8–12 mm.; pods pushed into the ground by the peduncles, surrounded by the reflexed enlarged calyces of the sterile flowers. **8. T. subterraneum**

  Flowers not as above. *6*

6 Stems prostrate and rooting at the nodes; heads all axillary; peduncles exceeding petioles. *7*

  Stems erect, or if prostrate not rooting at the nodes; all heads terminal, or some axillary; petioles usually exceeding the peduncles of axillary heads. *8*

7 Flowers pinkish, scentless; upper part of calyx becoming inflated in fruit, hairy and pinkish; petioles with persistent hairs; lateral veins of leaflets strongly curved. **11. T. fragiferum**

  Flowers white, rarely pink, scented; calyx not inflated in fruit; petioles glabrous at maturity; lateral veins of leaflets straight.
               10. T. repens

8 Heads all terminal. *9*

  Some heads axillary. *12*

9 Heads subtended by a single leaf; annual. **4. T. incarnatum**

  Heads subtended by a pair of ± opposite leaves; perennial. *10*

10 Flowers creamy-yellow. **2. T. ochroleucon**

  Flowers pink or red. *11*

11 Plant caespitose; free part of stipules triangular with a setaceous point; pod dehiscing by the thickened top falling off.
               1. T. pratense

  Plant with extensive slender creeping underground stems; free part of stipules subulate; pod dehiscing by a longitudinal slit.
               3. T. medium

12 Heads distinctly peduncled. *13*

  Heads sessile. *15*

13 Flowers ebracteate; heads usually cylindrical; stem and leaves usually softly hairy.　　　　　　　　　　**5. T. arvense**
Flowers subtended by bracts; heads ±globular; stem and leaves glabrous.　　　　　　　　　　　　　　　　*14*

14 Annual; flowers twisted so that the standard is below; calyx inflated in fruit, hairy.　　　　　　　　**12. T. resupinatum**
Short-lived perennial; flowers not twisted; calyx not inflated in fruit, glabrous.　　　　　　　　　　**9. T. hybridum**

15 Plant pubescent; flowers ebracteate.　　　　　　　*16*
Plant glabrous; flowers bracteate.　　　　　　　　*17*

16 Flowers pink; calyx tube ventricose in fruit, readily detaching itself from the peduncle when ripe; calyx teeth sub-erect; lateral veins of leaflets straight.　　　　　　　**6. T. striatum**
Flowers white; calyx tube not ventricose in fruit, persistent; calyx teeth recurved in fruit; lateral veins of leaflets strongly curved.　　　　　　　　　　　　　　**7. T. scabrum**

17 Internodes very short so that the heads are confluent; leaflets triangular (Breckland).　　　　　　　　**(T. suffocatum)**
Internodes longer so that the heads are not confluent; leaflets obovate (Breckland).　　　　　　　**(T. glomeratum)**

1. Trifolium pratense L.　　Red Clover　　　　Ray, 1660
Throughout the county in grassy and waste places, and by field margins. Cultivated plants, 'var. *sativum* (Crome) Schreb.', are more robust, less hairy and paler flowered than the wild one. All squares. General.

2. Trifolium ochroleucon Huds.　　　　　　Ray, 1660
Recorded from a number of localities; especially characteristic of roadsides and trackways on the boulder clay. *25*, 26, 34–6, *46*, 64, [45, 54–6, 58, 65, 67]. In grassy places mainly on boulder clay in eastern England from Essex to Lincs and west to Northants; probably introduced elsewhere. A continental species with a restricted English distribution. (See map, page 13, and plate 3.)

3. Trifolium medium L.　　　　　　　　Ray, 1660
Recorded from a few localities in grassland of which the only extant ones are Sawston Hall and Dernford Fen. 44, 45, [25, 44, 47, 65]. General.

4. Trifolium incarnatum L.
Recorded from a few localities as a relic of cultivation. 35, *45*, 48, [46, 55–7]. Cultivated and naturalized in many parts of Britain, especially in the south.

LEGUMINOSAE

5. Trifolium arvense L.   Hare's-foot   Ray, 1660

Occurs in grassy and waste places on the sands in the east of the county, and formerly grew at Gamlingay. Elsewhere probably only a casual. Common in the Breckland just over the Suffolk border. 26, 44, *45*, 54, 56, 65, 67, 76, [23, 25, 46, 66, 41]. **FH**. Scattered throughout the British Isles north to Inverness and mainly coastal in Ireland.

6. Trifolium striatum L.   Ray, 1660

Recorded from a few localities in sandy and gravelly places. 25, 39, *46*, 54, 66, 67, 76, [36, 45]. **FH**. Scattered throughout much of the British Isles north to Kincardine; local and only coastal in Ireland.

7. Trifolium scabrum L.   I. Lyons, 1763

Recorded from a few localities in sandy or gravelly places. Only recent records at Isleham and Furze Hills. 54, 67, [25, 36, 45, 66]. **FH**. Scattered throughout southern Britain northwards to Yorks, then on the east coast to Kincardine; also on S.E. Irish coast. Often on less acid soils than *T. striatum*.

8†. Trifolium subterraneum L.   Ray, 1663

Formerly occurred in sandy places at Gamlingay, where it was last seen with certainty in 1859 (CGE), though Evans (1939) says it still grows at the Cinques. Evans also records it for Hildersham and Chippenham but no specimens supporting these records have been seen. *25*, ?66, [?54]. Local from Cornwall and Kent to Cheshire and Lincs; also in Co. Wicklow.

9. Trifolium hybridum L.   Alsike Clover
    A. Fryer, end of nineteenth century

Naturalized on roadsides, trackways and in waste places in scattered localities. Often sown in rye-grass mixtures for temporary leys. 24, *25*, 26, 34, 35, 44–7, 66, 67, 20, 40, 50, [36, 38, 49]. Introduced; naturalized in grassy places in the British Isles; native of continental Europe.

10. Trifolium repens L.   White Clover   Ray, 1660

In pastures and on waste ground, roadsides and trackways throughout the county. Frequently sown with grasses for temporary leys. Very variable in leaf-marking, and robustness; wild plants are generally smaller than the many cultivars. All squares. General.

100

11. Trifolium fragiferum L. Strawberry Clover Ray, 1660
On trackways, roadsides and in other grassy places throughout much of the county, especially on heavy clay soils. 24, 25, 34–9, 44–8, 55–9, 67, 41, [26, 49, 65, 66, 40]. Scattered throughout the southern half of the British Isles, north to Co. Sligo, Isle of Man and Durham; a few isolated localities in S.E. Scotland.

12. Trifolium resupinatum L.
Casual. *45*.

13. Trifolium campestre Schreb. Hop Trefoil Ray, 1660
T. procumbens sensu Bab.
In grassy and waste places throughout most of the county, generally less abundant than *T. dubium*. 24–6, 33–6, 43–9, 54–6, 64–8, 76, *41*, [23, 38, 40]. General, but rare in the north and west.

14. Trifolium aureum Poll. A. Fryer, late nineteenth century
T. agrarium auct.
Recorded from Horselode Fen and clover fields near Chatteris by A. Fryer, and there is a specimen in CGE collected by C. E. Moss at Gamlingay in 1913. [25, 38, 48.] Naturalized in fields and waste places in many parts of the British Isles; native of continental Europe and Asia Minor.

15. Trifolium dubium Sibth. Ray, 1660
T. minus Sm.
Common in grassy and waste places, throughout the county. Similar in general appearance to *Medicago lupulina*; for distinguishing characters see under that species. All squares except *57*. General.

16. Trifolium micranthum Viv. J. Martyn in I. Lyons, 1763
T. filiforme L. nom. ambig. et sensu Bab., et Evans
Recorded from a few localities in sandy or gravelly places. Only seen recently in lawns in Cambridge. 45, [25, 38, 46, 48, 49, 54, 58]. In England, Wales and Ireland, but rarer in the north and only just reaching Scotland.

LEGUMINOSAE

ANTHYLLIS L.

1. Anthyllis vulneraria L.   Kidney Vetch   S. Corbyn, 1656

In dry grassy places on the chalk and sands, local. 23, 24, 33–6, 44–6, 54–6, 66, 67, 41. GG. General but more abundant on calcareous soils.

LOTUS L. (Birdsfoot Trefoil)

1. Lotus corniculatus L.   Ray, 1660

Throughout the county in grassy places. All squares except 68. General.

2. Lotus tenuis Waldst. & Kit. ex Willd.   Ray, 1660

A very local species of roadsides and other grassy places on chalk and boulder clay. *25, 34,* 35, 45, 59, 40, [26, 36, 44, 46, 66]. DF. Local, almost confined as a native to England south and east of a line from the Humber to Land's End.

3. Lotus uliginosus Schkuhr.   Ray, 1660

L. major sensu Bab.

A local species of marshy places, mainly in the Fenlands, but not characteristic of alkaline fens. Now rare at Wicken Fen though formerly apparently more abundant there. 25, 34, 43–6, 49, 54–7, 64–8, 20, [24, 38, 48, 58]. DF. General, except the extreme north.

GALEGA L.

1. Galega officinalis L.

Garden escape or casual. 45, 56.

ROBINIA L.

1. Robinia pseudoacacia L.

Planted tree. 25.

ASTRAGALUS L.

1. Astragalus danicus Retz.   Purple Milk-vetch   Ray, 1660

A. hypoglottis sensu Bab.

A local, characteristic species of old chalk grassland; occurring also on the Breckland sands. 24, *44,* 45, 54–6, 66, [23, 65]. F. Scattered over Britain from Wilts to Sutherland but mainly in the eastern half of the country; in Ireland only in the Aran Islands.

2. Astragalus glycyphyllos L.   Licorice   Ray, 1660
Recorded from several localities in hedgebanks and thickets or on
roadsides. *44*, 45, 54, 64, [25, 35, 46, 55, 56, 66]. CH. Scattered throughout Great Britain but very rare in the west and absent from Ireland.

ORNITHOPUS L.

1. Ornithopus perpusillus L.   Birdsfoot   Ray, 1660
Found only in grassy places on the sands at Gamlingay and in the east
of the county. Common in the Breckland just over the Suffolk border.
25, 54, 66, 67, 76. G. In Britain north to Moray, very rare in Ireland.

CORONILLA L.
1. Coronilla varia L.
Garden escape. 56, 67, [44, 46].

HIPPOCREPIS L.

1. Hippocrepis comosa L.   Horseshoe Vetch   Ray, 1685
A local species characteristic of old chalk grassland. 35, 44, 45, 54–6,
66, [23, 34]. Local in Britain north to Westmorland.

ONOBRYCHIS Mill.

1. Onobrychis viciifolia Scop.   Sainfoin   J. Gerarde, 1597
Probably native in old grassland on the chalk; elsewhere a relic of
cultivation; extensively grown as a fodder crop. 23, 24, 26, 33–6, 43–5,
54–7, 64–8, 76, [38]. GG. From Dorset and Kent northward to Durham,
introduced elsewhere.

VICIA L. (Vetch; Tare)

The first three species are similar in general appearance but can be distinguished as follows:
Calyx teeth subequal, somewhat exceeding tube; pod hairy, usually
  2-seeded; flowers 4–5 mm.   **1. V. hirsuta**
Calyx teeth unequal, upper two shorter than tube; pod glabrous, 4-
  seeded; flowers *c.* 4 mm.   **2. V. tetrasperma**
Calyx teeth unequal, upper two shorter than tube; pod glabrous, 5–8-
  seeded; flowers *c.* 8 mm.   **3. V. tenuissima**

1. Vicia hirsuta (L.) Gray   Hairy Tare   Ray, 1660
Scattered over the county in grassy places and on railway tracks. 23–5,
33, 36, 39, 43–7, 54, 56, 64, 65, 30, 40, [35, 38, 48, 49, 58]. Throughout
the British Isles; rare in Scotland and Ireland.

LEGUMINOSAE

2. Vicia tetrasperma (L.) Schreb.   Ray, 1660

In grassy and waste places in a number of scattered localities. 24, *25*, 29, 35, 36, 38, 45–8, [49, 54, 66]. Throughout England, rare in Wales and S. Scotland, naturalized in a few places in N. Scotland and Ireland.

3. Vicia tenuissima (Bieb.) Schinz & Thell.   Babington, 1860
V. gracilis Lois.

In a few localities in grassy places on the boulder clay. 25, 34–6, 45, 46. From Cornwall and Kent to Worcester, Cambs and Lincs; local.

4. Vicia cracca L.   Ray, 1660

Throughout the county in hedgerows and bushy and grassy places. All squares except 26, *54*, 20, 41. General.

5. Vicia orobus DC.

Casual. *45*.

6†. Vicia sylvatica L.   J. Hemsted, 1792

The only record is by J. Hemsted from Hall Wood near Wood Ditton in 1792. [65.] Scattered throughout the British Isles, local.

7. Vicia sepium L.   Ray, 1660

In grassy and bushy places, but not so common in Cambs as in most other English counties. 24, 25, 34–6, *38*, 43–5, *49*, 54–6, 58, 64–6, 76, [48]. General.

8. Vicia lutea L.

Casual. *35*.

9. Vicia hybrida L.

Casual. *45*.

10. Vicia sativa L. sens. lat.

The following taxa can be distinguished:

Leaflets obovate or oblong; flowers solitary or in pairs, 10–20 mm.; pods 50–70 mm.   **V. sativa** sens. strict.

Leaflets oblong; flowers usually in pairs, 10–15 mm.; pods 35–50 mm.
                                            V. angustifolia var. **angustifolia**

Leaflets linear; flowers usually solitary, 10–15 mm.; pods 25–35 mm.
                                            V. angustifolia var. **bobartii**

104

Vicia sativa L. sens. strict. Ray, 1660

Scattered over the county in grassy and waste places. Usually a relic of cultivation but sometimes casual. 25, 26, 33–6, 38, 39, 44–6, 54, 55, 57, 65–7, 76, [58, 40]. Introduced throughout the British Isles; probably native of W. Asia.

Vicia angustifolia L. Ray, 1690

V. sativa var. *angustifolia* (L.) Roth

On roadsides and in grassy and waste places throughout the county. Most plants are referable to var. **angustifolia**, but var. **bobartii** (Forst.) Koch is recorded from Gamlingay (1835 and 1836) and Kennett Heath (1954). All squares except 26, 38, 57, 59, 68, 20. General but local in the north.

11. Vicia lathyroides L. ?R. B. Smart in Babington, 1860.
First certain record H. Gilbert Carter and T. G. Tutin, 1932

Except for an unsubstantiated record from near Balsham by R. B. Smart this species is known only from a few localities on the sands in the east of the county. It is common just over the border in the Breckland of Suffolk. 67, 76, [?55]. K. Local, in dry grassy places.

12. Vicia faba L. Broad Bean, Horse Bean

Recorded as an escape from cultivation. It is widely grown as a crop (3600 acres in the Isle of Ely and 2600 in S. Cambs). 34, 36, 45.

LATHYRUS L.

1. Lathyrus aphaca L. Ray, 1660

Only seen recently in the Hardwick and Coton areas and at Cherry Hinton, but formerly in a number of scattered localities in grassy places. 34, 35, 45, 54, [44, 55, 56, 58, 65, 66, 40]. From Devon and Kent to Worcester and Norfolk, also Glamorgan and Denbigh; very local.

2. Lathyrus nissolia L. Ray, 1663

Rather rare, in grassy places. 25, 26, 35, 36, 45, 56, [47, 58]. GG. From Cornwall and Kent to Cheshire and S.E. Yorks; very local.

3. Lathyrus pratensis L. Fingers-and-thumbs Ray, 1660

Common in grassy and waste places throughout the county. All squares except 49, 59, 30. General.

LEGUMINOSAE

4. Lathyrus sylvestris L.   S. Corbyn, 1656

Rare in woods and copses mainly on the boulder clay. 25, 35, 36, 43, 45, [54, 56, 64, 66]. H. Scattered throughout Great Britain, local.

5. Lathyrus latifolius L.   Everlasting Pea

Garden escape or casual. 48, 58, [46, 55].

6. Lathyrus palustris L.   Relhan, 1820

Formerly occurred in several places in the Fens but now only to be found at Wicken, where it is plentiful. The record by J. Martyn from Little Eversden is almost certainly referable to *L. sylvestris*. 56, [58, 31]. W. Scattered over England, Wales and Ireland; rare.

(The Pea, **Pisum sativum** L., is commonly grown as a crop (7300 acres in the Isle of Ely and 800 acres in S. Cambs) and often occurs as a relic of cultivation.)

ROSACEAE

FILIPENDULA Mill.

1. Filipendula vulgaris Moench.   Dropwort   Ray, 1660

Spiraea filipendula L.; *F. hexapetala* Gilib.

In scattered localities in dry grassland, mainly on the chalk. *25*, 26, 35, 36, 44–6, 54–6, 65, 66, [23, 24, 34, 58]. GG. Widespread in England but rather local; rare elsewhere.

2. Filipendula ulmaria (L.) Maxim.   Meadow-sweet   Ray, 1660

Spiraea ulmaria L.

By ditches, rivers, pits and other damp places throughout the county, though local north of Wisbech. All squares except 41, [29]. General.

RUBUS L. (Blackberry)

This account of the Cambridgeshire *Rubi* is based mainly on the work of W. H. Mills. He first took a serious interest in them when he showed H. J. Riddelsdell round the county in 1930. The results of this joint study were published by Riddelsdell (1931). During the next 30 years he examined *Rubi* in all parts of the county and found that the only areas that held any number of species were the Gamlingay greensands and some of the boulder-clay woods. P. D. Sell has seen nearly all the recorded species in the county since 1950 and is responsible for a large number of the field records for *R. conjungens* and *R. ulmifolius*. Anyone wishing to study the genus should consult the monograph by W. C. R. Watson (1958).

ROSACEAE

1. Rubus idaeus L.   Raspberry   Relhan, 1793

In a few localities in woods, copses and fens, sometimes if not always of garden origin. 44, 45, 56, 31, [24, 25, 48, 49, 65]. General, but rare in the Cambridge area and the extreme south-west of Britain and Ireland.

2. Rubus caesius L.   Dewberry   Ray, 1660

Common on ditchsides and in hedgerows, copses, woods and scrub throughout the county. All squares except 29, 39, 47, 59, 68, 30, [38]. Common in England, Wales and S. Ireland, local elsewhere.

× **ulmifolius** Schott f. This hybrid was said to be common by W. H. Mills but he mentioned no particular localities.

3. Rubus conjungens (Bab.) W. C. R. Watson   Babington, 1860

R. corylifolius var. *conjungens* Bab.

In hedgerows and scrub throughout most of the county. 25, 29, 34, 35, 38, 45, 48, *54*, 55–8, 68, 20, 31, 40, 50, [36, 46]. General.

4. Rubus sublustris Lees   Babington, 1860

R. corylifolius var. *sublustris* (Lees) Bab.

Recorded by Babington at Cambridge and still to be found at Gamlingay. 25, [45]. **G.** General.

5. Rubus balfourianus Bloxam ex Bab.   Babington, 1848

Babington (1860) recorded this species in several localities, but it has not been found since, so the identifications may be wrong. [35, 45, 57, 65.] General.

6. Rubus warrenii Sudre   W. H. Mills, 1948

Known only from Gamlingay and Wicken Fen. 25, 57. **W.** England, local.

7. Rubus halsteadensis W. C. R. Watson

H. J. Riddelsdell and W. H. Mills, 1930

R. raduliformis (A. Ley) W. C. R. Watson, non Sudre

Known only from Over and Borley Woods. 37, 54. Frequent in England and Wales.

8. Rubus purpureicaulis W. C. R. Watson   Babington, 1860

R. corylifolius var. *purpureus* Bab.

In hedgerows in a few places mostly on the boulder clay. 35, 37, [65, 40]. England and Wales.

107

ROSACEAE

9. Rubus tuberculatus Bab.   Babington, 1851

R. dumetorum var. *tuberculatus* (Bab.) Rogers

Said by Mills to be widely distributed, but there are few actual records, which are mainly on the boulder clay. It seems to be exceptionally common around Dry Drayton. 35, 36, 56, [45, 57, 65]. Common in England.

10. Rubus babingtonianus W. C. R. Watson   W. W. Newbould, 1847

R. altheaefolius sensu Bab.

In hedges and woods in a few places on the boulder clay. Some of the old records by Babington may not be correct. 35, 45, 54, [34, 44, 46, 47, 56, 66]. H. Scattered over Great Britain and in one locality in Ireland.

11. Rubus scabrosus P. J. Muell.   W. H. Mills, 1952

On the boulder clay at Kingston, Linton, Hardwick and Bourn. 35, 54. H. Common in England and Wales.

12. Rubus myriacanthus Focke   Babington, 1855

R. diversifolius sensu Bab.; *R. dumetorum* var. *myriacanthus* (Focke) Evans

Mainly on the boulder clay, at Willingham, Gog Magogs, Hardwick, Hildersham, Caldecote, Kingston and Cambridge. 35, 45, 47, [54]. H. General.

13. Rubus nemoralis P. J. Muell. (*R. selmeri* Lindeb.)

Only known from Wilberforce Road, Cambridge, where it is almost certainly introduced. 45. General.

14. Rubus laciniatus Willd.

Garden escape. 57.

15. Rubus lindleyanus Lees   H. J. Riddelsdell, 1930

Recorded from Doddington Wood and Gamlingay (in Evans (1939)) and from Hayley Wood where it was last seen by W. H. Mills in 1945. *25, 49.* General.

16. Rubus poliodes W. C. R. Watson   W. H. Mills, 1953

R. mercicus auct.

Known only from Borley Wood. 54. England and Wales, local.

17. Rubus pyramidalis Kalt. Moss, 1912
Known only at Gamlingay. 25. G. General.

18. Rubus polyanthemus Lindeb. W. H. Mills in Evans, 1939
Occurs at Gamlingay and Borley Wood. 25, *54*. G. General.

19. Rubus rhombifolius Weihe ex Boenn. H. J. Riddelsdell, 1930
R. subcarpinifolius (Rogers & Riddelsd.) Riddelsd.
Known only at Gamlingay. 25. G. General.

20. Rubus cardiophyllus Muell. & Lefèv. W. H. Mills, 1930
R. rhamnifolius auct.
Known only at Gamlingay. 25. G. General but commonest in England and Wales.

21. Rubus ulmifolius Schott Ray, 1660
R. discolor sensu Bab.; *R. rusticanus* Merc.
Common in hedgerows, copses and scrub throughout the county. One of the few brambles that grows in plenty on the chalk. This is also one of the few sexual diploid species of the *R. fruticosus* aggregate; most other species are apomictic tetraploids. It can be easily recognized by its rather small dark green leaves which are white-felted beneath, its glabrous pruinose stem, and by its usually rose-pink flowers. 24–6, 29, 34–8, 44–6, 48, 49, 54–8, 65, 68, 76, 20, 31, 40, 50. General.

22. Rubus procerus P. J. Muell.
Commonly grown in gardens and probably escaping more frequently than records suggest. 34, 45, 46.

23. Rubus falcatus Kalt. ?Babington, 1856
First certain record, C. E. Moss, 1911
R. thyrsoideus sensu Evans, et ?Bab.
Now known only at Gamlingay, Babington's (1860) localities of *R. thyrsoideus* may be referable to this species. 25, [?34, ?35, ?45]. G. England especially the Midlands.

24. Rubus vestitus Weihe & Nees
H. J. Riddelsdell & W. H. Mills, 1930
Frequent in woods on the eastern boulder clay and introduced round Cambridge. 43, 45, 46, 54, 55, 65. DPW. Common in England and Wales.

ROSACEAE

25. Rubus conspicuus P. J. Muell.   W. H. Mills, 1945
Known only in Lt. Chishill Wood (v.c. 19). 43. England, very local.

26. Rubus criniger (E. F. Linton) Rogers   W. H. Mills, 1930
Known only at Gamlingay. 25. G. England, common.

27. Rubus radula Wiehe ex Boenn.   W. Mathews jun., 1851
In woods mostly on the boulder clay. 24, *25*, 35, 43, 54. General.

28. Rubus discerptus P. J. Muell.
                        C. D. Pigott, P. D. Sell and S. M. Walters, 1951
Known only from Morden Grange Plantation. 24. General.

29. Rubus echinatoides (Rogers) Sudre   H. J. Riddelsdell, 1930
The only record is by H. J. Riddelsdell from Gamlingay in 1930. *25*.
General.

30. Rubus foliosus Weihe & Nees   H. J. Riddelsdell, 1930
R. flexuosus Muell. & Lefev.
Known only at Gamlingay. 25. G. Common in England, rare elsewhere.

31. Rubus insectifolius Muell. & Lefèv.   W. H. Mills, 1930
R. fuscus var. *nutans* Rogers
Known only from Lt. Chishill Wood (v.c. 19). 43. S. England, frequent.

32. Rubus rufescens Muell. & Lefèv.
                        H. J. Riddelsdell & W. H. Mills, 1930
R. hystrix var. *infecundus* Rogers
In Hildersham and Borley Woods and Sparrows Grove on the eastern
boulder clay. 54, 65. General.

33. Rubus dasyphyllus (Rogers) Rogers   Babington, 1854
R. koehleri var. *pallidus* sensu Bab.
Recorded by Babington (1860) from Balsham and Caldecote but it has
not been seen since. [35, 55.] General.

34. Rubus bellardii Weihe & Nees   W. H. Mills, 1947
Known only at Longstowe Wood. 35. Rare in England and Wales.

POTENTILLA L. (Cinquefoil)

1†. Potentilla palustris (L.) Scop.   Ray, 1660
Comarum palustre L.
Formerly occurred in marshy places at Gamlingay, Wicken Fen and
Wimblington. Last seen at Wicken Fen in 1886. [25, 49, 57.] General
on peaty soils.

2. Potentilla sterilis (L.) Garcke   Barren Strawberry   Ray, 1660
P. fragariastrum Pers.
Restricted to the boulder-clay woods where it is locally abundant.
(For vegetative distinction from *Fragaria vesca*, see that species.) 25, 26,
35, 36, 43, 54, 55, 64, 65, [49, 56, 66]. B. General, but rare in N.
Scotland.

3. Potentilla anserina L.   Silverweed   Ray, 1660
On roadsides and grassy and waste places throughout the county. All
squares. General.

4. Potentilla argentea L.   Ray, 1685
A rare species found only at Hildersham Furze Hills, Chippenham and
Kentford Heath. There are old records for Gamlingay and Cambridge.
54, 66, 76, [25, 45]. FH. In Britain from Moray and Cumberland
southwards but local, especially in the west.

5. Potentilla canescens Bess.
P. inclinata M. & K.
Casual. [45.]

6. Potentilla recta L.
Garden escape. *45.*

7. Potentilla supina L.
Casual. [45.]

8. Potentilla norvegica L.
Casual or garden escape. 45, [56].

9. Potentilla tabernaemontani Aschers.   Relhan, 1785
P. verna sensu Bab., et Evans
A rare species found only in a few places in chalk grassland. Also

ROSACEAE

recorded from Gamlingay by Relhan. 45, 55, [25, 66]. **GG.** Very local from Banff and Cumberland to Suffolk, Hants and Somerset.

10. Potentilla erecta (L.) Räusch.   Tormentil   Ray, 1660
P. tormentilla Stokes

Scattered over the county on damp peaty ground, but very local. 24, 25, 35, 36, 44, 45, 49, 54–6, 65, 66, 40, [34, 46–8, 58]. **W.** General, but preferring neutral to acid soils.

× **reptans.** Grows with the parents at 'Sawston Moor' in the grounds of Sawston Hall. 44.

11. Potentilla anglica Laichard. ?W. W. Newbould, in Babington, 1860
         First certain record H. Gilbert Carter and S. M. Walters, 1945

P. procumbens Sibth.

Recorded from several localities. The only record supported by a herbarium sheet is from Sutton Meadlands in 1945 (CGE). The species behaves as a calcifuge at least in East Anglia (cf. *Montia fontana*). 37, [?25, ?47, ?49, ?56]. Widespread but local in the British Isles and very rare in N. Scotland.

× **reptans.** Growing with parents at Sutton Meadlands in 1945 (det. S. M. Walters conf. D. H. Valentine). 37.

12. Potentilla reptans L.   Creeping Cinquefoil   Ray, 1660

Common throughout the county. The var. **microphylla** Trattinick has been recorded from Chippenham. All squares. General, but rare in N. Scotland.

FRAGARIA L.

1. Fragaria vesca L.   Wild Strawberry   Ray, 1660

Common in woods and copses especially on the boulder clay. Easily confused with *Potentilla sterilis* when not in fruit. It can readily be distinguished from it, however, by the silky appressed (not spreading) hairs on the back of the leaflets. 24–6, 35, 36, 39, 43–5, 54–6, 64–6, 30, 40, [23, 49]. General.

2. Fragaria ananassa Duchesne   Garden Strawberry
                         E. A. George and E. F. Warburg, 1939
F. chiloensis auct.

112

Naturalized in a number of localities especially on railway banks. About 2000 acres are grown yearly in the Fens, mainly around Wisbech. 23, 25, 33, 38, 46. This commonly cultivated strawberry originated in France as a hybrid between the two octoploid species, *F. virginiana* Duchesne from eastern N. America and *F. chiloensis* Duchesne from Chile.

GEUM L.

1. Geum urbanum L.   Wood Avens   Ray, 1660
Throughout the county in hedgebanks, woods, thickets and streamsides. All squares except 37, 59. General, but rare in N. Scotland.

2. Geum rivale L.   Water Avens   Relhan, 1785
Occurs in a few localities, mainly in woods on the eastern boulder clay, where, however, it is locally abundant. 44, 45, 54, 65, [25, 64, 67]. **DPW.** Rather common in Scotland, N. England and Wales, local in S. England; widespread but local in Ireland.

× **urbanum** (*G.* × *intermedium* Ehrh.). Occurs with the parents in a few woods on the eastern boulder clay. It is fertile and forms hybrid swarms showing a large range of variation. 65. **DPW.**

AGRIMONIA L.

1. Agrimonia eupatoria L.   Agrimony   Ray, 1660
In hedgebanks, roadsides, waste places and field margins throughout the county. All squares except 39, 30, [49, 40]. General.

2. Agrimonia odorata (Gouan) Mill.   C. E. Moss, *c.* 1913
Recorded by C. E. Moss from Weston Colville and Gamlingay (*c.* 1913) and by E. W. Jones from Brinkley in 1932. There are no herbarium specimens to support these records. This is a species to be looked for on the sands at Gamlingay and in the east of the county. It is distinguished from *A. eupatoria* by the leaves having numerous, not few or no, glands beneath, and by the basal spines of the fruit being deflexed, not spreading laterally. 65, [25]. In Britain from Inverness southwards; widespread but local in Ireland; usually absent from calcareous soils.

ALCHEMILLA L.

1. Alchemilla vestita (Buser) Raunk.   Lady's Mantle   Ray, 1685
A. vulgaris sensu Bab., et Evans
The locality known to Ray and R. B. Smart at Balsham was rediscovered

ROSACEAE

by J. Rishbeth in 1953. The plant grows in damp grazed pasture near the wood. The records of 'A. vulgaris' from Linton and Gamlingay cannot be determined in the absence of herbarium specimens, but they were probably *A. vestita*. 54, [?25]. General; the commonest *A. vulgaris* species in lowland England.

APHANES L.

1. Aphanes arvensis L. sens. lat.   Parsley Piert   Ray, 1660
Alchemilla arvensis (L.) Scop.

The aggregate can be divided as follows:
Lobes of stipules surrounding inflorescence triangular-ovate; fruit 2·2–2·6 mm., with diverging calyx-teeth.   **A. arvensis** sens. strict.
Lobes of stipules surrounding inflorescence oblong; fruit 1·4–1·8 mm., with converging calyx-teeth.   **A. microcarpa**

Aphanes arvensis L. sens. strict.   Henslow, 1826
Scattered over the county as a weed of arable land and waste places. 23–5, 34, 37, 39, 43–5, 54, 55, 65–7, 76, 40. General.

Aphanes microcarpa (Boiss. & Reut.) Rothm.   S. M. Walters, 1949
A rather rare weed of arable land and waste places, mainly on the sands. 25, 39, 45, 54, 65, 66, 76. G. General.

SANGUISORBA L.

1. Sanguisorba officinalis L.   Great Burnet   Ray, 1660
Very local and now much reduced by drainage and reseeding of old pastures; formerly abundant near Cottenham. 25, *35*, 36, 47, [29, 37, 45, 46, 55]. From Hants, Surrey and Suffolk westwards and northwards to Ayr and Berwick; in Ireland in W. Mayo, Down, Antrim and Londonderry.

POTERIUM L.

1. Poterium sanguisorba L.   Salad Burnet   Ray, 1660
A characteristic species of chalk grassland but not confined to it. 23–6, 33–6, 44–6, 54–7, 64–6, 76, [47, 49, 58]. **GG.** Widespread and common on calcareous grassland in England and Wales; very local in Scotland and introduced north of Dunbarton and Angus; widespread in S.E. Ireland to mid-Cork, E. Galway and Meath, also Donegal and Antrim.

2. Poterium polygamum Waldst. & Kit.

W. Mathews jun. & W. W. Newbould, 1849

P. muricatum Spach

Recorded from a number of localities by the margins of fields and in waste places. 34, 38, 45, *46*, 55, 65, 66, [35, 67]. Formerly introduced as a fodder crop and still to be found naturalized in many places in S. England and Wales, and occasionally elsewhere. Native of the Mediterranean region extending to C. Asia.

ROSA L. (Rose)

This account of the genus *Rosa* in Cambs is based mainly on the work done by W. H. Mills. This was first carried out in the 1930's and the results published in Evans (1939), but most if not all the localities were visited again after 1950, and all Mills's records are here recorded as recent. Many of Mills's identifications have been confirmed by R. Melville. Anyone wishing to make a study of this complex group should consult A. H. Wolley-Dod (1930–31).

1 Styles united into a column, which equals at least the shorter stamens; flowers white; trailing shrub.   **1. R. arvensis**
   Styles free or united into a short column; flowers pink or white; not trailing.   *2*

2 Flowers always solitary, without bracts; leaflets small, 3–5 pairs; stems densely prickly and bristly.   **(R. pimpinellifolia)**
   Flower 1 or more, bracts resembling the stipules; leaflets 2–3(–4) pairs; stems not bristly or with few bristles.   *3*

3 Styles united into a column in flower, becoming free in fruit; disc conical, prominent.   **2. R. stylosa**
   Styles free throughout; disc flat or nearly so.   *4*

4 Leaves ±densely covered over the whole surface beneath with conspicuous brownish viscid fruity-scented glands, glabrous or somewhat pubescent but not tomentose; prickles hooked.   *5*
   Leaves not glandular beneath, or with glands confined to the main veins, or if with glands scattered over the whole surface then usually ±densely tomentose; the glands smaller, not fruity-scented; prickles hooked to straight.   *7*

5 Leaflets cuneate at base; pedicels glabrous.   **(R. elliptica)**
   Leaflets rounded at base; pedicels glandular-hispid.   *6*

6 Stems erect; prickles usually unequal; styles pilose; sepals persistent at least till the fruit reddens, erect or spreading.   **8. R. rubiginosa**
   Stems arching; prickles ±equal; styles glabrous; sepals soon falling, usually reflexed.   **9. R. micrantha**

ROSACEAE

7 Prickles straight or slightly curved; leaves always pubescent,
usually very tomentose, always doubly serrate.   8
Prickles hooked or strongly curved; leaves glabrous or pubescent,
simply or doubly serrate.   9

8 Stems flexuous (often zigzag) but scarcely arching; leaves bluish
green; pedicels relatively short; sepals erect or ascending, persistent till the fruit ripens; styles usually villous; disc not more
than 3·5 times diam. of orifice.   **7. R. sherardii**
Stems arching; leaves pale green, not bluish; pedicels relatively
long; sepals erect to reflexed, falling before the fruit is ripe,
though usually persistent till it reddens; styles pilose or glabrous,
rarely villous; disc 4–6 times diam. of orifice.   **6. R. tomentosa**

9 Leaves never pubescent or glandular beneath.   **3. R. canina**
Leaves pubescent at least on the veins beneath and often with
glands on the main veins.   10

10 Outer sepals with narrow usually entire lobes.   **4. R. dumetorum**
Outer sepals with broad, usually toothed or lobed lobes; prickles
more strongly hooked and with stouter bases.   **5. R. obtusifolia**

1. Rosa arvensis Huds.   T. Martyn, 1763
Common on the boulder clay in hedgerows, local elsewhere and absent
from the northern Fens. The var. **gallicoides** (Bak.) Crep. is recorded
from a number of places on the boulder clay. *24*, 25, 26, 34–6, 43, 44, 54,
64, 65, [23, 38, 45, 46, 48, 49, 55–8]. Common in S. England and Wales,
becoming local in N. England and rare in Scotland; throughout Ireland
but local.

(**R. pimpinellifolia** L. (*R. spinosissima* sensu Bab.). Babington (1860) gives
this species from Swaffham Prior on the authority of Henslow (1826) and in
his notes records it for Burwell. There are no specimens in CGE to support
these records. [?56.] General but especially near the sea.)

2. Rosa stylosa Desv.   Babington, 1855
The var. **systyla** (Bast.) Bak. occurs in a number of localities mainly on
the boulder clay. 25, 34–6, 45, 65, [38]. Local in S. England, Wales and
S. Ireland.

3. Rosa canina L.   Ray, 1660
A very variable species common throughout the county in hedgerows,
woods and scrub. Five varieties and one form are recorded by Mills:
var. **lutetiana** (Leman) Bak. is frequent; var. **senticosa** (Ach.) Bak. is at
Madingley; var. **spuria** (Pug.) W.-Dod is frequent and its forma

116

syntrichostyla (Rip.) Rouy occurs in several places; var. **carioti** (Chab.) Rouy is frequent in chalky places and occasional elsewhere; and var. **verticillacantha** (Mérat) Bak. is in a few places. Other varieties were recorded by Babington but his specimens were not seen by Mills and may have been wrongly identified. All squares except 37, 65, 67, 30. General but rare in the north.

4. Rosa dumetorum Thuill.   W. H. Mills in Evans, 1939
This species is often placed under *R. canina* but Mills followed Wolley-Dod in keeping it distinct. Mills recorded var. **typica** forma **urbica** (Leman) W.-Dod (*R. forsteri* Sm.) and forma **semiglabra** (Rip.) W.-Dod as frequent; var. **inserta** (Deseg.) W.-Dod from near Coton; var. **mercica** W.-Dod from several places; and var. **seticaulis** W.-Dod as rather frequent in the district west of Cambridge. 34–6, 45–7, 58. General.

5. Rosa obtusifolia Desv.   W. H. Mills, 1940
The var. **tomentella** (Leman) Bak. occurs in a few places on the boulder clay. According to Mills, Cambs plants invariably lack subfoliar glands and presumably come under forma **canescens** (Bak.) W.-Dod. 25, 45, 46, 54. Local in England, Wales and Ireland, but distribution uncertain.

6. Rosa tomentosa Sm.   T. Martyn, 1763
R. villosa sensu Bab.
Frequent on the boulder clay in woods and hedgerows, rare elsewhere. Mills recorded var. **pseudocuspidata** (Crep.) Rouy from several places. 24–6, 35, 36, 43, 45, *46*, 54, 56, 64, 65, [38, 58, 66]. Common in England, Wales and Ireland, rare in Scotland.

7. Rosa sherardii Davies   W. H. Mills in Evans, 1939
Mills found the var. **omissa** (Desegl.) W.-Dod by the Beck Brook near Girton in the 1930's but it can no longer be found. *46*. General in Great Britain but commoner in north, rare in Ireland.

8. Rosa rubiginosa L.   Ray, 1670
Scattered about the county in hedgerows and woods, especially on the boulder clay. 25, 35, 38, 44, 45, 48, 49, 54, 55, 64–7, [36, 46, 41]. **FH.** Widespread in England and Wales, rarer in Scotland and Ireland.

117

ROSACEAE

9. Rosa micrantha Borrer ex Sm.   R. Ross, 1934–37

Frequent in woods on the boulder clay, rare elsewhere. *24, 25, 26, 35, 36, 45, 54, 55, 64–6.* H. Common in England and Wales, rare in Scotland and Ireland.

(**R. elliptica** Tausch (*R. inodora* sensu Bab.). Babington (1860) recorded this species from Hinton, Swavesey, Madingley and Snailwell. There are no specimens in CGE so the species must be regarded as unconfirmed for the county. [?36, ?45, ?66.])

PRUNUS L.

1. Prunus spinosa L.   Blackthorn; Sloe   Ray, 1660

P. communis sensu Bab. pro parte

Common throughout the county in hedgerows, woods, copses and scrub. All squares except 29, 59, 68, 30. General, but rare in the far north.

2. Prunus domestica L.   Ray, 1685

P. communis sensu. Bab. pro parte

Through much of the county in hedgerows, woods and copses. The ssp. **domestica** ('Plum') with a large (4–8 cm.) variously coloured fruit is frequent in hedgerows, and is widely cultivated. The ssp. **insititia** (L.) C. K. Schneid. (*P. insititia* L.) ('Bullace') with smaller (2–4 cm.) blue-black fruit is said by Evans (1939) to be frequent in the boulder-clay woods but the records do not support this. The 'Bullace' does not now appear to be much grown but the 'Damson', which is usually regarded as a form of it, is more widely cultivated, as is also the ssp. **italica** (Borkh.) Hegi ('Greengage') with a large green fruit. 24, 25, 34, 35, 38, 43, 45, 49, 54–6, 58, 66, 20, 31, 41, 50, [36, 46–8]. General, but rare in the north.

3. Prunus cerasifera Ehrh.   Cherry-Plum

                H. Gilbert Carter & T. G. Tutin, 1932

Introduced in hedgerows, woods and copses in a number of scattered localities. *25, 34, 45,* 46, *49,* 54, *55,* 65, 68. Planted in many parts of England; native of Asia.

4. Prunus avium (L.) L.   Wild Cherry; Gean   Ray, 1685

Rather rare in copses and hedgerows, mainly on the chalk and clay, usually as single trees. An old tree known to Babington in Lime Kiln Close, Cherry Hinton, is now dead, but many young saplings are growing from root suckers. Babington states (1860) that Ray recorded the Wild Cherry from here, but the record is not in the obvious Ray sources, and the first certain record for Hinton is in J. Martyn (1727) annotated by T. Martyn (c. 1760). References to cherries grown at Hinton go back to Elizabethan times. 25, 34, 35, 45, 55, 65, 66, 67, [36, 47, 40]. Rather common in England, Wales and Ireland, becoming rare in N. Scotland.

5. Prunus cerasus L.   Sour Cherry   P. Dent in Ray, 1685

Recorded from Teversham by P. Dent and J. Fisher, from Gamlingay Wood by R. S. Adamson, and from Chippenham by A. Fryer. None of these records is substantiated by herbarium specimens. [25, 45, 66.] Introduced. Frequent in S. England, Wales and Ireland but local; rare elsewhere. The garden Sour and Morello Cherries belong to this species and all our wild plants are probably descended from them.

6. Prunus padus L.   Bird-Cherry   Relhan, 1786

Now only to be found in the Chippenham area but formerly recorded in a few other places. 45, 65, 66, 67, [25, 35, 46, 40]. C. Throughout much of the British Isles but most frequent in the north and probably not native in the south.

7. Prunus laurocerasus L.

Planted tree. 35, 36.

8. Prunus lusitanica L.

Planted tree. [25.]

9. Prunus serotina Ehrh.

Planted tree. [25.]

COTONEASTER Medic.

1. Cotoneaster simonsii Bak.

Garden escape. 55.

2. Cotoneaster horizontalis Decne   J. Rishbeth, 1940

Recorded from walls at Trumpington and Cambridge and well naturalized (presumably bird-sown) at Cherry Hinton chalk pit. Commonly cultivated in gardens. 45. Widely naturalized in Britain; native of W. China.

CRATAEGUS L. (Hawthorn; Whitethorn)

1. Crataegus oxyacanthoides Thuill.   ?Ray, 1660
                                                 First certain record, Babington, 1835

C. oxyacantha sensu Bab. pro parte et sensu Evans

A characteristic species of boulder-clay woods where it appears to be

ROSACEAE

native. This tree has a much looser habit of growth than *C. monogyna* and is typically much less thorny. *24*, 25, 26, 34–7, 43–6, 49, *54–6*, 58, 64, 65, 76, 20, [57]. H. Local in England and mainly in the south-east.

2. Crataegus monogyna Jacq.   ?Ray, 1660.
First certain record, Babington, 1840
C. oxyacantha sensu Bab. pro parte
Common throughout the county as a hedgerow bush; common also as a small tree in woods and copses. All squares. General.

× oxyacanthoides
Frequent where the parents are found together, and occurring sometimes in their absence. *24*, 25, 35, *45*, *56*, *64*, 65, [38]. H.

(**Mespilus germanica** L., Medlar, is occasionally planted for its fruit.)

SORBUS L.

1. Sorbus aucuparia L.   Rowan; Mountain Ash   S. Corbyn, 1656
Pyrus aucuparia (L.) Ehrh.
In a number of scattered localities, where it was very probably originally planted; effectively naturalized or native at Chippenham and Gamlingay. 25, 33, 34, *45*, 56, 66, [38, 65, 41]. G. General, commonest in the north and west.

2. Sorbus torminalis (L.) Crantz   Service Tree   A. M. Smith, 1909
Known only from Gamlingay Wood where there is at least one fine tree and numerous saplings from root suckers. The Gransden locality given by Ray and later authors is presumably in Hunts. 25. G. In Britain from Westmorland and Lincs southwards.

PYRUS L.

1. Pyrus communis L.   Wild Pear   Ray, 1670
There are a number of records of single trees, none of which are likely to be native. 35, 45, *54*, [36, 37, 48, 55]. ?Introduced. Widespread but local in England and Wales, rare in Scotland and Ireland.

120

MALUS Mill.

1. Malus sylvestris Mill. Crab Apple Ray, 1660
Pyrus malus L.

Scattered over the county in hedgerows, woods and copses. Both the sspp. occur: ssp. **sylvestris** with a nearly glabrous pedicel, receptacle and calyx, and a small sour apple; and ssp. **mitis** (Wallr.) Mansf., with a tomentose pedicel, receptacle and calyx, and a larger, often sweet apple. The former is the wild plant, whilst the latter is apparently always descended from cultivated trees. Most garden apples can be referred to the latter. Over 4000 acres are grown in the county, especially around Histon, Cottenham, Haddenham, Wilburton and Wisbech. 25, 26, 34–6, 44–6, 48, 49, 54–6, 58, 59, *64*, 65, [38, 47]. General, but rare in the north.

CRASSULACEAE

SEDUM L. (Stonecrop)

1. Sedum telephium L. Orpine Ray, 1660
A rare species of woods and hedgebanks on the boulder clay. Cambs specimens in CGE have been determined as ssp. **telephium** (ssp. *purpurascens* (Koch) Syme) by Professor J. Jalas. 25, 54, *64*, 65, [35, 44, 45]. DPW. General, but local.

2. Sedum dasyphyllum L. Relhan, 1786
Recorded from Trumpington in 1857 (where it was still to be found according to Evans (1939)), Grantchester (1907), and Fulbourn (Babington, 1860) where it was refound in 1958. All localities are on walls. *45*, 55. ?Introduced. Possibly native on limestone rocks in Cork. Naturalized elsewhere on old walls, mainly in S. and C. England.

3. Sedum album L. White Stonecrop Ray, 1660
On walls, rooftops and hedgebanks in a number of localities. Both the ssp. **album** with oblong leaves and petals 3–4 mm., and ssp. **micranthum** (DC.) Syme with oblong-ovate to subglobose leaves, and petals 2–3 mm., have been recorded. 36, 44, 45, *55*, 56, 65, [38, 47, 49]. Possibly native in the Malverns, Mendips and in Devon, naturalized elsewhere.

4. Sedum acre L. Wall-pepper Ray, 1660
Throughout the county on walls, rooftops and in dry sandy places. 23, *24*, 25, 33–5, *36*, 38, 39, 44–8, 54–6, 58, 65, *66*, 67, 76, 20, 41, [49, 40]. General.

121

CRASSULACEAE

5. Sedum sexangulare L.   I. Lyons, 1763

Recorded from Cambridge by I. Lyons, Chippenham gravel pit by Relhan, Ely Cathedral by T. Martyn, and Tydd St Giles by Skrimshire. [45, 58, 67, 41.] Naturalized on old walls in a few places in England and Wales; native of continental Europe.

6. Sedum reflexum L.

Garden escape. [45, 56, 58.]

SEMPERVIVUM L.

1. Sempervivum tectorum L.   Houseleek   I. Lyons, 1763

Recorded from a number of localities on walls. 35, 36, *45*, *54*, *56*, 20, [44, 46, 40]. Planted for centuries on old walls and roofs; origin unknown.

CRASSULA L.

1. Crassula tillaea L.–Garland   W. H. Mills, 1930

Tillaea muscosa L.

Found by W. H. Mills between Chippenham and Kennett in 1930 but not recorded since. It is plentiful in the Breckland just over the Suffolk border. *66.* In N. Somerset, Dorset, S. Wilts, Hants, Norfolk, Suffolk, Notts and the Channel Islands; formerly in S. Devon, Kent and Middlesex.

SAXIFRAGACEAE

SAXIFRAGA L. (Saxifrage)

1. Saxifraga tridactylites L.   Ray, 1660

Scattered over the county on walls and in dry, usually sandy places. 25, 44, 45, 54–6, 58, 67, 76, [34–6, 38, 46, 47, 49, 57, 65, 66, 40]. K. Throughout England and Wales; E. Scotland north to Caithness and in Inner and Outer Hebrides; almost throughout Ireland.

2. Saxifraga granulata L.   Ray, 1660

Occasional in old pasture in the south of the county especially on the sands; absent from the Fens. 25, 45, *46*, 54, *55*, 56, 66, 67, 76, [35, 57]. G. In Britain from E. Ross and Renfrew southwards; in Ireland native round Dublin, and probably introduced elsewhere.

PARNASSIACEAE

PARNASSIACEAE

PARNASSIA L.

1. Parnassia palustris L.   Grass of Parnassus   Ray, 1660

Formerly in a number of fens, but now only at Chippenham and Sawston and there very rare. 44, 66, [25, 45, 54, 56]. C. Widespread in the British Isles, but rare and decreasing in the south.

GROSSULARIACEAE

RIBES L.

1. Ribes sylvestre (Lam.) Mert. & Koch   Red Currant

I. Lyons, 1763

R. rubrum sensu Bab.

Recorded from a number of localities, some of those in the Fens perhaps native. 25, 36, 43–5, 54, 58, 59, 65, *67*, 20, [47, 49, 56]. **CH.** ?Native. Widespread in Great Britain but much cultivated and often an escape. Very rare and only an introduction in Ireland.

2. Ribes nigrum L.   Black Currant   Ray, 1660

Recorded from a number of localities in damp places, mostly in the Fens, where it is probably native, as at Chippenham. *25*, 34, 45, 54, 56, 59, 66, [48, 57]. C. Widespread in the British Isles, though local in Wales and Ireland, but commonly cultivated for its fruit and often an escape.

3. Ribes uva-crispa L.   Gooseberry   I. Lyons, 1763

R. grossularia L.

Occurs throughout the county in hedges and thickets but doubtfully native. About 1000 acres are grown in the county, mainly in the northern Fens. 23, 34, 35, 44, *45*, 55, 56, 59, 66, 75, 76, 20, [36, 46, 48, 49, 40]. Throughout the British Isles but commonly cultivated and often an escape.

DROSERACEAE

DROSERA L. (Sundew)

1†. Drosera rotundifolia L.   Ray, 1660

Recorded from Hinton, Teversham and Fulbourn Moors, Melbourn and Fowlmere Commons, Gamlingay and Elm, and Wicken and

123

DROSERACEAE

Chippenham Fens. At all these places there must have been acid *Sphagnum* bog that has now gone. Last seen at Chippenham Fen by N. D. Simpson, *c*. 1913. [25, 34, 44, 45, 55, 56, 66, 40.] General in suitable habitats.

2†. Drosera anglica Huds.   Relhan, 1802

Known only from Relhan's records from Sawston and Hinton Moors. [44, 45.] A very local species in England and Wales but very frequent in W. Scotland and Ireland. Usually amongst *Sphagnum* and often associated with *Rhynchospora alba*.

3†. Drosera intermedia Hayne   Ray, 1660

D. longifolia sensu Evans

Recorded only from Hinton, Teversham and Sawston Moors. It was last seen by Relhan. [44, 45.] Rather local, but more frequent in England and Wales than the preceding species.

LYTHRACEAE

LYTHRUM L.

1. Lythrum salicaria L.   Purple Loosestrife   S. Corbyn, 1656

Throughout most of the county by ditches, rivers and lodes, and in marshy places, commonest in the Fens. All squares except 24, *25*, 26, 54, 64, 65. Throughout much of the British Isles but less frequent in Scotland and absent from the extreme north.

2. Lythrum hyssopifolia L.   Grass Poly   Ray, 1660

Very rare and sporadic in seasonally wet hollows, particularly at the edge of arable fields; found in quantity near Thriplow in September 1958 by Mrs G. Crompton. 44, *66*, [36, 45, 46, 48, 58]. Probably native in some southern and eastern counties; elsewhere casual.

PEPLIS L.

1†. Peplis portula L.   Ray, 1685

Known to the early botanists at Gamlingay where Babington (1860) still knew it, but it has not been recorded there since. It was also found at Ely by W. J. Cross in 1878. Specimens from both these localities in CGE have the calyx teeth short and are referable to var. **portula**. [25, 58.] Scattered throughout the British Isles.

THYMELAEACEAE

DAPHNE L.

(Daphne mezereum L.
Recorded from the Gog Magog Hills (CGE) by Shrubbs in 1879 where it was presumably planted. [45.] Native on calcareous soils in England from Sussex to Yorks, but local and rare. Often planted in gardens.)

1. Daphne laureola L.   Spurge Laurel   Ray, 1660
A characteristic species of boulder-clay woods flowering from February to April; rare elsewhere. 25, 34–6, 44, 45, 54, 55, *41*, [47, 56, 57, 65, 40].
B. Widespread and common in England, local in Wales, alien in Scotland, and Ireland.

ONAGRACEAE

EPILOBIUM L. (Willow-herb)

The species of *Epilobium* can be distinguished in the field with care and a hand-lens. Hybrids occur rather frequently and can be recognized by their failure to set good seed. The mixtures of species and hybrids which sometimes grow where woodland has been felled present a special difficulty, but even here the majority of plants can usually be identified.

1 Stigma 4-lobed.   *2*
     Stigma entire.   *5*

2 Flowers large, petals more than 12 mm. (wet places).
                                                    1. E. hirsutum
     Flowers medium or small, petals not over 10 mm.   *3*

3 Stem clothed with soft, spreading hairs.   **2. E. parviflorum**
     Stem subglabrous with rather sparse curved or crisped hairs.   *4*

4 Lower leaves ovate or ovate-lanceolate, short-stalked (usually
        c. 2 mm.); petals 8–10 mm. (common).   **3. E. montanum**
     Lower leaves elliptic-lanceolate, with obvious stalks (usually 4–8
        mm.), petals paler pink, 6–8 mm. (rare).   **4. E. lanceolatum**

5 Lower leaves with an obvious petiole, 2 mm. or more; stem with
        spreading glandular hairs in upper part.   *6*
     Lower leaves ±sessile; stem eglandular.   *7*

6 Petiole of lower leaves usually more than 5 mm.; petals at first al-
        most white, later pale pink.   **5. E. roseum**
     Petiole of lower leaves 2–3 mm.; petals pale pink, later purplish.
                                                    6. E. adenocaulon

125

ONAGRACEAE

7 Stem smooth with no raised lines, though usually with 2 lines of
   hairs; stem-base with very slender white stolons terminating in a
   bulbil (rare in Cambs).   **10. E. palustre**
   Stem with 2–4 raised lines or ridges; stolons if present thick.   *8*

8 Capsule 4–6 cm.; glandular hairs usually present at base of calyx
   (sometimes on capsule also); late summer stolons usually present.
                                                            9. E. obscurum
   Capsule 6–9 cm.; inflorescence wholly eglandular; autumn rosettes
   subsessile.   **7. E. adnatum** (includ. *E. lamyi*)

1. Epilobium hirsutum L.   Ray, 1660
Abundant by streams, rivers, lodes and in marshy places throughout the
county. All squares. General, but rare in the north and west.

× **parviflorum**. There are specimens in CGE from Bourn Brook near
Toft (1910) and Cherry Hinton Hall (1941). *45*, [35].

2. Epilobium parviflorum Schreb.   Ray, 1660
Occurs throughout the county by water and in marshy places. All
squares except 24, 29, 67. General.

3. Epilobium montanum L.   Ray, 1660
Common on the clays and gravels in waste places, hedgerows and
especially woodland clearings; locally rare or absent on parts of the
chalk and in the Fens. 24, 25, 34–6, 38, 39, 43–6, 49, 54, 58, 64–6, 20, 30,
40, 41, 50, [47, 56]. General.

× **obscurum**. There is a specimen in CGE collected at Gamlingay in 1952
and determined by G. M. Ash. 25.

× **parviflorum**. There is a specimen in CGE collected at Eversden Wood
in 1910 and determined by E. S. Marshall. [35.]

4. Epilobium lanceolatum Seb. & Mauri   S. M. Walters, 1953
Recorded twice, on disturbed ground in woodland at Morden Grange
and Hardwick (CGE). 24, 35. S. and S.W. England; apparently
spreading in recent years and recorded since 1950 in Cambs, S. Essex,
W. Suffolk and Norfolk.

5. Epilobium roseum Schreb.   G. Goode, 1912
Occurs as a roadside and garden weed in Cambridge. 45. **Cam.**
Throughout lowland Britain north to Sutherland, and in N.E. Ireland.

6. Epilobium adenocaulon Hausskn.   J. N. Mills, 1946
Now widespread in the county especially in disturbed woodland. It is
not known when the spread took place, as it was not recognized by
Cambs botanists until after 1946. 25, 35, 37, 38, 43, 45, 49, 54, 55, 57,
65, 20, 50. First recorded in 1891 since when it has spread rapidly
especially in the south-east; native of N. America.

× **montanum.** Recorded from Ditton Park Wood in 1952, the specimen
in CGE being confirmed by G. M. Ash. 65. DPW.

× **parviflorum.** Recorded from Hardwick Wood and Ditton Park
Wood; a specimen from the latter locality is in CGE and was determined by G. M. Ash. 35, 65.

7. Epilobium adnatum Griseb.   Ray, 1660
E. tetragonum sensu Bab., et Evans
Common throughout the county in woodland clearings and waysides.
15, 24, 25, 35, *36*, 39, 43, 45–7, 49, 54, 56–9, *64*, 65, [38, 48, 56, 40].
Mainly a species of England and Wales south of the Humber; probably
only casual in Scotland and Ireland.

8. Epilobium lamyi F. W. Schultz   S. M. Walters, 1951
There are specimens in CGE from Milton (1951) and Cherry Hinton
(1952). Both were determined by G. M. Ash. The taxonomic rank of
E. lamyi is much disputed; many authorities, finding the diagnostic
characters relatively unsatisfactory, prefer to treat it as a subspecies of
E. adnatum. 45, 46. Local in S. England and S. Wales.

9. Epilobium obscurum Schreb.   W. W. Newbould, 1859
Known with certainty only from Gamlingay and Cambridge. There is
an unsubstantiated record from Wimblington Fire Lots. 25, 45, (?49).
G. General, very common in the west and probably preferring acid
soils.

10†. Epilobium palustre L.   Ray, 1660
A number of old records exist for fens and marshy places. Said by
Evans to be 'common' at Wicken Fen, but not recently seen there.
Probably somewhat calcifuge, and intolerant of highly calcareous fen
water. Last seen at Milton Fen in 1940. *46*, [25, 44, 45, 55, 58, 40]. General.

ONAGRACEAE

CHAMAENERION Adans.

1. Chamaenerion angustifolium (L.) Scop.   Rosebay Willow-herb
J. N. Bullock, 1884

Epilobium angustifolium L.

Only known as a casual until the end of the last century, after which it spread rapidly and is now rather common on waste land and in woodland clearings throughout the county. There are specimens in CGE of both var. **macrocarpum** (Leighton) Clapham with thick, very wavy margined leaves and long (4–8 cm.) capsules, and var. **brachycarpum** (Leighton) Clapham with thinner, straighter-margined leaves and shorter (2·5–4 cm.) capsules. The status of these variants is not clear. All squares except 39. General. A century ago this species was local, but there has been a phenomenal spread in the last few decades.

OENOTHERA L. (Evening Primrose)

1. Oenothera biennis L.

Recorded from a cornfield at Soham by C. E. Moss, and Bottisham in 1959 by C. Turner (Herb. ipse). 56, [57]. A widespread casual in England, Wales and S. Scotland; absent from Ireland. Native of N. America.

2. Oenothera erythrosepala Borbás   Evans, 1939

O. lamarkiana sensu Evans

A frequent casual of disturbed roadsides and waste places. 35, 38, 45, 49, 58, 66, 40. A widespread casual of southern England, often naturalized; believed to have arisen in Europe but probably native of N. America.

CIRCAEA L.

1. Circaea lutetiana L.   Enchanter's Nightshade   S. Corbyn, 1656

In hedgebanks and woody places, especially characteristic of the boulder-clay woods. 25, 26, 35, 36, 43–6, 49, 54, 55, 58, 64–6, 20, 40, 50, [34]. General, but rare in the north.

HALORAGACEAE

MYRIOPHYLLUM L.

1. Myriophyllum verticillatum L.   Ray, 1660

Scattered throughout the county in ponds, pits and lodes, though more frequent in the Fens. 29, 37–9, 47, 48, 55, 56, 59, 68, 40, [25, 34, 44–6,

49, 57, 58, 66]. **W.** Widespread in England, except the south-west; in central Ireland and very rare in Wales.

2. Myriophyllum spicatum L. Ray, 1660

In ponds, pits and lodes throughout the county, but more frequent in the Fens. 25, 29, 34, 37–9, 46–9, 55, 58, 59, 67, 68, 20, 30, 31, 40, 41, [36, 45, 57, 66]. General.

3†. Myriophyllum alterniflorum DC. Babington, 1848

Recorded from Gamlingay, where it was probably last seen by Evans (1939) early this century, and Wicken Fen in 1919. [25, 56.] General, calcifuge.

HIPPURIDACEAE

HIPPURIS L.

1. Hippuris vulgaris L. Mare's-tail S. Corbyn, 1656

Scattered throughout the county in ponds, pits and lodes, frequent in the Fens. 25, 29, 34, 38, 39, 44, 45, *46*, 47–9, 55–7, 59, 68, 20, 30, 31, 40, 41, [35, 58]. General, but often local.

CALLITRICHACEAE

CALLITRICHE L. (Starwort)

Until recent years the genus *Callitriche* has not received very much attention in the county. The taxonomic difficulty of the genus makes it impossible to accept any of the old records, unless they are supported by herbarium sheets. Anyone seriously interested in the group should consult the paper by Schotsman (1954), abstracted by Kent (1955). The three Cambs species can be distinguished as follows:

Rosette leaves broadly ovate to ± circular; submerged leaves ± similar, never linear; fruit 1·6 × 2 mm., somewhat broader than long, conspicuously keeled, its lateral grooves deep; stigma bases curving downwards in fruit, persistent. **1. C. stagnalis**

Rosette leaves broadly elliptical, gradually narrowed into the stalk; submerged leaves linear or narrowly elliptical; fruit *c.* 1·5 mm., somewhat longer than broad, lateral lobes keeled with fairly deep grooves; stigma-bases erect, sometimes persistent. **2. C. platycarpa**

Rosette leaves ± rhomboidal; submerged leaves narrower, sometimes linear; fruit *c.* 1·5 × 1 mm., longer than broad, its lobes with rounded margins and with a very shallow groove between members of a parallel pair; stigma bases erect, persistent. **3. C. obtusangula**

CALLITRICHACEAE

1. Callitriche stagnalis Scop.   F. R. Tennant, 1900
C. platycarpa sensu Bab. pro parte
Throughout most of the county. Some of the field records especially from the Fens may be referable to *C. platycarpa*. 25, 29, 34–7, 39, 44, 45, 47–9, 55, 56, 58, 59, 64, 65, 67, 68, 20, 30, 31, 40, [46]. General.

2. Callitriche platycarpa Kutz.   G. B. Jermyn, 1825
C. polymorpha Lonnr.
Herbarium specimens exist (CGE) from Swaffham Prior (1825) and Gamlingay (1855), and another from Quy (1898) is doubtful. A. O. Chater has seen it in several places in Cambridge during the last few years. The species may prove to be of more frequent occurrence in Cambs. 45, [25, ?55, 56]. **Cam.** Distribution imperfectly known but probably scattered over the British Isles and perhaps frequent in lowland Britain.

3. Callitriche obtusangula Le Gall   Babington, 1830
C. platycarpa sensu Bab. pro parte; *C. verna* sensu Bab. pro parte
Throughout much of the county but commonest in the Fens. 29, 34–9, 45–7, 49, 54–6, 58, 59, *65*, 67–9, 20, 30, 31, 40, 41, 50, [48]. Great Britain northwards to Lancs, and Yorks; C. Ireland.

LORANTHACEAE

Viscum L.

1. Viscum album L.   Mistletoe   Ray, 1660
Parasitic on branches of trees; recorded from a few localities on *Malus*, *Crataegus*, *Quercus*, *Fraxinus* and *Tilia*. Often grown in gardens and in orchards on apple trees. *25*, 34, 35, *45*, *46*, [54, 56]. Frequent from Cornwall and Kent to Denbigh and mid-west Yorks, rare elsewhere.

SANTALACEAE

Thesium L.

1. Thesium humifusum DC.   Bastard Toadflax   Ray, 1660
A rare hemiparasitic species of old chalk grassland. *44*, 45, 55, 56, 66, [23, 54, 67]. **D.** Chalk and limestone in S. and E. England, north to Lincolnshire, and in the Channel Isles.

130

CORNACEAE
CORNUS L.

1. Cornus sanguinea L.　Dogwood　　　　Ray, 1660
Thelycrania sanguinea (L.) Fourr.
In hedges, woods and copses throughout most of the county except the
northern Fens, where it is very rare. 23–6, 33–6, 38, 43–7, *49*, 54–8, 64,
65, *66*, 67, 75, [48, 40]. In Britain from Durham and Cumberland
southwards; probably introduced farther north; local in Ireland.

ARALIACEAE
HEDERA L.
1. Hedera helix L.　　　　Ivy　　　　　Ray, 1660
Throughout the county, climbing on trees and buildings or creeping
along the ground or on banks. Old 'trees' of Ivy, with trunks almost as
thick as the Ash (*Fraxinus*) on which they are growing, may be seen at
Lime Kiln Close, Cherry Hinton. The vigour with which this Atlantic
species grows even in Cambridge emphasizes that the climate here is
only comparatively continental, and is far from the extremes of
continental Europe. All squares except 37. General.

UMBELLIFERAE
HYDROCOTYLE L.

1. Hydrocotyle vulgaris L.　Pennywort　　　S. Corbyn, 1656
In a number of localities by the edges of ponds or pits and in marshy
places mainly in the Fens. 25, 44, 45, 48, *49*, 55–7, 59, 66, 30, [38, 47,
58, 40]. W. General.

SANICULA L.
1. Sanicula europaea L.　Sanicle　　　　Ray, 1660
Frequent in woods and copses but especially characteristic of the
boulder-clay woods, and absent from the Fenlands. 23–6, 33–6, 43–5,
54, 55, 64, 65, 75, [56, 66]. General.

CHAEROPHYLLUM L.

1. Chaerophyllum temulentum L.　Rough Chervil　Ray, 1660
C. 'temulum' L.
Throughout the county in grassy and waste places. The second of the

UMBELLIFERAE

three common white-flowered roadside Umbellifers to flower. All squares. General, but rare in N. Scotland and W. Ireland.

ANTHRISCUS Pers.

1. Anthriscus caucalis Bieb. var. **scandix** (Scop.) Hyl.   Ray, 1660
A. vulgaris Pers., non Bernh.; *A. neglecta* Boiss. & Reut.

A weed of waste places, frequent on the sands in the east of the county, usually casual elsewhere. 29, 39, 45, 48, 49, 54, 65–7, 76, [23, 25, 38, 46, 56, 57, 68, 40]. FH. Scattered throughout the British Isles except the extreme north.

2. Anthriscus sylvestris (L.) Hoffm.   Hedge Parsley   Ray, 1660
Throughout the county in grassy and waste places. The earliest to flower, and the most profuse, of the three common white-flowered roadside Umbellifers. All squares. General.

SCANDIX L.

1. Scandix pecten-veneris L.   Shepherd's Needle   Ray, 1660
A weed of arable land scattered over the county and sometimes locally abundant, especially on the chalk. *25*, 26, 34–6, 39, 43, *44*, 45, 46, 55, 56, 65, 66, [38, 47–9, 58, 67, 40, 41]. General, but rare in Wales and the north, and only common now in south and east England.

TORILIS Adans.

1. Torilis japonica (Houtt.) DC.   Ray, 1660
T. anthriscus (L.) C. C. Gmel., non Gaertn.; *Caucalis anthriscus* (L.) Huds.

Throughout the county in grassy and waste places. The latest to flower, and the least conspicuous, of the three common white-flowered roadside Umbellifers. All squares except 37, 43. General, except the far north.

2. Torilis arvensis (Huds.) Link.   S. Corbyn, 1656
T. infesta Spreng; *Caucalis arvensis* Huds.

A weed of arable land in several scattered localities, formerly more frequent. 26, 34–6, 45, 46, 55, 56, 65, *66*, [38, 39, 44, 47, 49, 54, 57, 58, 67, 40]. Scattered throughout S. and E. England north to Yorkshire.

3. Torilis nodosa (L.) Gaertn. Ray, 1660
Caucalis nodosa (L.) Scop.
Scattered throughout the county on bare banks, dry fields and waste places. 26, 29, 34, 39, 44–8, *49*, 54, *55*, 57, 65, *66*, 68, 75, 30, 31, 41, [35, 36, 38, 58, 40]. Widespread, but only common in eastern England.

CAUCALIS L.
1. Caucalis platycarpos L. Ray, 1660
C. daucoides L., 1767, non L. 1753; *C. lappula* Grande
Recorded from a number of localities as a weed of cornfields and waste places. Last seen at Bourn airfield in 1945. *35*, *54*, [23, 25, 34, 36, 45, 56]. Casual or ± naturalized in arable fields and waste places in S. and E. England north to Durham; rather rare and less frequent than formerly. Native of continental Europe.

2†. Caucalis latifolia L. Ray, 1660
A casual found in a number of localities by the early botanists, but not seen since 1833. [35, 36, 45, 46, 56.] Native of C. and S. Europe, N. Africa and W. Asia.

CORIANDRUM L.
1. Coriandrum sativum L. Coriander Relhan, 1793
Recorded from a few localities where it is a casual or perhaps an escape from cultivation. 45, 65, [47, 40]. Probably native of the east Mediterranean region.

SMYRNIUM L.
1. Smyrnium olusatrum L. Alexanders Ray, 1660
Occurs in a number of scattered roadside localities especially in the neighbourhood of Cherry Hinton and Fulbourn. Has spread in recent years and seems to be still increasing. 26, *36*, 44–6, 55, 67, [34]. CH. Introduced. Extensively naturalized in hedges, waste places and on cliffs in Britain north to Dunbarton and Banff, especially near the sea, and almost throughout Ireland. Native of S.W. Europe and Mediterranean region.

CONIUM L.
1. Conium maculatum L. Hemlock Ray, 1660
Very poisonous herb; frequent throughout the county in hedgerows and waste ground, though more abundant in the Fenland, where it replaces *Anthriscus sylvestris* to some extent. All squares except *24*. General, except the north and west.

UMBELLIFERAE

BUPLEURUM L.

1. Bupleurum rotundifolium L.   Thorow-wax (Through-wax)

Ray, 1660

Formerly a widespread weed of arable land, last seen by Shrubbs near Abington in 1910. [23, 25, 34–6, 38, 45, 47, 54–7.] In cornfields from Devon and Kent to N. Yorks; casual elsewhere.

2. Bupleurum tenuissimum L.   Ray, 1663

Formerly in a number of localities mostly in the Fens, but only seen recently at Foul Anchor. 41, [25, 45, 47, 56, 58]. A rare plant of salt marshes and waste places from Dorset and Kent to Cheshire and Durham, usually near the coast.

APIUM L.

1. Apium graveolens L.   Wild Celery   Ray, 1660

Formerly frequent by lodes, ditches and in marshy places especially in the Fenlands, now rare. *44*, 45, 49, *55*, *40*, 41, 50, [25, 34–6, 38, 39, 46–8, 56–8, 67]. In most maritime counties of the British Isles north to Perth and the Clyde Isles.

(About 1750 acres of cultivated Celery, **A. dulce** Mill., are grown annually in the Fens, especially in the Littleport–Ely area.)

2. Apium nodiflorum (L.) Lag.   Fool's Water-cress   Ray, 1660
Helosciadium nodiflorum (L.) Koch

Common in ditches, streams and lodes throughout the county. For distinction from *Berula erecta*, see note under that species. All squares except 29, 64, 68, 30, 40, [38]. General, but rare in Scotland.

(**Apium repens** (Jacq.) Lag. (*Helosciadium nodiflorum* var. *repens* (Jacq.) Bab.). H. J. Riddelsdell and E. G. Baker (1906) identified material of '*A. repens*' collected by Babington (CGE) from Ely and Upware as varieties of *A. nodiflorum*. A plant from Stourbridge Fair Green, Cambridge (1850), they thought 'came very near to true *A. repens*', but the specimen lacks ripe fruit (CGE). There has been no recent investigation of the Cambs material.)

3. Apium inundatum (L.) Reichb. f.   C. Miller in I. Lyons, 1763
Helosciadium inundatum (L.) Koch

Formerly in a number of localities mainly in the Fens, but only seen recently in Wicken Lode. 56, [25, 29, 35, 44, 45, 47, 48, 54, 58]. W. Widely distributed, but local.

PETROSELINUM Hill

1. Petroselinum crispum (Mill.) Nyman   Parsley
P. sativum Hoffm.
Escape from cultivation. *46*, [48, 58].

2. Petroselinum segetum (L.) Koch   Ray, 1663
Carum segetum (L.) Benth. ex Hook. f.
A rather rare species of hedgerows and grassy places. Very similar in general appearance to *Sison amomum*, but leaves with more numerous pinnae and bracts of inflorescence very short (never half as long as any of the rays). *35, 36, 37,* 44, 46, *57*, [25, 45, 47, 48, 55, 58]. From Cornwall and Kent to Staffs and S. Yorks; rare in Wales.

SISON L.

1. Sison amomum L.   Stone Parsley   Ray, 1660
Frequent in hedgerows and grassy places in the south of the county, local in the southern Fens and absent from the north. 24–6, 34–6, *44*, 45–7, 57, [38, 48, 49, 56, 58]. Throughout most of England south of the Humber; rare in Wales.

CICUTA L.

1†. Cicuta virosa L.   Water Hemlock   J. Martyn in I. Lyons, 1763
Twice found in the eighteenth century in the Prickwillow area but not seen since. [58.] Scattered throughout the British Isles but mainly in E. England, S. Scotland and N. Ireland, decreasing.

FALCARIA Bernh.

1. Falcaria vulgaris Bernh.   D. E. Coombe and C. D. Pigott, 1949
Naturalized at Cherry Hinton. 45. **CH.** In a few localities in E. England; native of continental Europe and W. Asia.

CARUM L.

1. Carum carvi L.   Caraway   Ray, 1685
Naturalized or casual in a few places. *23,* 44, *41,* [45, 56, 58, 40]. Generally assumed to be an introduction in scattered localities over the British Isles; native of continental Europe and Asia. Cultivated for its seeds, used for flavouring.

135

UMBELLIFERAE

BUNIUM L.

1. Bunium bulbocastanum L.   W. H. Coleman, 1839
Carum bulbocastanum (L.) Koch

One of the classical Cambridgeshire rarities, long known at Cherry Hinton, and formerly occurring in one or two other places on the chalk. Resembles *Conopodium* but can easily be distinguished by the general involucre (absent in *Conopodium*). Flowers in May and early June, and ripens fruit in early July. By late July the stems are withered and very inconspicuous. 45, [23, 34, 55]. **CH.** Very local in Herts, Bucks, Cambs, and Beds.

CONOPODIUM Koch

1. Conopodium majus (Gouan) Loret   Pignut   Ray, 1660
Bunium flexuosum With.

Rare; in a few localities, mainly in boulder-clay woods; at Gamlingay in grassland in damp sandy soil, and in small quantity on the chalk under trees at Wandlebury. 25, 35, 36, 45, *46*, [65, 40]. **G.** General. Common, except on chalk and in fens.

PIMPINELLA L.

1. Pimpinella saxifraga L.   Burnet Saxifrage   Ray, 1660

Common in dry grassland on the chalk and sands, rare elsewhere. A variable species particularly in leaf shape, but easily recognized on Cambridgeshire roadsides by its late flowering (July–August) and drooping young inflorescence. 23–6, 33–6, 43–6, 49, 54–7, *58*, *64*, 65–7, 75, 76, [47]. Scattered throughout the British Isles; rarer in the north.

2. Pimpinella major (L.) Huds.   Ray, 1660
P. magna L.

Local in hedgerows, woods and grassy places, especially characteristic of the edges and rides of boulder-clay woods. 25, 35, 36, 43, 47, 54, 64, 65, [49, 55]. Throughout England, probably absent from Scotland and Wales, and in Ireland only in the south and west.

AEGOPODIUM L.

1. Aegopodium podagraria L.   Goutweed   Ray, 1660
Throughout the county in hedgerows, ditch banks and waste places, and as a weed of gardens. All squares except 24, 37. General. Said to have been introduced as a pot-herb, but perhaps native in some northern woods.

SIUM L.

1. Sium latifolium L.   Water Parsnip   Ray, 1660
A local species occurring by the sides of lodes, rivers and pits mainly in the Fens. 35, 37, *38*, 45, *47*, 48, 49, 55–7, *20*, 30, [29, 46, 58, 40]. Scattered throughout S. and E. England, very rare in Scotland; C. Ireland.

BERULA Koch

1. Berula erecta (Huds.) Coville   Ray, 1660
Sium angustifolium L.; *Sium erectum* Huds.
By ditches and lodes and in marshy places mainly in the Fenlands. It is difficult to find a leaf character which will serve to distinguish with certainty this species from the rather variable *Apium nodiflorum*, although normally *Berula* has 4 or more pairs of leaflets with a ring of purple towards the base of the leaf, whilst *Apium* has fewer than 4 and no ring of purple. In flower the presence of a general involucre to the umbel in *Berula* provides an easy diagnostic character. 29, 37, 44, 45, 47–9, 54–8, *67*, 20, 30, [25, 38, 66]. General, but uncommon in the north and west.

SESELI L.

1. Seseli libanotis (L.) Koch   Ray, 1690
One of the classical Cambridgeshire rarities, known at Cherry Hinton since the time of Ray. This handsome plant is still frequent in and around the chalk-pits at Cherry Hinton, where it regenerates abundantly on chalk rubble. It is characteristic of open calcareous ground in its very few British stations, which represent the north-western European limit for this continental species. 45. CH. Otherwise only in Sussex, Herts and Beds.

OENANTHE L. (Dropwort)

1. Oenanthe fistulosa L.   Water Dropwort   Ray, 1660
Occasional by rivers and lodes in the Fens, rare elsewhere. 25, 29, 36, 37, 45, 46, 48, *49*, 55–9, 67, 20, 30, [38, 44, 47, 40]. **W.** Throughout

137

UMBELLIFERAE

England and S. Scotland, local in Wales; in Ireland mainly in east and centre.

2. Oenanthe silaifolia Bieb.   Relhan, 1802

A very local species recorded from the water-meadows of the Ouse and Cam. Occurs frequently in Hunts in the water-meadows around Houghton, and the only recent Cambs records from near Over are in this area. 37, *47*, [45, 57]. From Dorset and Kent to Worcester and Notts.

3. Oenanthe lachenalii C. C. Gmel.   Relhan, 1786

A very local species of fens and marshes, formerly more frequent. *45*, 55–7, 41, [26, 35, 38, 44, 46, 49, 58, 40]. W. Scattered throughout the British Isles except the extreme north.

4. Oenanthe aquatica (L.) Poir.   Ray, 1660

O. phellandrium Lam.

Scattered throughout most of the county in ponds, rivers and especially Fen lodes. 29, 36–9, 44–9, 57, 59, 67, 30, 31, 40, 41, 50, [35, 56, 58]. W. From Somerset and Kent to Cumberland and the Lothians; Ireland, particularly the centre.

5. Oenanthe fluviatilis (Bab.) Colem.   Babington, 1843

In lodes and rivers, especially in the Fenlands. Closely allied to *O. aquatica*, from which it can be distinguished by the segments of its submerged leaves being cuneate and cut at ends, not capillary, and by the fruit (5–6 mm. long and at least 3 times as long as the styles in *O. fluviatilis*, 3–4 mm. and about twice as long as styles in *O. aquatica*). 29, 34, 37, 39, 45, 46, *47*, 48, 49, 56, 57, 20, 30, 31, [35, 38, 44, 58]. W. From Somerset and Kent north to mid-Yorks; and in C. Ireland.

AETHUSA L.

1. Aethusa cynapium L.   Fool's Parsley   Ray, 1660

A common weed of arable land throughout the county. Varies in habit; dwarf cornfield ecotypes, only a few inches tall, contrast with erect arable weeds of gardens up to 3 ft. in height. All squares except 37, 39. General, but rare in the north.

138

FOENICULUM Mill.

1. Foeniculum vulgare Mill.   Fennel   I. Lyons, 1763

F. officinale All.

Naturalized in a few localities in waste places or on roadsides. 35, 36, 45, *46*, 56, 41, [44]. Perhaps native on sea-cliffs in England, Wales and Ireland. The plant is cultivated for the leaves which are used for culinary purposes.

SILAUM Mill.

1. Silaum silaus (L.) Schinz & Thell.   Ray, 1660

Silaus pratensis Bess; *Silaus flavescens* Bernh.

In hedgerows and grassy places, almost entirely confined to the boulder clay; also in wet peaty meadows. 24, 25, 26, 35, 36, 44–6, 55–7, 66, 67, [38, 47–9, 54, 65, 40]. From Devon and Kent to Fife, rare in Wales and Scotland.

SELINUM L.

1. Selinum carvifolia (L.) L.   W. J. Cross and W. Marshall, 1882

Very local at Chippenham Fen and in damp places in the grounds of Fordham Abbey and Sawston Hall. First recorded as a British plant at Chippenham in 1882, where it is (in spite of early rumours of introduction!) apparently native. Resembles *Peucedanum palustre* and *Silaum silaus* in its foliage but is easily distinguished by its very late flowering (late August–September) and the deeply ridged stem below the white umbel. 44, 66, 67, [58]. C. A rare species of Cambs, N. Lincs and Notts; now extinct through drainage in all but the Cambs localities.

ANGELICA L.

1. Angelica sylvestris L.   Ray, 1660

Throughout most of the county in fens, damp meadows and woods. All squares except 24, *67*, 68, [26]. General.

PEUCEDANUM L.

1. Peucedanum palustre (L.) Moench   Relhan, 1786

Formerly in several fens but now only to be found at Wicken where it is still fairly common. The food plant of the Swallowtail butterfly (*Papilio machaon* L.). 56, [58, 66, 68]. In fens and marshes in Somerset, Sussex, Hunts, Cambs, Suffolk, Norfolk, Lincs and S. Yorks.

UMBELLIFERAE

PASTINACA L.

1. Pastinaca sativa L.   Wild Parsnip   Ray, 1660
Peucedanum sativum (L.) Benth. ex Hook. f.
Throughout the county in grassy and waste places; especially common on roadsides. About 300 acres are grown in the county yearly, especially in the Chatteris area; it is also commonly cultivated in gardens. All squares except 26, 20. Scattered throughout England and Wales; only an escape from cultivation in Scotland; S. and E. Ireland.

HERACLEUM L.

1. Heracleum sphondylium L.   Hogweed   Ray, 1660
Throughout the county in grassy and waste places, hedgebanks and especially on roadsides. The var. **angustifolium** Huds., with linear-lanceolate leaf segments, is recorded. All squares. General.

2. Heracleum mantegazzianum Somm. & Levier sens. lat.
Babington, 1828
Naturalized in a number of localities. The nomenclature and taxonomy of these introduced 'Giant Hogweeds' is very confused. In the Botanic Garden, Cambridge, and at Milton there is evidence of hybridization with *H. sphondylium*. 25, 26, 36, 45, 46. Native of the Caucasus.

LASER Borkh.

1†. Laser trilobum (L.) Borkh.   J. C. Melvill, 1867
Siler trilobum (L.) Crantz
Naturalized at Cherry Hinton and recorded frequently there between 1867 and 1910, but not recently seen. [45.] Native of continental Europe.

DAUCUS L.

1. Daucus carota L. ssp. **carota**   Wild Carrot   Ray, 1660
Throughout the county in dry grassland, and most common on the chalk. Over 3000 acres of cultivated Carrots, ssp. **sativus** (Hoffm.) Thell., are grown in the county, especially around Chatteris. All squares except 37, 39, 59, 68, [40]. General, but mainly coastal in the north.

140

CUCURBITACEAE

BRYONIA L.

1. Bryonia dioica Jacq.   White Bryony   Ray, 1660
Throughout the county in hedges, woods and copses. All squares except 30. England north to N.W. Yorks, and Northumberland, rare in Wales, introduced locally in S. Scotland and S. and E. Ireland.

ARISTOLOCHIACEAE

ARISTOLOCHIA L.

1. Aristolochia clematitis L.   Birthwort   Ray, 1685
Formerly occurred at Ickleton and Milton and still to be found at Whittlesford, Waterbeach and Cambridge. 44–6. Long cultivated as a medicinal plant and naturalized in a number of places mainly in S. England. Native of C. and S. Europe and the Orient.

EUPHORBIACEAE

MERCURIALIS L.

1. Mercurialis perennis L.   Dog's Mercury   Ray, 1660
A characteristic species of the herb layer of woods and copses on the boulder clay and often dominating large areas. 25, 26, 34–6, 43–5, 54–6, 64–6, 75, 76, 20, [48]. Common over most of Great Britain; local in Ireland, where it may be an introduction.

2. Mercurialis annua L.   Ray, 1685
A local weed of arable land mostly in the southern Fens. 45, 48, 55–8, 65, [66, 67]. ?Native. Widespread in S. England; rare in Wales, N. England and S. Scotland; in S. Ireland north to Dublin and Clare.

EUPHORBIA L. (Spurge)

1. Euphorbia lathyrus L.   Caper Spurge
L. Jenyns, first half 19th century
Garden escape naturalized in a number of localities. 35, 37, 45, 46, 49, 54, 56, 58, 50, [44]. Perhaps native in woods in a few places in England and Wales; more common as a garden escape.

EUPHORBIACEAE

2. Euphorbia platyphyllos L.   Ray, 1685
Rare cornfield weed in a few places on the boulder clay. 26, 35, 36, 45, *46*, [25, 38, 49]. Widespread but local in S. England; rare in N. England; Glamorgan.

3. Euphorbia helioscopia L.   Sun Spurge   Ray, 1660
A common weed of arable land throughout the county. All squares except 24, 26, 54. General.

4. Euphorbia peplus L.   Petty Spurge   Ray, 1663
A common weed of arable land, especially gardens, throughout the county. All squares except 37, 54, [66]. General, but rarer in the north.

5. Euphorbia exigua L.   Ray, 1660
A common weed of arable land throughout most of the county. 23–6, 33–6, 38, 43–6, 48, 49, 54–8, 64–6, 75, 20, [29, 39, 47, 41, 50]. Common in England, Wales, S. Scotland and E. Ireland, rare elsewhere.

6. Euphorbia uralensis Fisch. ex Link   W. H. Mills, 1917
E. virgata Waldst. & Kit., non Desf.; *E. esula* auct.
Scattered over the southern half of the county in waste and grassy places. *E. esula* L. sens. strict. is apparently not in Cambs, but the taxonomy of the group, including *E. cyparissias*, needs revision. 24, 34, 35, *44*, 45, *46*, *54*, 56, [66]. Naturalized in England and E. Scotland; native of E. Europe.

7. Euphorbia cyparissias L.   Shrubbs, 1910
Known only from Newmarket Heath and the Gogs. 45, 66. **GG**. Possibly native in a few scattered localities; more common as a garden escape.

8. Euphorbia amygdaloides L.   Wood Spurge   Ray, 1660
A rare species found only in a few woods on the eastern boulder clay. Also formerly occurred at Gamlingay. 35, 65, [25, 66]. **DPW**. Throughout the southern half of England and Wales; in Ireland only in Cork.

POLYGONACEAE

POLYGONUM L.

1. Polygonum aviculare L. sens. lat.   Knot-grass   Ray, 1660
An abundant weed of arable land and waste places throughout the county. All squares. General.

POLYGONACEAE

The following three segregates have been recorded. The herbarium material has been checked by B. T. Styles (cf. Styles, 1962):

P. aviculare L. sens. strict. Babington, 1835

P. heterophyllum Lindm.

Specimens in CGE from Cambridge (1835, 1836, 1863 and 1899), Madingley (1912), Fordham (1912), Chesterton (1940) and Barrington (1957). The segregate is probably common throughout the county especially in cultivated ground.

P. rurivagum Jord. ex Bor. C. E. Moss, 1912

Specimens in CGE from between Fulbourn and Gt Wilbraham (1912) and Babraham (1955).

P. arenastrum Jord. ex Bor. Babington, 1835

P. aequale Lindm.

Specimens in CGE from Cambridge (1835), Swavesey (1882), Toft (1912) and Barton (1941). This segregate is probably common in the county, especially on paths and in waste places.

2†. Polygonum bistorta L. Snakeweed Ray, 1660

Formerly recorded in a few localities in damp meadows. Last seen at Fulbourn by Henslow, in 1830. [45, 46, 55, 56.] Scattered throughout the British Isles.

3. Polygonum amphibium L. Ray, 1660

Throughout the county in streams, rivers and lodes and on banks. Plants growing in water develop floating leaves with glabrous shiny upper surfaces; on land the stems are suberect and the leaves are roughly hairy, with a dull colour. These two forms can be developed on the same plant, as first noted by Ray (see Raven (1950), p. 59). All squares except 24, 26, 54, 64, 65. General.

4. Polygonum persicaria L. Ray, 1660

A common weed of arable land and waste places throughout the county. All squares except 54, 64. General.

5. Polygonum lapathifolium L. I. Lyons, 1763

A common weed of arable land throughout most of the county. 29, 34, 36–8, 39, 44–9, 55–9, 66–9, 20, 31, 40, 41, 50, [25]. General, but rarer in the north.

6. Polygonum nodosum Pers. Babington, 1836

P. laxum sensu Bab.

A common weed of arable land in the Fens, less frequent elsewhere.

143

POLYGONACEAE

This species with its dirty pink, not greenish-white, flowers, and strongly glandular inflorescence is easily distinguished from *P. lapathifolium* in Cambs. 38, 44, 45, 47–9, 54, 56, 58, 59, 68, 69, 20, 31, 40, 41, 50, [25, 46, 57]. Scattered throughout much of the British Isles, but very rare in Scotland.

7. Polygonum hydropiper L.   Water-pepper   Ray, 1660

Occasional in wet places in the Fens, rare elsewhere. 29, 36, 37, 39, 44, 45, *46*, 48, 59, 67, 20, 30, [25, 35, 38, 47, 56, 66]. General except N. Scotland.

8. Polygonum mite Schrank   Babington, 1836

P. laxiflorum Weihe

A rather rare species of wet places in the Fens. 37, 45, 48, 59, 67, 68, [46, 47, 49, 57, 40]. Rare; throughout the southern part of Great Britain; N. and W. Ireland.

9. Polygonum minus Huds.   Ray, 1660

A very rare species of wet places in the Fens, only seen recently on the Welney Washes. 59, [46–8]. Scattered throughout the British Isles, but always very local, and absent from N. Scotland.

10. Polygonum convolvulus L.   Black Bindweed   Ray, 1660

A common weed of arable land and waste places throughout the county. Var. **subalatum** Lej. & Court. with the exterior perianth segments narrowly winged is recorded from several localities. It should not be confused with *P. dumetorum* L. (not recorded for the county), which has longer fruiting pedicels (up to 8 mm., not 1–2 mm.) jointed at or below, not above, the middle, and shining, not dull fruits. All squares. General.

11. Polygonum baldschuanicum Regel

Naturalized alien. 44, 45.

12. Polygonum cuspidatum Sieb. & Zucc.   T. G. Tutin, 1946.

Naturalized in a number of localities. 29, 34, 44–6, 49, 58, 40, 41, 50. Sometimes cultivated and commonly naturalized, especially in W. Britain; native of Japan.

13. Polygonum sachalinense F. Schmidt   F. H. Perring, 1959

A large clump near Thorney in 1959. 20. Sometimes cultivated and occasionally naturalized; native of Sakhalin.

FAGOPYRUM Mill.

1. Fagopyrum esculentum Moench   Buckwheat   Ray, 1660
Casual and relic of cultivation in a number of localities. 34–6, 45, *46*, 55–8, 66, 68, [54]. Widespread casual or relic of cultivation, native of C. Asia.

RUMEX L. (Dock)

1. Rumex acetosella L. sens. lat.   Sheep's Sorrel   Ray, 1660
Scattered throughout the county in dry grassy or waste places, but not abundant. A specimen from Gamlingay in CGE collected by C. E. Moss in 1912 has the leaves 3–4 times as long as broad with the perianth segments closely applied to the fruit, and has been determined by J. E. Lousley as **R. angiocarpus** Murb. This is probably the common species in the county. Specimens in CGE from Kennett (1954) and Hildersham (1957) have the leaves 7–10 times as long as broad and the perianth segments not closely applied to the fruit, and have been determined by Lousley as **R. tenuifolius** (Wallr.) Love. 25, 29, 35–9, *44*, 45, *46*, 48, 49, 54–6, 58, 59, 65–7, 76, 20, 30, 31, 40, 41, 50, [23, 47]. General.

2. Rumex acetosa L.   Sorrel   Ray, 1660
Throughout the county in meadows and grassy places. All squares except *57*, [24]. General.

3. Rumex hydrolapathum Huds.   Great Water Dock   Ray, 1660
Frequent in ditches and lodes and by pits and rivers in the Fens, local elsewhere. 29, 36–9, 44, *45*, 46–8, *49*, 56–9, 67–9, 20, 30, 31, 40, 41, 50. General, but rare in Scotland.

4. Rumex crispus L.   Curled Dock   Ray, 1685
A variable species common throughout the county in cultivated and waste places. All squares. General.

× **obtusifolius** (*R. pratensis* Mert. & Koch). A number of old records, but not recorded since 1938. *24*, *46*, *57*, *58*, [25, 34–6, 38, 44, 47].

× **pulcher.** Recorded by C. E. Moss from Chippenham in 1913 and by J. S. L. Gilmour at Soham, Stretham and Wicken. 56, 57, [66].

5. Rumex obtusifolius L. ssp. **obtusifolius**   Ray, 1660
Abundant on roadsides and waste ground throughout the county. All squares except *24*, 26, *35*. General.

POLYGONACEAE

6. Rumex pulcher L.   Fiddle Dock   Ray, 1685

Scattered over the Fens in grassy places, rare elsewhere. 39, 45, 46, 56, 58, 67, 68, 41, [25, 35, 36, 38, 44, 47–9, 55, 65, 66, 40]. In England south of a line from the Humber to the Severn, in Wales in coastal districts of the south and north-west, and on the S. coast of Ireland, where it is never more than a casual.

7. Rumex sanguineus L.   Relhan, 1786

Throughout the county in woods, copses, waste places and ditch-sides; common in boulder-clay woods, rarer in the Fens. Most if not all our plants are green or occasionally with rust-red veins to the leaves, and are referable to var. **viridis** Sibth. (*R. condylodes* Bieb.). All squares except 24, 29, 59, 68, 30, 40. Common in S. England, Wales and Ireland, less frequent in the north.

8. Rumex conglomeratus Murr.   Ray, 1660

By ditches, in woods and grassy places throughout the county. All squares except 26, 67, 41, [39]. General, but rare in Scotland.

× **crispus.** Recorded from Chesterton (1930), between Stretham and Wicken (1931), and from Earith bridge (1938). The last record was that of J. E. Lousley. *37, 46, 57.*

× **sanguineus.** This hybrid occurs and sets fertile seed. In this respect it differs from most other *Rumex* hybrids, which are highly sterile.

9. Rumex palustris Sm.   Marsh Dock   Ray, 1660

Recorded in the Fenlands on open usually peaty ground, by ditches, ponds and lodes. 29, 37, 38, 45, *46*, 48, 49, 56, 57, 59, 40, 41, 50, [39, 47, 58]. W. Local in S. and E. England north to the Tees.

10. Rumex maritimus L.   Golden Dock   Ray, 1685

A number of old records exist for the Fenlands, but the plant has only been seen recently at Wicken Fen, where it occurs at the water's edge of the Mere and on disturbed peaty ground by ditches in apparently similar habitats to *R. palustris. 46*, 56, [25, 38, 39, 45, 47–9, 57, 58, 40]. W. Scattered throughout England, very rare in Wales, Scotland and Ireland.

× **obtusifolius.** Recorded from Roswell Pits, Ely, by J. E. Lousley in 1938. *58.*

URTICACEAE
PARIETARIA L.

1. Parietaria diffusa Mert. & Koch   Pellitory-of-the-wall Ray, 1660
P. ramiflora sensu Evans; *P. erecta* Koch
Occasional on old walls and buildings throughout the county. 29, 34, 36, 44, 45, *46*, 47, 49, *54*, 55–8, 66, 67, 20, 40, [38, 65, 50]. Throughout England, Wales and Ireland but rather local; rare in Scotland. Native at least on rocks in the west.

HELXINE Req.
1. Helxine soleirolii Req.
Garden escape. 34, 45, *46*.

URTICA L.

1. Urtica urens L.   Small Nettle   Ray, 1660
A weed of arable land and waste places throughout the county. All squares except 24. General.

2. Urtica dioica L.   Stinging Nettle   Ray, 1660
Abundant throughout the county. At Wicken and Chippenham Fens the native plant is narrow-leaved and practically stingless, and is probably referable to forma **angustifolia** (Wimm. & Grab.) Moss. The forma **microphylla** (Hausm.) Moss with smaller leaves also occurs in the county. All squares. General.

3†. Urtica pilulifera L.   Roman Nettle
                    Dr Jermyn, early nineteenth century
Recorded by Dr Jermyn from Wisbech. According to Babington it was the plant with nearly entire leaves referable to var. **dodartii** (L.) Aschers. [40.] Extinct in Britain; never native.

CANNABIACEAE
HUMULUS L.

1. Humulus lupulus L.   Hop   Ray, 1660
In hedgerows and woods throughout the county. 24–6, 34–8, 44–9, 54–8, *64*, 30, 40, 50, [66, 67]. Widely distributed; apparently native in wet woodland, but doubtless often an escape from cultivation. A valuable constituent of the best beers.

147

CANNABIACEAE

CANNABIS L.

1. Cannabis sativa L. Hemp

Casual from bird-seed. *36, 45, 46, 49.*

ULMACEAE

ULMUS L. (Elm)

Elms are amongst the most common trees to be found in the county and are especially frequent on the boulder clay and sands. It is probable that the great majority if not all of our present stock is of planted origin. They are extremely variable, which makes the taxonomy very difficult. Hybrids are said to be frequent. *U. glabra* and *U. procera* are fairly easy to recognize, but the variation in leaves and branching of the smooth-leaved Elms and the supposed hybrids of these with *U. glabra* is very great. It is highly probable that this group is made up of a large number of clones, and some of these have received names. The three species here recognized can be distinguished as follows:

Leaves large, 8–16 cm., broadly ovate to obovate or elliptic, long-cuspidate, scabrid above and pubescent beneath, base of long side forming a rounded lobe almost overlapping the short petiole. **1. U. glabra**

Leaves 4·5–9 cm., suborbicular to ovate, acute, ± scabrid above, pubescent beneath, the rounded to subcuneate base unequal.

2. U. procera

Leaves very variable in size and shape, nearly or quite glabrous except for axillary tufts beneath; bases subequal or unequal.

3. U. carpinifolia

The so-called hybrids between *U. glabra* and *carpinifolia* are variable but usually have the leaves smooth like *U. carpinifolia*. An account of Cambridgeshire elms has been published by R. H. Richens (1958 and 1960).

1. Ulmus glabra Huds. Wych Elm Ray, 1660

U. montana Stokes

Throughout the county in woods, copses and hedgerows. 23–6, 33–6, 38, 39, 43–6, 48, 49, 54–6, 58, 64–6, 30, 31, [47, 57, 41]. General.

2. Ulmus procera Salisb. English Elm Ray, 1660

U. suberosa sensu Bab.; *U. campestris* sensu Evans

Scattered throughout the county in woods, copses and hedgerows and planted in meadows and other places. Most frequent in the south-west. 24, 26, 34–6, 38, 39, 43–6, 49, 54–6, 58, *64*, 65, 20, 31, 40, 41, 50, [48, 66]. General, but less frequent in the north.

148

ULMACEAE

3. Ulmus carpinifolia Gled. sens. lat.   Ray, 1660

U. suberosa var. *glabra* sensu Bab.; *U. nitens* sensu Evans; *U. sativa* sensu Evans; *U. coritana* Melville; *U. plotii* Druce; *U. diversifolia* Melville

Throughout the county mainly in copses and hedgerows and occasionally in woods. A very variable species which has been divided into several segregates, of which **U. coritana** Melville and **U. plotii** Druce have been recorded. The Jersey Elm, **U. × sarniensis** (Loud.) Melville (*U. stricta* var. *sarniensis* (Loud.) Moss), is occasionally planted. 24–6, 34–7, 43–7, 49, 54–8, 64–7, 76. Common in eastern England; west to Cornwall if *U. angustifolia* (Weston) Weston is included in the aggregate.

× **glabra.** This hybrid is recorded from a number of localities, especially the forms known as **U. × hollandica** Mill. (the Huntingdon Elm) and **U. × vegeta** (Loud.) A. Ley (the Dutch Elm). *U. vegeta* is frequently planted in Cambridge and is easily recognized by its short trunk and ascending branches which spread fan-wise to form a rounded head. It produces few or no suckers. *24*, 34, 36, 44, 45, 56, 57, [25, 48].

MORACEAE

FICUS L.

1. Ficus carica L.   Fig

Casual. 45.

JUGLANDACEAE

JUGLANS L.

1. Juglans regia L.   Walnut

Often planted for its fruit. 34, 45, 58, 31.

MYRICACEAE

MYRICA L.

1. Myrica gale L.   Bog Myrtle; Sweet Gale   S. Corbyn, 1656

Although said by Ray to be abundant in many parts of the Isle of Ely, and known at Wimblington Firelots until 1895, it is now only to be found at Wicken Fen. 56, [49, 58]. W. Common in bogs, wet heaths and fens in the north and west of the British Isles, widely scattered but local elsewhere.

149

PLATANACEAE

PLATANACEAE

PLATANUS L.

1. Platanus × hybrida Brot. (*P. acerifolia* (Ait.) Willd.)
Planted tree. 45.

BETULACEAE

BETULA L.

1. Betula pendula Roth   Silver Birch   Babington, 1860
B. alba sensu Bab., et Evans; *B. verrucosa* Ehrh.
May be native in some woods; commonly planted over much of the county, occasionally self-sown. 23–6, 33–5, 44, 45, 49, 54–7, 64–6, 68, 30, 40, 41, [36, 38]. Throughout most of the British Isles, but common in S. England and rare in the north.

× **pubescens.** Recorded from Gamlingay Heath Wood in 1916 and Basefield Wood in 1932. *65*, [25].

2. Betula pubescens Ehrh. ssp. **pubescens**   Birch   Ray, 1660
B. glutinosa Fries
May be native at Gamlingay and Chippenham, but elsewhere widely planted. 24, 25, 34, 35, 43, 45, *54*, 57, 66, 68, 30, 40, [56]. General.

ALNUS Mill.

1. Alnus glutinosa (L.) Gaertn.   Alder   Ray, 1660
Scattered over the county in fens, by streams and in other wet places. Typical alder carr is not common in Cambs, but is developed at Chippenham, Hildersham, Dernford and Whittlesford. The var. **macrocarpa** Loudon with large leaves and pistillate catkins occurs at Chippenham Fen. *23*, 34–6, 44, 45, 54–8, 65, 66, 68, [25, 38, 48]. General.

2. Alnus incana (L.) Moench
Planted tree. 45, *55*.

CORYLACEAE

CARPINUS L.

1. Carpinus betulus L.   Hornbeam   Ray, 1660
Scattered over the county in a number of localities, mostly in boulder-clay woods, but often planted. 25, 33–5, *36*, 38, 43–5, *54*, *64*, 65, [24, 76]. **B.** Native in S.E. England, planted elsewhere.

CORYLUS L.

1. Corylus avellana L.   Hazel   Ray, 1660
Throughout most of the county in woods, copses and hedgerows. One of the dominant shrubs in the boulder-clay woods. 23, *24*, 25, 26, 34–6, 43–6, 49, 54–6, 58, 64–8, 75, 31, 41, [38, 47, 48, 57, 40]. General.

FAGACEAE

FAGUS L.

1. Fagus sylvatica L.   Beech   Relhan, 1820
Not known to Ray, so all Cambridgeshire Beech may be of planted origin. Occurs throughout much of the county, but most frequent in plantations on the chalk. 23–5, 33–6, 39, 43–6, 48, 49, 54, 55, 59, 64–7, 20, 31, 40, 41. Native of southern England, planted elsewhere.

CASTANEA Mill.

1. Castanea sativa Mill.   Sweet Chestnut   C. E. Moss, 1909
Frequent at Gamlingay; a few records elsewhere. 24, 25, 44, 65, *40*. **G.** Commonly planted throughout the British Isles, though rare on limestone. Native of S. Europe, N. Africa and W. Asia.

QUERCUS L.

1. Quercus cerris L.   Turkey Oak   C. E. Moss, *c.* 1912
Frequent and regenerating in the Heath Wood, Gamlingay; scattered records elsewhere. 25, 26, 34, 36, 65, 40. **G.** Commonly planted, native of S. Europe and S.W. Asia.

FAGACEAE

2. Quercus robur L.   Oak   Ray, 1660

Throughout the county in woods, copses and hedgerows. The dominant standard tree of the 'coppice with standards' woods on the boulder clay. All squares except 37. General, on heavy especially basic soils, thus the dominant species of S., E. and C. England, rarer in the west.

3. Quercus petraea (Mattuschka) Liebl.   Durmast Oak

C. E. Moss, 1908

Q. sessiliflora Salisb.

Recorded by Moss from White Wood, Gamlingay, and illustrated in his *Camb. Brit. Fl.*, plate 73, from Cambs material. Also recorded as rare in the Heath Wood, Gamlingay, in 1916. 25. Throughout most of the British Isles but mainly on acid soils and more dominant in the north and west.

× **robur.** (*Q.* × *rosacea* Bechst.; *Q.* × *intermedia* Boenn. ex Reich.). There are specimens in CGE from Gamlingay in 1909 and Hildersham Wood (a single tree) in 1956. 25, 54.

SALICACEAE

POPULUS L.

1. Populus alba L.   White Poplar   Ray, 1660

Planted in a number of scattered localities. 25, 34, 45–7, 57, 58, 41, 50, [38, 48, 49, 54, 56, 40]. Frequently planted especially in S. England; native of S. and C. Europe, N. Africa and W. Asia.

2. Populus canescens (Ait.) Sm.   Grey Poplar   Relhan, 1820

Scattered throughout the county in woods, copses and hedgerows. 25, 26, 34–6, 43, 44, *45*, *46*, 47, 49, 55, *58*, 66, *67*, 20, 50, [38, 48, 56]. Probably native in damp woods in S., E. and C. England; planted elsewhere.

× **tremula.** Recorded from Fulbourn (CGE) in 1932. *55*.

3. Populus tremula L.   Aspen   Ray, 1660

In numerous localities, mostly boulder-clay woods; occasionally planted. The var. **villosa** (Lang) Wesm., with the young leaves covered with silky hairs, is not uncommon. 24, 25, 34–6, 43–6, 48, 49, 54, 64, 65, 41. General, but common only in north and west.

152

4. Populus nigra L.   Black Poplar   Ray, 1660

Recorded from a number of scattered localities, but some of these records may be referable to *P.* × *canadensis*. Several fine male trees grow at Madingley. Characterized by the large bosses on the trunk and main branches, and by the branches spreading, and arching downwards, to form a wide crown. 34–*36*, 38, 44, *45*, *46*, 48, 40, [54, 55, 57, 67]. Possibly native in E. and C. England, planted elsewhere.

5. Populus italica (Duroi) Moench   Lombardy Poplar
E. A. George, 1939

Frequently planted. 25, 26, 34–6, 44–6, *56*. Native of C. Asia.

6. Populus × **canadensis** Moench   C. E. Moss, 1911

P. serotina Hartig.

Planted in many parts of the county. Said to be a hybrid between the American *P. deltoidea* Marshall and *P. nigra* L.; very variable in habit and leaf shape but differing from *P. nigra* by the branches ascending and curving upwards to form a wide fan-like crown, and by lacking bosses on trunk and main branches. 25, 34, 36, 39, 44, 45, *46*, 47, *55*, 57, 58, 20, 30, 31, 41, [65]. Planted generally.

SALIX L.

1. Salix pentandra L.   Bay-leaved Willow

Planted at Wicken Fen and Cambridge. 45, 56.

2. Salix alba   White Willow   Ray, 1660

Common throughout the county though often of planted origin. The var. **vitellina** Stokes is recorded from several localities; it has the twigs bright orange or yellow in their first year, the leaves nearly glabrous above and only thinly hairy below. All squares except *65*. General.

× **fragilis**. According to R. D. Meikle, *S. alba* var. *coerulea* (Sm.) Sm. (*S. coerulea* Sm.), which is grown to provide wood for cricket bats, almost certainly originates from a cross between *S. alba* and *fragilis*. There are three records of this, two of which are supported by specimens in CGE. (See also *S. fragilis* var. *russelliana*.) 44, 45, [48].

× **pentandra** (*S.* × *ehrhartiana* Sm.). C. E. Moss (1914) gives this hybrid from Cambs. No precise locality is given and there are no specimens in CGE.

153

SALICACEAE

× **triandra.** There are a number of records from Cambs of *S. undulata* Ehrh. (*S. lanceolata* Sm.) which has been thought by some authors to have originated from a hybrid between *S. triandra* and *viminalis* and by others from *S. alba* and *triandra.* Specimens in CGE from Stretham (1833 and 1932), Cambridge (1940) and Isleham (1955) seem to be intermediate between *S. alba* and *S. triandra.* 45, 57, 67, [?38, ?46–9].

3. Salix babylonica L.   Weeping Willow
This tree and its hybrids are often planted, especially in the Cambridge area. 45.

4. Salix fragilis L.   Crack Willow   Ray, 1660
Scattered over the county, and often if not always of planted origin. Three varieties are recorded and can be distinguished as follows: var. **russelliana** (Sm.) Koch (*S. russelliana* Sm.; *S. alba × fragilis* auct.) has the leaves 7·5–14 × 1·2–1·5 cm. and the ovaries longer than the subtending scale. This is the common variety in the county and according to Professor K. H. Rechinger fil. and other authors is a *S. alba × fragilis* hybrid. Var. **latifolia** Anderss. with the leaves 7·5–12 × 2–4 cm. is of rare occurrence. Var. **decipiens** (Hoffm.) Koch (*S. decipiens* Koch) with the leaves small 5–9 × 1·5–2·5 cm. is probably frequent in the Fens. It is doubtful if var. **fragilis** with the leaves 6·5–10·5 × 1–2 cm. and the ovaries shorter than the subtending scale is found in the county. All squares except 26, 55, 57, 64, 67. General, but not native in N. Scotland and doubtfully so in Ireland.

× **pentandra** (*S. × meyeriana* Rostk.). There are specimens in CGE from Ely (1864), Cambridge (1942) and Wendy (1953), and several other field records. All are presumably of planted origin. 34, 44, 45, 56, [58]. **Cam.**

× **triandra** (*S. × speciosa* Host.; *S. alopecuroides* (Reichb.) A. Kerner). There are specimens in CGE from Cottenham (1893), Fen Ditton (1912), Lord's Bridge (1912) and Waterbeach (1912). [45, 46, 56].

5. Salix triandra L.   Almond-leaved Willow   Ray, 1660
A local species mostly by pits, streams and in marshy places in the Fens. Two varieties occur: var. **triandra** with the leaves linear-lanceolate and the stipules small and not very persistent, and var. **hoffmanniana** (Sm.) Bab. (*S. hoffmanniana* Sm.) with the leaves narrowly ovate and the stipules large, ear-shaped and persistent. The var. **amygdalina** (L.) Bab.

with the leaves ovate and ± glaucous beneath has been recorded but there are no supporting specimens in CGE. This species and *S. viminalis* are commonly planted in Osier beds in the Fens. *25*, 35–7, 44–6, 48, *55*, 56, *57*, 59, 64, 65, [34, 38, 47, 49, 54, 58, 66, 40]. Widespread and rather common in England, less so in Wales, very local in S.E. Scotland and S.E. Ireland.

6. Salix purpurea L.   Purple Osier   Ray, 1660

Scattered throughout the county by ditches, pits and rivers, commonest in the Fenlands. 25, 34, 35, 37, 44, 45, *46*, 48, 49, 55–8, *65*, 66, 68, 20, [38, 47, 67]. **W.** General, but rare in the north.

× **viminalis** (*S.* × *rubra* Huds.). There are a number of records of this hybrid but they are unsupported by herbarium specimens and need confirmation. *45*, *57*, [?35, 46–8, 58, 68].

7. Salix viminalis L.   Osier   Ray, 1660

By ditches, rivers and pits throughout most of the county. The var. **linearifolia** Wimm. & Grab., with more slender leaves, branches and catkins is recorded. 24, 25, 34, 35, 38, 39, 44–9, 56–9, 66, 68, 75, 20, 31, 40, 50, [36, 67]. General.

S. viminalis hybrids with either *S. cinerea* sens. lat. or *S. caprea* are probably frequent. 34, *35*, *45*, *55*, 57, [38, 47–9].

8. Salix calodendron Wimm.   C. C. Townsend, 1959

This species grows on 'Paradise', Cambridge, and at Nine Wells, Trumpington. If it is a hybrid, as some authorities believe, it is probably either *S. caprea* × *viminalis* or *S. caprea* × *cinerea* × *viminalis*. 45. Local in E. England and S. Scotland, probably introduced.

9. Salix caprea L. ssp. caprea   Goat Willow   Ray, 1660

Scattered throughout the county by ditches and pits and in hedgerows. The characteristic Willow of boulder-clay woods. 25, 29, 34–6, 43–7, 49, 54–6, 64–6, [38, 48, 57, 41]. General.

× **cinerea** L. (*S. reichardtii* A. Kerner). There are specimens in CGE from Bottisham Fen (1838), Langley Wood (1932), and Hildersham Wood (1956). 54, *64*, [56].

10. Salix cinerea L.   Sallow   Ray, 1660

By streams, rivers and pits and in hedgerows, fens and marshy places throughout the county. Very variable. The most frequent subspecies is

155

SALICACEAE

probably ssp. **atrocinerea** (Brot.) Silva & Solr. (*S. oleifolia* Sm., non Vill.; *S. atrocinerea* Brot.) which has the surface of the leaves slightly glaucous beneath where it is clothed with a varying number of short rusty red hairs, especially on the nerves and veins. The mature twigs are almost glabrous. Ssp. **cinerea** (*S. aquatica* Sm.) with the leaves and twigs densely grey-felted is characteristic of the Fens, as at Wicken. All squares except 59, 20. The ssp. *atrocinerea* is general, but ssp. *cinerea* is probably confined to the Fenlands of Eastern England.

× **purpurea** × **viminalis** (*S.* × *forbiana* Sm.; *S. rubra* var. *forbiana* (Sm.) Bab.). There are several records of this hybrid as *S. forbiana* Sm., but the only specimen in CGE was collected by E. F. Warburg at Stretham in 1932. ?48, *57*, [?35, ?58].

× **repens** (*S* × *subsericea* Doell.). This hybrid occurs at Wicken Fen where both parents are abundant. 56. **W.**

11. Salix aurita L.   Eared Sallow
First certain record, E. F. Warburg, 1935
The only certain records for this species are Hayley Wood (1935) and Gamlingay Wood (1957) which are supported by specimens in CGE. Other field records must be regarded as doubtful. 25, [?35, ?44–6, ?55, ?65, ?66]. **G.** General but usually on acid soils.

× **cinerea**. A specimen in CGE collected by E. F. Warburg at Hayley Wood in 1941 seems to be referable to this hybrid. *25*.

(According to Professor K. H. Rechinger two sterile specimens in CGE from Chippenham are probably referable to **S. nigricans** Sm. Catkin-bearing specimens are needed, however, before this northern species can be recorded with certainty for the county.)

12. Salix repens L. ssp. **repens**   Ray, 1685
The var. **fusca** (L.) Wimm. & Grab. with a short rhizome and erect stems is still plentiful at Wicken Fen, and it was probably this variety that formerly occurred in other localities in the Fens. To judge by herbarium material, the plant which used to grow at Gamlingay was probably var. **repens** with long rhizomes and prostrate or ascending stems. This variety has recently been found on the edge of Hayley Wood (1962). 25, *49*, 56, 57, [44, 48, 66]. **W.** General, but local in some areas.

ERICACEAE

ERICACEAE

RHODODENDRON L.

1. Rhododendron ponticum L.

Planted evergreen shrub. Not suited to Cambs soils. 25, 54.

CALLUNA Salisb.

1. Calluna vulgaris (L.) Hull   Heather   Ray, 1660

Formerly in a number of heathy places. Now known only at Gamlingay, Kennett and in small quantity on Newmarket Heath. Abundant on the Breckland heaths in Suffolk and Norfolk. 25, 56, 66, [44, 45, 54, 55, 76]. G. General on acid soils.

ERICA L.

1†. Erica tetralix L.   Cross-leaved Heath   Ray, 1685

Formerly occurred at Gamlingay where it was last seen in 1920. [25.] General in wet heaths and bogs.

2†. Erica cinerea L.   Bell-heather   Relhan, 1785

Formerly occurred at Gamlingay where it was last seen in 1914. [25.] General on heaths.

VACCINIUM L.

1†. Vaccinium oxycoccos L.   Cranberry   Ray, 1685
Oxycoccus palustris Pers.

Formerly occurred at Gamlingay where it was last seen in 1859. [25.] Widespread but local in bogs.

MONOTROPACEAE

MONOTROPA L.

1. Monotropa hypopitys L. sens. lat.   Yellow Bird's-nest
                                                            Relhan, 1802

This saprophyte has been recorded from a number of localities mostly in boulder-clay woods, and was last seen at Hardwick Wood in 1946. The species has been divided into two segregates, both of which have been recorded in Cambs. **M. hypopitys** L. sens. strict., with the filaments, style and the inside of the 9–12 mm. long petals covered with rather stiff hairs, is represented in CGE by specimens from Madingley (1835

157

MONOTROPACEAE

and 1890), Wimpole (1836) and Eversden Wood (1849). **M. hypophegea** Wallr., with the filaments, style and inside of the 8–10 mm. long petals glabrous, is represented from near Coton (1886), Madingley (1878), Shepreth (1890) and a wood on Wort's Causeway (1940). *24, 35, 45, 55,* [34, 36]. Widespread but local in England; rare in Wales, Scotland and Ireland.

PLUMBAGINACEAE

LIMONIUM Mill. (Sea Lavender)

1. Limonium vulgare Mill.   Skrimshire in Relhan, 1802
Statice limonium L.
Only to be found by the tidal banks of the River Nene near Wisbech. 41. FA. Round the coasts of Great Britain north to Fife and Kirkcudbright.

2†. Limonium bellidifolium (Gouan) Dumort.
                                    J. Hempsted in J. E. Smith, 1796
Statice caspia sensu Bab.
Formerly found by the tidal River Nene near Wisbech where it was last seen in 1860. [41.] Coasts of Norfolk and Lincs.

ARMERIA Willd.

1. Armeria maritima (Mill.) Willd.   Thrift   I. Lyons, 1763
Formerly occurred by the banks of the tidal River Nene and its tributaries at Parsons Drove and near Wisbech, where it was last seen in 1930. *41,* [30]. General on coasts of British Isles and on mountains inland.

PRIMULACEAE

PRIMULA L.

1. Primula veris L.   Cowslip; Paigle   Ray, 1660
In meadows and other grassy places throughout most of the county, but less common than formerly after ploughing and re-seeding of old pasture. 23–6, 33–7, 43–6, 54–7, 64–7, 75, 76, 41, [38, 48, 49, 58, 40]. General but rare in north.

× **vulgaris.** Recorded from a number of places on the boulder clay where the two species grow near together. *25,* 35, 36, 43, 54, [45].

PRIMULACEAE

2. Primula elatior (L.) Hill   Oxlip   Ray, 1660
A characteristic spring-flowering species of the boulder-clay woods,
where, alone or with *P. vulgaris* and their hybrid, it covers large areas.
25, 35, 36, 54, 55, 64, 65, 75. **HA.** Woods, or rarely wet meadows, on
boulder clay in Essex, Herts, Suffolk, E. Norfolk, Cambs, Beds and
Hunts.

× **veris.** No certain records for this hybrid have been traced, though it
has been reported on several occasions. ?24, ?36, ?65.

× **vulgaris.** In hybrid swarms wherever the parents are found growing
together. 25, 35, *36*, 54. **B.**

3. Primula vulgaris Huds.   Primrose   Ray, 1660
Frequent in boulder-clay woods. Elsewhere occurs as a garden escape.
25, 26, 34–6, 43–6, 54, 56, 66, [24, 38, 49, 55, 65]. Throughout the
British Isles but rare in the north. Appears to be absent from some of
the Oxlip woods.

HOTTONIA L.

1. Hottonia palustris L.   Water Violet   Ray, 1660
Common in Fen ditches and lodes, rare elsewhere. 29, 34, 37–9, *45*, *46*,
47–9, 55–7, 59, *66*, 68, 20, 30, 40, [44]. Widely distributed in England
but local and rare except in the east. Also recorded from N. Wales,
Inverness and N.E. Ireland. (See plate 4.)

LYSIMACHIA L.

1. Lysimachia nemorum L.   S. Corbyn, 1656
Known only from woods at West Wickham, and Ditton Park Wood,
on the eastern boulder clay. 64, 65. **DPW.** General; probably rarer in
Cambs than most other counties.

2. Lysimachia nummularia L.   Creeping Jenny   Ray, 1660
In moist places throughout the county. All squares except *24*, 66, 68,
[26, 38]. General, but absent from the north.

3. Lysimachia vulgaris L.   Yellow Loosestrife   Ray, 1660
Scattered through the Fenlands in marshy places. 35, 37, 38, 44, 45,
46, 48, 49, 55–7, 59, 66, 20, [47, 58, 65, 40]. **W.** General, but rare in the north.

4. Lysimachia punctata L.
Casual. [46.]

159

PRIMULACEAE

ANAGALLIS L.

1. Anagallis tenella (L.) L.   Bog Pimpernel   S. Corbyn, 1656
Formerly in a number of boggy places. Still occurs at Quy, Sawston,
Thriplow and Chippenham. 44, 55, 56, 66, [25, 34, 38, 45, 58]. **T.** In
bogs and damp peaty and grassy places throughout most of the British
Isles.

2. Anagallis arvensis L.   Scarlet Pimpernel   Ray, 1660
A common weed of arable land throughout the county. The forma
azurea Hyl. (*A. arvensis* var. *caerulea* sensu Bab.) with blue flowers is
recorded. The ssp. **foemina** (Mill.) Schinz & Thell. is also recorded;
this has blue flowers but differs in having the margins of the petals
sparingly, not densely, fringed with 4-celled glandular hairs, the corolla
lobes not overlapping, and the calyx teeth concealing the corolla in
bud. All squares except 29, 37, 39, 59. General, but rare in the north.

3†. Anagallis minima (L.) E. H. L. Krause   Relhan, 1820
Centunculus minimus L.
Recorded from Gamlingay by Relhan. [25.] Local in suitable habitats
north to Lewis.

GLAUX L.

1. Glaux maritima L.   I. Lyons, 1763
Only to be found by the banks of the tidal River Nene and its tributaries. 30, 41. **FA.** All round the coasts of the British Isles.

SAMOLUS L.

1. Samolus valerandi L.   Brookweed   S. Corbyn, 1656
Usually found on bare, muddy or peaty ground; common in the Fens,
local elsewhere. 34, 37–9, 44–9, 55–7, 59, 66, 67, 20, 30, 31, 40, 41, 50,
[25, 35, 58]. **W.** General, but mainly coastal in the west.

BUDDLEJACEAE

BUDDLEJA L.

1. Buddleja davidii Franch.
Planted shrub, but not freely self-sown. 38, 45, 58, 67, 40.

OLEACEAE
FRAXINUS L.

1. Fraxinus excelsior L.   Ash   Ray, 1660
Common in woods, copses and hedgerows throughout the county, often planted. Co-dominant with *Quercus robur* in the boulder-clay woods and regenerating freely; also on peat (as at Chippenham) and chalk (as at Cherry Hinton). All squares. General.

SYRINGA L.

1. Syringa vulgaris L.   Lilac
Planted or self-sown. 23, 45, 58, 65, *66*, [40].

LIGUSTRUM L.

1. Ligustrum vulgare L.   Privet   Ray, 1660
In woods, copses and hedgerows throughout the county. The plant more commonly grown in gardens with ovate, more nearly evergreen, leaves is **L. ovalifolium** Hassk. All squares except 37, 39, 30. Native on calcareous soils in England, probably naturalized elsewhere.

APOCYNACEAE
VINCA L.

1. Vinca minor L.   Lesser Periwinkle   Ray, 1660
In copses and hedgerows, often, perhaps always, an escape from cultivation. *25*, 34–6, 44, 45, *46*, 49, 54, 55, *40*, 41, [56, 64]. General, but rare in Ireland; doubtfully native.

2. Vinca major L.   Greater Periwinkle   I. Lyons, 1763
An escape from cultivation, in hedgerows and copses, and often naturalized. 23, 25, 29, 34–6, 44–6, 54, 56, 58, [*47*, 49]. Naturalized garden escape in the southern half of the British Isles; native of C. and S. Europe and N. Africa.

GENTIANACEAE

GENTIANACEAE

CENTAURIUM Hill (Centaury)

1. Centaurium pulchellum (Sw.) Druce　　　　　Relhan, 1802
Erythraea pulchella (Sw.) Fries
Formerly occurred in a number of localities, but only seen recently in a ride in Ditton Park Wood, at Chippenham and on Green Hills, Soham. 65–7, [24, 35, 36, 45, 56–8, 41]. **DPW.** Common near the sea in S. and C. England, more local inland, rarer northwards; also in S. and E. Ireland.

2. Centaurium erythraea Rafn.　　　Common Centaury　　　Ray, 1660
Erythraea centaurium sensu Bab., et Evans; *C. minus* auct.
Occasional in grassland, though nearly absent from the Fens. *25*, 26, 34, 35, *36*, 43–5, 54–6, 65–7, [24, 46, 49, 57, 40]. General, but rare in Scotland.

BLACKSTONIA Huds.

1. Blackstonia perfoliata (L.) Huds.　　Yellow-wort　　Ray, 1660
Chlora perfoliata (L.) L.
A local species of grassland on the chalk and boulder clay. *24*, 26, 34, 35, *36*, 44, 45, 55, 56, 64–6, [25, 38, 46, 58, 41]. Base-rich soils in England, Wales and S. Ireland.

GENTIANELLA Moench

1. Gentianella amarella (L.) Börner ssp. **amarella**　Felwort　Ray, 1660
Gentiana amarella L.
A local species of open grassland on the chalk and sands. 24, *25*, 34, 35, 44, 45, 54–6, 66, [23, 36]. **GG.** General in suitable habitats.

MENYANTHACEAE

MENYANTHES L.

1. Menyanthes trifoliata L.　　Buckbean　　　Ray, 1660
Formerly in a number of boggy places in the Fens, but only seen recently at Wicken and Chippenham. *44*, 56, 66, [25, 45, 46, 48, 58, 59]. **C.** General.

MENYANTHACEAE

NYMPHOIDES Hill

1. Nymphoides peltata (S. G. Gmel.) Kuntze   Ray, 1660
Villarsia nymphaeoides Vent.
Aquatic perennial herb with floating leaves intermediate in size between *Hydrocharis morsus-ranae* and *Nuphar lutea*. The handsome yellow flowers are heterostylous; it is interesting that only the 'pin' form has been found in the county. Characteristic of rivers and lodes in the Fens, particularly abundant in the Old West River. 29, 37–9, 47–9, *56*, 57–9, 68, 20, 30, 31, 40, [36, 46]. Local in E. and C. England, occasional as an introduction elsewhere.

BORAGINACEAE

CYNOGLOSSUM L.

1. Cynoglossum officinale L.   Hound's-tongue   Ray, 1660
Frequent in disturbed places on the sands in the east of the county and on the chalk; also frequent on lode banks in the Fenlands. 24, 35, 43–5, 47, 48, 54–7, 64–7, 76, 40, [46, 49]. In Britain north to Angus but local; in Ireland almost confined to the coast.

ASPERUGO L.

1. Asperugo procumbens L.   Madwort   Ray, 1660
Recorded by Ray from Newmarket, and seen at Melbourn *c*. 1906. Found at Cherry Hinton in 1925 where it has been recorded sporadically until 1953. 45, [34, 66]. A rare casual occasionally naturalized by the margins of fields and in waste places; native of continental Europe, W. Asia and N. Africa.

LAPPULA Fabr.

1. Lappula myosotis Moench (*L. echinata* Gilib.)
Casual. *46*, [45].

SYMPHYTUM L.

1. Symphytum officinale L.   Comfrey   Ray, 1660
Throughout most of the county, common in the Fenlands, rare on the chalk. The cream-flowered form occurs in native habitats by stream and lode banks or in fens and marshy places, while the purple-flowered form

BORAGINACEAE

is more often on roadsides and railway banks. All squares except 24, 26, 29, 54, 64, [25]. General, though less common in the north and not native in Ireland.

S. asperum × officinale   A. J. Crosfield, 1923
S. peregrinum sensu Evans; *S. × uplandicum* Nyman
Occurs in a number of scattered localities. One of the parents, *S. asperum* Lepech., is not recorded for the county. 23, 34, 35, 37, 44, *46*, 56, 68, 41, [45]. General.

2. Symphytum orientale L.   Babington, 1849
S. tauricum sensu Bab.
Well established in a number of localities, especially around Cambridge. 34, *35*, 45, 46, 64, 65. **Cam.** Naturalized and spreading particularly in eastern England; native of Turkey.

3. Symphytum armeniacum Bucknall
Casual or garden escape. *35*.

4. Symphytum caucasicum Bieb.
Garden escape. 45.

5. Symphytum grandiflorum DC.
Garden escape. 65.

BORAGO L.

1. Borago officinalis L.   Borage
Casual or garden escape. 35, 44, *45*, 58, [34, 46, 41].

PENTAGLOTTIS Tausch

1. Pentaglottis sempervirens (L.) Tausch
Anchusa sempervirens L.
Garden escape, rarely naturalized. 34, 43–5, [46].

LYCOPSIS L.

1. Lycopsis arvensis L.   Bugloss   Ray, 1660
A common weed of arable and waste land on the sands, very local elsewhere. 25, 38, 39, 44, 45, *46*, 54, 55, 65–7, 76, [58]. General in Great Britain; in Ireland mainly in the east and near the sea.

NONEA Medic.

1. Nonea rosea (Bieb.) Link
Garden escape. 45, 46.

2. Nonea lutea DC.
Garden escape. 45.

MYOSOTIS L. (Forget-me-not)

1 Hairs on calyx-tube appressed, rarely almost 0.   *2*
At least some hairs on calyx-tube short, stiff, hooked or crisped,
  not appressed.   *3*

2 Style equalling or exceeding calyx-tube; corolla often more than
    6 mm. in diam. (common).   **1. M. scorpioides**
Style *c.* ½ length of calyx-tube; corolla usually not over 6 mm. in
    diam. (rare).   **2. M. caespitosa**

3 Fruit pedicels as long as or longer than calyx; cymes in fruit shorter
    or not much longer than leafy part of stem.   *4*
Fruit pedicels shorter than calyx, but if as long then cymes in fruit
    much longer than leafy part of stem.   *5*

4 Style shorter than calyx-tube; corolla not over 5 mm. diam., lobes
    concave, tube shorter than calyx (common).   **4. M. arvensis**
Style longer than calyx-tube; corolla 6–10 mm. diam., lobes flat,
    tube equalling or exceeding calyx (rare).   **3. M. sylvatica**

5 Corolla at first yellow or white; tube finally about twice as long as
    calyx; calyx teeth ultimately nearly erect; nutlets rounded at base,
    dark brown; cymes in fruit not much longer than leafy part of
    stem (rare).   **5. M. discolor**
Corolla never yellow; tube shorter than calyx; calyx teeth spread-
    ing; nutlets truncate at base, pale brown; cymes in fruit much
    longer than leafy part of stem (occasional).   **6. M. ramosissima**

1. Myosotis scorpioides L.   Ray, 1660
M. palustris (L.) Hill

Throughout the county by ditches, streams and pits. The var. **strigulosa**
(Reichb.) Schinz & Keller with the hairs of the stem appressed is
recorded. All squares except 24, 26, 41. General but in relatively base-
rich habitats.

BORAGINACEAE

2. Myosotis caespitosa K. F. Schultz.   Henslow, 1829
In a few localities by streams and in marshes. Formerly may have been
more frequent in the Fens but records need substantiating (cf. Welch
(1961)). 25, 44, 45, ?55, 56, [?29, ?35, ?38, ?46–9, 57, ?58, ?40]. General.

3. Myosotis sylvatica Hoffm.   Shrubbs, 1900
Abundant in wet woodland in grounds of Hildersham Hall with
Silene dioica, and possibly native there. Otherwise a rare garden
escape. 45, 46, 54, 65. Scattered throughout Great Britain, mainly in
the east.

4. Myosotis arvensis (L.) Hill   Common Forget-me-not Ray, 1660
In arable and waste land and grassy places throughout the county. A
robust plant with larger flowers referable to var. **silvestris** Schlecht.
(var. *umbrosa* Bab.) has been recorded. All squares except 37. General.

5. Myosotis discolor Pers.   Ray, 1670
M. versicolor Sm.
Occurs on the sands in the east of the county and at Gamlingay, very
rare elsewhere. 25, *34*, 39, *67*, 40, [38, 44, 45, 48, 54, 66, 76]. General.

6. Myosotis ramosissima Rochel   Babington, 1835
M. collina sensu Bab., et Evans; *M. hispida* Schlecht.
In a few dry sandy or gravelly localities, mostly in the east of the county.
25, 45, 54–6, 65–7, 76, [38, 46]. **FH.** General in drier regions.

LITHOSPERMUM L.

1. Lithospermum purpurocaeruleum L.
Garden escape. *45*, 56.

2. Lithospermum officinale L.   Gromwell   Ray, 1660
Occasional in wood margins, hedgerows, arable land and waste places.
25, 26, 34, 36, 44–6, 55, 56, 66, *67*, [24, 54, 57, 58, 40]. General in
England and Wales, rare in Scotland and Ireland.

3. Lithospermum arvense L.   Corn Gromwell   Ray, 1660
Recorded for most areas of the county as a weed of arable land but
especially characteristic of the boulder clay. 23–6, 33–7, 39, 44–6, *49*,
55, 56, 64, 65, 20, 31, 40, [38, 47, 58, 66, 41]. General in England, rare
elsewhere.

166

ECHIUM L.

1. Echium vulgare L.   Vipers' Bugloss   Ray, 1660
Babington noted this species as common in the Cambridge, Royston and Wimpole districts, but it is now found frequently only on the sands in the east of the county, and is elsewhere only occasional. 24, 29, 44, 45, *46*, 55, 56, 66, 67, 76, 40, [26, 34–6, 38, 47, 48, 54, 57]. Scattered throughout England and Wales, rare and perhaps not native in Scotland; in Ireland mainly coastal and native only in the east.

2. Echium lycopsis L.
E. plantagineum L.
Garden escape. 35.

CONVOLVULACEAE

CONVOLVULUS L.

1. Convolvulus arvensis L.   Bindweed   Ray, 1660
An exceedingly persistent weed of arable land and waste places throughout the county. A small-leaved variant, var. **minimus** Bab., has been known near Eversden since the time of Ray. All squares. General but rare in north-west.

CALYSTEGIA R.Br.

The three species can be distinguished as follows:

1   Bracteoles not overlapping, flat or somewhat keeled at base, acute at apex; corolla not over 50 mm. long; anthers 4·5–5·5 mm.
1. C. sepium
Bracteoles overlapping, saccate at the base, obtuse to truncate at apex; corolla more than 50 mm. long; anthers 6–7 mm.   2

2   Plant with sparse hairs on stem and usually on petiole and peduncle; corolla pink with paler veins outside.   **2. C. pulchra**
Plant practically glabrous; corolla white.   **3. C. silvatica**

1. Calystegia sepium (L.) R.Br.   Bellbine; Great Bindweed
Ray, 1660
Convolvulus sepium L.
Throughout the county in hedgerows and copses; certainly native in fen carr. A pale pink-flowered form occurs with the type at Wicken Fen. All squares except 26, *64*, 65, *40*. General; less frequent northwards.

CONVOLVULACEAE

2. Calystegia pulchra Brummitt & Heywood   Babington, 1850
C. dahurica auct.

In a few scattered localities in hedgerows near gardens. 35, 45, 57, [36, 44, 58]. Garden escape scattered over Great Britain and Ireland, and in W. Europe; origin unknown.

3. Calystegia silvatica (Kit.) Griseb.   C. D. Pigott, 1947
C. sylvestris (Willd.) Roem. & Schult.

Locally frequent in hedgerows as an escape from gardens. Plants intermediate between this and *C. sepium* occur at Cherry Hinton and elsewhere; they are presumably of hybrid origin. 24, 25, 34–6, 44–7, 49, 56, 67, 30, 41, 50. Commonly naturalized in many parts of the British Isles; native of S.E. Europe.

CUSCUTA L.

1. Cuscuta europaea L.   Great Dodder   Ray, 1660

Formerly occurred in a number of localities on a variety of hosts, mainly in the Fenlands. The only recent record is from the Botanic Garden at Cambridge. 45, [35, 36, 44, 46, 47, 56, 58, 66]. Rare in S. England; very rare and introduced elsewhere.

2. Cuscuta epilinum Wiehe

Introduced with flax seed at Ely in 1853 and Chatteris in 1876, but it disappeared with that crop. [38, 58.]

3. Cuscuta epithymum (L.) L.   Small Dodder   Relhan, 1786
C. trifolii Bab.

Formerly recorded in a number of localities on a great variety of hosts, but only seen recently at Eversden and Abington. 35, *45*, 54, [25, 34, 36, 38, 44, 55]. Locally common in England and Wales; very local in S. Scotland and Ireland.

SOLANACEAE

LYCIUM L.

1. Lycium halimifolium Mill.   Duke of Argyll's Tea-plant
                                               Babington, 1848

Naturalized in hedgerows in many parts of the county. 35, 37, 45, *46*, 49, *54*, 56–8, 65, 67, 40, 41, 50, [25, 34, 55]. Long cultivated and widely naturalized in the British Isles; probably native of S.E. Europe and W. Asia.

SOLANACEAE

2. Lycium chinense Mill.   C. E. Moss, 1912

L. rhombifolium Dippel

Like the last species naturalized in hedgerows but not as common. 44, 55, 58, 30, 41, [35, 45]. Frequently naturalized, native of E. Asia.

ATROPA L.

1. Atropa bella-donna L.   Deadly Nightshade   S. Corbyn, 1656

In a number of scattered localities in woods, hedgerows and waste places, certainly native in some places on the chalk. A powerful narcotic and very poisonous, formerly grown for medicinal purposes. 34, 35, 39, 44, 45, *46*, *54*, 56, [38, 47, 55, 57, 65, 67, 40, 41, 50]. Native in woods and thickets on calcareous soils in England and Wales, local. Naturalized elsewhere.

HYOSCYAMUS L.

1. Hyoscyamus niger L.   Henbane   Ray, 1660

An occasional casual of disturbed ground especially around farmyards. 26, *35*, *38*, 39, 44–6, 55, 56, [34, 47, 48, 57, 58, 66, 40]. Widely scattered throughout the British Isles as a casual; perhaps native in sandy places near the sea in S. England.

SOLANUM L.

1. Solanum dulcamara L.   Bittersweet; Woody Nightshade

Ray, 1660

Common throughout the county in hedgerows, ditches, woods and copses. All squares except [38]. General, except for the extreme north.

2. Solanum nigrum L.   Black Nightshade   Ray, 1660

Weed of arable land and waste places throughout the county. All squares except 24, 26, 64. Throughout England but rare in the north, very local in Wales, S. Scotland and Ireland.

3. Solanum rostratum Dunal

Casual. *48*.

(**Solanum tuberosum** L., the Potato, is widely cultivated in the county (36,000 acres in the Isle of Ely and 6100 in S. Cambs). It is often to be found as a relic of cultivation by the edges of fields and in farmyards.)

(**Lycopersicum esculentum** Mill., Tomato, is much cultivated in gardens and sometimes appears as a casual on rubbish tips, sewage farms and other waste places.)

169

SOLANACEAE

DATURA L.

1. Datura stramonium L.   Thorn-apple   Relhan, 1802
A casual of waste land found in a number of scattered localities.
Narcotic and very poisonous. 34, 35, *36*, 45, *46*, 49, *55*, 56, [38, 48, 58,
40, 41]. Casual, sometimes naturalized; native of many of the temperate and subtropical parts of the northern hemisphere.

SCROPHULARIACEAE

VERBASCUM L. (Mullein)

1. Verbascum thapsus L.   Ray, 1660
Scattered throughout the county in waste and grassy places. 23, 34, 35,
38, 44–6, 48, 54–8, 66, 67, 20, 41, [25, 26, 36, 49, 65, 50]. General but
rare in the extreme north.

2. Verbascum phlomoides L.
Casual or garden escape. 34.

(**Verbascum lychnitis** L. was recorded by Lyons (1763) and included by
Relhan on this authority. Either the plant was a casual or there was a mistake
in identification, as no other records exist.)

3. Verbascum nigrum L.   Black Mullein   Ray, 1660
A number of records mainly in the east of the county. The species is
common in the Breckland just over the Suffolk border. *24*, 44, 45, *46*,
54, *55*, 66, [25, 58, 65, 67]. Fairly common in S. England, extending
north to Caernarvon and Nottingham, naturalized farther north.

× **thapsus** (*V.* × *semialbum* Chaub.). Recorded from Whittlesford by
C. E. Raven in 1942. *44*.

4. Verbascum blattaria L.
Casual or garden escape. 45, [38, 49, 65].

5. Verbascum virgatum Stokes
Casual or garden escape. 45.

SCROPHULARIACEAE

MISOPATES Raf.

1. Misopates orontium (L.) Raf.   Skrimshire, 1819
Antirrhinum orontium L.
Formerly a cornfield weed around Odsey and near Wisbech, where it
may have been native. There are a few other records as a casual. *55*,
[23, 45, 67, 40]. Native from Cumberland and Yorks southwards and in
S. Ireland; local.

ANTIRRHINUM L.

1. Antirrhinum majus L.   Snapdragon   I. Lyons, 1763
A garden escape scattered over the county on old walls and in waste
places. 29, 33, 44, 45, 55, 58, 20, 40, [46, 54, 56, 65]. Naturalized in
many parts of England and Ireland; native of the Mediterranean region.
Commonly cultivated.

LINARIA Mill.

1. Linaria purpurea (L.) Mill.
Garden escape. 34, 44, 45, 20.

2. Linaria vulgaris Mill.   Toadflax   Ray, 1660
A common weed of arable and waste land throughout the county. All
squares except 26, 57. Common in England and S. Scotland; widespread
but perhaps not native in Ireland.

CHAENORHINUM (DC.) Reichb.

1. Chaenorhinum minus (L.) Lange   Ray, 1660
Linaria minor (L.) Desf.
Not uncommon in arable land but also common on railway tracks.
24–6, 34–9, 44–6, 48, 49, 54–6, 59, 66, 20, 40, 41, [23, 65, 67]. Common
in England and Wales, rarer in Scotland; widespread in Ireland.

KICKXIA Dumort. (Fluellen)

1. Kickxia spuria (L.) Dumort.   Ray, 1660
Linaria spuria (L.) Mill.
Widespread weed of arable land, commonest on the boulder clay and
rare in the Fens. 24–6, 34–6, 44–6, 54–6, 58, 64, [23, 47, 57, 65, 66].
S. England and Wales north to Pembroke, Nottingham and Lincoln.

SCROPHULARIACEAE

2. Kickxia elatine (L.) Dumort.   Ray, 1660
Linaria elatine (L.) Mill.

A weed of arable land, distributed like the last species and usually growing with it. Distinguished from *K. spuria* by its hastate, not ovate, upper leaves and glabrous, not villous, pedicels. 24–6, 34, 35, 44, 45, 47, 48, 54–6, 66, [23, 36, 38, 39, 46, 49, 57, 58, 65]. S. England and Wales extending north to Cumberland and Isle of Man; S. Ireland north to S.E. Galway and Wexford.

CYMBALARIA Hill

1. Cymbalaria muralis Gaertn., Mey. & Scherb.
Ivy-leaved Toadflax   I. Lyons, 1763
Linaria cymbalaria (L.) Mill.

Frequent on old walls. 29, 34, 36, 38, 44–6, 49, 54, 55, 57, 58, 65, 20, 30, 40, [47]. Native of S. Europe; first recorded in 1640 and now naturalized throughout the British Isles.

SCROPHULARIA L. (Figwort)

1. Scrophularia nodosa L.   Ray, 1660

Frequent in the boulder-clay woods, rare elsewhere. 24–6, 34–7, 39, 43, 45, *46*, 54, 55, 57, 64, 65, 20, [38, 44, 47, 49, 66]. General.

2. Scrophularia aquatica L.   Ray, 1660

Common by the edges of ponds and streams and other wet places throughout the county. All squares except 26, 54, 59, 64, 65, 30, [38]. Britain north to Ayr and Fife but rare in Scotland, local in Ireland.

3. Scrophularia vernalis L.

Garden escape. Naturalized at Fen Ditton where it was known from 1908 until 1940. 45, *46*, 56. An alien sometimes well established.

LIMOSELLA L.

1†. Limosella aquatica L.   Mudwort   Ray, 1685

Formerly occurred in muddy places at Gamlingay and several localities in the Fens. Last seen by A. Fryer about 1877. [25, 38, 46–8.] Local in England and Wales, rare in Scotland, and in Clare, Fermanagh and S. Galway in Ireland.

ERINUS L.

1. Erinus alpinus L.
Garden escape. *45*, [55].

DIGITALIS L.

1. Digitalis purpurea L.   Foxglove
R. S. Adamson and C. E. Moss, 1909
Found by Moss and Adamson in White Wood, Great Heath Wood and
Gamlingay Wood on the greensand at Gamlingay. It is still to be
found at White Wood. Occurs elsewhere as a garden escape, except
perhaps at Chrishall Grange where it may be native. 25, 44, 46. G. On
acid soils throughout the British Isles.

VERONICA L.

1. Veronica beccabunga L.   Ray, 1660
Common by streams and ponds and in marshes throughout the county.
All squares except 24, 26, 38, 68, 40, 41. General.

2. Veronica anagallis-aquatica L.
Ray, 1660 (including *V. catenata* Pennell)
V. ? anagallis sensu Bab. pro parte
In wet places throughout the county. 29, 34, 36, 38, 43–9, 55–9, 67, 69,
20, 30, 40, [66]. General.

3. Veronica catenata Pennell
Ray, 1660 (including *V. anagallis-aquatica* L.)
V. anagallis sensu Bab. pro parte; *V. anagallis* Bernh., non Gray
Frequent in wet places in the Fens; occasional elsewhere. This and the
preceding species occur occasionally as submerged aquatics in streams
and rivers, in which condition they may not produce aerial shoots and
flowers. Swollen, distorted fruits are often produced by the attacks of a
gall-insect in both species. The sterile hybrid with *V. anagallis-aquatica*
has been recorded in the county. 25, 34, 37, 38, 39, 44–9, 55, 56, 58, 59,
65–8, 75, 20, 30, 31, 41, 50. General, but very rare in Scotland and
N. Ireland.

4. Veronica scutellata L.   Ray, 1660
Formerly in wet places especially in the Fens. Only seen recently at
Childerley and Monk's Hole Wood near Great Chishill. The plant at the

173

SCROPHULARIACEAE

latter locality has glandular pedicels and is referable to var. **villosa** Schum. 36, 43, [25, 35, 44–9, 55, 56, 58, 66]. General, but usually calcifuge.

5. Veronica officinalis L.   Ray 1660

A local species of grassy and heathy places known only in a few localities. 25, 34, *37*, 65, 66, 76, 41, [23, 36, 45, 46, 54, 55]. G. General.

6. Veronica montana L.   Ray, 1685

A rare species of wet rides in boulder-clay woodland only seen recently in Cadge's Wood, Brinkley Wood and in the grounds of Kirtling Towers. 64, 65, [25, 35, 49, 54, 55]. Local in Britain and Ireland; rare in the far north.

7. Veronica chamaedrys L.   Germander Speedwell   Ray, 1660

Common in hedgerows, woodlands and grassy places throughout the county. All squares except *24*, 59, 30. General.

8. Veronica spicata L. ssp. **spicata**   Ray, 1660

This rare species was known to Ray at Horseheath and 'in great plenty' on Newmarket Heath. It survived in unmown grassland at the end of the Beacon Course until ploughed up in 1954 and probably still grows elsewhere on Newmarket Heath within the county. 56, [55, 64]. Very rare in the East Anglian Breckland. The ssp. *hybrida* (L.) E. F. Warburg occurs on limestone cliffs in W. England and Wales.

9. Veronica longifolia L.

Garden escape. *36*, 66.

10. Veronica serpyllifolia L. ssp. **serpyllifolia**   Ray, 1660

Occasional in damp grassy places throughout the county. 23, 25, 34–6, 39, 43–7, 49, 54, 55, 65, 66, 20, [38, 48, 40]. General.

11. Veronica peregrina L.

Casual garden weed. 34, 45.

12. Veronica arvensis L.   Ray, 1660

Common throughout the county in dry open places, and on walltops. All squares except 57, *58*, 68, 30. General.

13. Veronica hederifolia L.   Ray, 1660

Common weed of arable land throughout the county. All squares except 24, 59, 68, [48, 49]. General, but rare in the north and west.

14. Veronica persica Poir.   Henslow, 1826

V. buxbaumii Ten., non Schmidt

An abundant weed of arable land and waste places throughout the county. All squares. First recorded in Britain in 1825; now everywhere a common weed of arable land; native of W. Asia but now widespread in continental Europe.

15. Veronica polita Fr.   Henslow, 1835

A weed of arable land frequent throughout the south of the county but rare in the Fens. The var. **grandiflora** Bab. with larger flowers is recorded. 23–6, 35, 36, 44, 45, 54, 56, 65–7, *40*, 50, [34, 38, 46–9, 57, 41]. General, but rare in Scotland.

16. Veronica agrestis L.   Ray, 1660

A characteristic weed of arable land in the Fens, rare elsewhere and mainly in old gardens. 35, 38, 39, 45–7, 49, 56–9, 68, 69, 76, 30, 31, 40, 41, 50, [25, 36, 48, 55, 65–7]. General.

17. Veronica filiformis Sm.   T. G. Tutin, 1948

Has recently become well established as a weed in lawns and grassland in a number of places, especially in and around Cambridge. Normally sets no seed, apparently because of the self-sterility of the vegetatively spread clones. Fruiting material was, however, found at Girton in 1958. 35, 45, 46, 65. **Cam.** Naturalized throughout the British Isles; native of Asia Minor and the Caucasus.

PEDICULARIS L.

1. Pedicularis palustris L.   Ray, 1670

Formerly in a number of fens but only seen recently at Wicken. 56, [25, 44, 45, 66]. **W.** General, but rare in S.E. England.

2. Pedicularis sylvatica L.   Ray, 1660

Formerly recorded in several localities in marshy places. Last seen at Gamlingay by C. E. Moss and Evans around 1912. [25, 44, 65, 66.] General, but rare in the east.

RHINANTHUS L.

1. Rhinanthus minor L. sens. lat.   Yellow Rattle   Ray, 1660

Grassy places in a number of scattered localities mainly in the Fenlands. The Wicken Fen plants are much branched with the branches flowering

SCROPHULARIACEAE

freely, the flowers yellow and usually with 2 pairs of intercalary leaves, and are referable to ssp. **stenophyllus** (Schur) O. Schwarz. *25, 35, 45, 46, 49, 55, 56,* [24, 34, 44, 47, 48, 66, 40]. **W.** General.

MELAMPYRUM L.

1. Melampyrum cristatum L.   Crested Cow-wheat
S. Corbyn, 1656
A local species of wood margins on the boulder clay. *25, 35, 36, 54, 65,* [43]. **H.** Very local in C. and E. England.

2†. Melampyrum arvense L.
G. S. Gibson (1862) records this species from a field of tares at Heydon in 1849 and there is a specimen in CGE from this locality. This locality is presumably within the present county boundary but is in the botanical v.c. 19. There is also a specimen in CGE collected by Messrs Bowman, Bloxham and Babington labelled 'Cambridgeshire' without further information. Fordham's record from Ashwell, near the point where Sandon, Morden, Kelshall and Ashwell parishes meet, given in Pryor (1887), p. 314, must have been within a stone's throw of the Cambs border. [?23, 44.] A rare plant of S. and E. England from Dorset to E. Yorks.

3. Melampyrum pratense L.   Common Cow-wheat   Ray, 1660
Formerly occurred in a few woods on the boulder clay and sands, where it was last seen at West Wickham Wood by T. G. Tutin around 1930. There is a specimen in CGE collected at Gamlingay in 1930. It was recently reported in small quantity by the edge of Dernford Fen by the Bishop's Stortford College Natural History Society, who have a specimen in their herbarium. *45, 65,* [25, 55]. General.

EUPHRASIA L. (Eyebright)

All the older authors placed the Eyebrights under one species, *E. officinalis* L. It is fairly certain from the localities given 'In pascuis, as upon the moors, and Gogmagog hills, and almost in all meadows' that Ray must have seen both the species, *E. nemorosa* and *E. pseudokerneri*, which are found in the county today. It was however left to Babington (1860) to recognize the segregate *E. nemorosa* and for Pugsley (1929) to describe *E. pseudokerneri*. Pugsley (1933) determined a specimen from the Woolstreet as *E. micrantha* Reichb. and one from Chippenham as *E. brevipila* Burn. & Gremli. No specimens have been seen, but the Woolstreet is a most unlikely locality for *E. micrantha*

176

and the Chippenham plant was probably a large form of *E. nemorosa*. A specimen in CGE from Hayley Wood was determined by Pugsley as ?*E. brevipila × pseudokerneri*. This is most unlikely and the plant seems to be a large specimen of *E. nemorosa*. *E. pseudokerneri* can usually be distinguished from *E. nemorosa* by its larger flowers (7–11 not 5–7 mm. long and broad), bushier habit with more branches near the base, and finer leaf-teeth.

1. Euphrasia nemorosa (Pers.) Wallr.   Ray, 1660
In undisturbed grassland, trackways and by chalk-pits in a number of localities mainly on the chalk and boulder clay. 23, 24, 34, 35, 43, 44, *45*, 55, 65, 66, [25, 36]. C. Common in England and Wales, local in Scotland and Ireland.

2. Euphrasia pseudokerneri Pugsl.   Ray, 1660
A rare species occurring in a few localities on undisturbed chalk grassland. Where it grows with *E. nemorosa* hybrids occur. 44, 45, 55, 66, [65]. D. Grasslands on the chalk and oolite from S. Lincs and Kent to Wilts and Dorset.

ODONTITES Ludw.

1. Odontites verna (Bellardi) Dumort.   Red Bartsia   Ray, 1660
Bartsia odontites (L.) Huds.; *O. rubra* Gilib.
Frequent throughout the county in grassy and waste places, and particularly common on trackways. Two subspecies occur: ssp. **verna** with the branches coming off at an angle of less than 45°, the leaves lanceolate, and the bracts longer than the flowers, and ssp. **serotina** (Wettst.) E. F. Warburg with the branches spreading at a wide angle, the leaves linear-lanceolate, and the bracts shorter than or equalling the flowers. 23, 25, 26, 34–6, 43–7, 54–6, 64–6, 75, [24, 38, 39, 48, 49, 40, 41]. General.

OROBANCHACEAE

LATHRAEA L.

1. Lathraea squamaria L.   Toothwort   Shrubbs, 1889
There are two records for the county. A specimen in CGE was collected at Madingley Wood by Shrubbs in 1889, and the Bishop's Stortford College Natural History Society has a colour slide of a plant growing on roots of hazel, in alder carr at Dernford Fen in 1954. 45, [36]. Great Britain north to Perth and Inverness; Ireland.

OROBANCHACEAE

2. Lathraea clandestina L.   B. Reynolds, 1908

Grown in the Botanic Garden and introduced from there on to Coe Fen; it is now naturalized on roots of Willows along the banks of the Cam and Hobson's Brook in Cambridge (see plate 5). 45. Native of W. Mediterranean region.

OROBANCHE (Broomrape)

1†. Orobanche ramosa L.   Relhan, 1802

In Skrimshire's time this species was frequent in Hemp fields around Wisbech and, according to Babington (1860), also around Upware. There are no other records. [57, 40.] Formerly introduced into some eastern and southern counties with Hemp; no recent records.

2†. Orobanche rapum-genistae Thuill.   Ray, 1660

O. major sensu Evans

Parasite on *Sarothamnus scoparius* or other shrubby Leguminosae; formerly recorded from a few scattered localities. Last seen at Gamlingay in 1913. [25, 45, 46, 65.] Throughout England and Wales, S.W. Scotland and S. and S.E. Ireland, but now a very rare plant.

3. Orobanche elatior Sutton   Ray, 1660

The commonest *Orobanche* species in the county; parasitic on *Centaurea scabiosa* and therefore most frequent on the chalk and absent from the Fens. 24, 34, 35, 43–5, 54–6, 64, [65]. GG. S. and E. England and Wales, rare in the north; common in Cambs and Wilts.

4. Orobanche minor Sm.   Lesser Broomrape   Henslow, 1829

Parasite mainly on clover crops; recorded from a number of localities. A specimen with the back of the corolla tube gently curved throughout and the filaments very hairy was collected at Swaffham Bulbeck in 1957 and was apparently growing on *Crepis*. This specimen is referable to var. **compositarum** Pugsl. 35, 44, 45, 54–7, [34, 38, 48, 66, 76]. England, Wales, Ireland, and Fife in Scotland.

5. Orobanche picridis F. W. Schultz ex Koch
                W. W. Newbould, 1848

Parasitic on *Picris* and *Crepis*; formerly recorded from a few scattered localities. Last seen at Haslingfield in 1935 (Herb. Kew). *45*, [25, 35]. In England from Devon and Kent to Cambs, Bucks and Worcester, and from Pembroke in Wales.

178

6. Orobanche hederae Duby　　　　　　H. C. Gilson, *c.* 1930
Parasite on *Hedera helix*; reported at least twice from the grounds of
Girton College between 1930 and 1940, and recorded in the Botanic
Garden in 1939; presumably introduced. *45, 46.* Local species of S.
England, Wales and Ireland.

LENTIBULARIACEAE

PINGUICULA L.

1. Pinguicula vulgaris L.　　　Butterwort　　　　　S. Corbyn, 1656
Formerly in a number of fens and bogs, but only seen recently at
Chippenham Fen where it still persists and flowers in small quantity
annually. 66, [25, 34, 44, 45, 55, 56, 58]. General but rare in southern
England.

UTRICULARIA

1. Utricularia vulgaris L.　　　Bladderwort　　　　S. Corbyn, 1656
Scattered through the Fenlands in lodes and pits but apparently much
less common than formerly. Still abundant in the old brick-pits at
Wicken Fen. The submerged shoots bear small bladders which trap
Daphnia and similar small aquatic animals. 29, 36, 47, 48, 56, 57, 20,
[37, 38, 44–6, 58, 66, 40, 41]. **W.** Britain north to Dunbarton and
Angus, rare in Ireland.

2. Utricularia minor L.　　　　　　　　　　　　　　　Ray, 1685
Formerly occurred in ditches, pits and lodes in a number of localities
in the Fenlands. Last seen at Wicken Fen about 1948. *56,* [25, 44–6,
55, 66, 67]. General.

　　(There is a specimen of U. **intermedia** Hayne in CGE collected at
Chippenham Fen in 1898 by Shrubbs and a field record for Welches Dam
by C. E. Moss. According to A. Bennett (1899), U. **neglecta** Lehm may have
occurred in Burwell Fen, but in the absence of flowering specimens this is
not certain.)

VERBENACEAE

VERBENA L.

1. Verbena officinalis L.　　　Vervain　　　　　　　Ray, 1660
Occasional throughout the county on roadsides and in grassy and waste
places. 34, *35,* 38, 44–6, *47,* 54–8, 66, 67, 69, *41,* [23–6, 36, 48, 49, 64,
65, 40]. England, Wales, Ireland, and Fife in Scotland.

179　　　　　　　　　　　　　　　　I2-2

LABIATAE

MENTHA L.

This genus is taxonomically difficult because of the great variation within species, the profusion of hybrids, and the free vegetative spread which gives rise to uniform clones. All the specimens in CGE were determined by R. A. Graham in 1950 and have been used as a basis for the following account.

1 Whorls of flowers all axillary, the axis terminated by leaves, or
   with very few flowers in the axils of the uppermost pair.   2
   Whorls of flowers forming terminal spikes or heads.   5

2 Calyx teeth scarcely longer than broad, calyx campanulate,
   stamens normally exserted.   **2. M. arvensis**
   Calyx teeth much longer than broad; either calyx tubular or
   stamens included.   3

3 Pedicels and base of calyx hairy; stamens included.
        M. aquatica × arvensis
   Pedicels and base of calyx glabrous.   4

4 Leaves usually more than twice as long as broad, ±hairy; calyx
   campanulate; stamens included.   **M. arvensis × spicata**
   Leaves usually less than twice as long as broad, glabrous or thinly
   hairy; calyx usually tubular; stamens ±exserted.
        M. aquatica × arvensis × spicata

5 Flowers in a head often with axillary whorls below.   **3. M. aquatica**
   Flower in a spike.   6

6 Leaves stalked.   **M. aquatica × spicata**
   Leaves sessile or subsessile (petioles not over 3 mm.).   7

7 Pedicels and calyx tube glabrous; leaves glabrous or very thinly
   hairy.   8
   Pedicels and calyx tube hairy; leaves densely hairy at least below.   9

8 Leaves more than twice as long as broad, not normally rugose.
        4. M. spicata
   Leaves not more than twice as long as broad, ±rugose.
        M. rotundifolia × spicata

9 Leaves lanceolate, not or scarcely rugose.   **5. M. longifolia**
   Leaves oblong, ovate or suborbicular, ±rugose.   10

10 Leaves not over 4 cm., often a few suborbicular, rounded at apex
   or minutely cuspidate, densely white-tomentose below, crenate
   or dentate.   **(M. rotundifolia)**
   Leaves 3–10 cm., never suborbicular, cuspidate or acuminate,
   varying from green and hairy to white-tomentose below, serrate.
        M. longifolia × rotundifolia

1†. Mentha pulegium L.   Penny-royal   Ray, 1660
Recorded from Cambridge by Ray, from Harlton, Denny Abbey and
Wisbech by Relhan, and on the way to the Gog Magog Hills by Mr Vernon
(in Babington's notes). [35, 45, 46, 40.] Rare plant of winter-wet
habitats, mainly in S. England.

2. Mentha arvensis L.   Ray, 1660
A very variable weed of arable land and waste places throughout the
county. A number of specimens in CGE with broadly elliptic, obtuse
leaves with a convex margin and thin scattering of hairs are var.
obtusifolia Briq. All squares except 26, 29, 39, 59, 67, 30, [57]. General.

× **spicata** (*M.* × *gentilis* L.; *M. pratensis* Sole). Recorded from Shelford,
Hauxton and Horningsea by Relhan, and recently by M. J. d'Alton at
Cambridge, the latter being confirmed by R. A. Graham. 45, [46].

3. Mentha aquatica L.   Water Mint   Ray, 1660
A very variable herb found in ditches, by lodes, rivers and pits, and in
marshy places throughout the county. Many varieties of this species
have been described, but R. A. Graham (1954) recommends that these
varieties should not be retained, as all combinations of characters exist.
All squares except 24, 26, [38]. General.

× **arvensis** (*M.* × *verticillata* L.; *M.* × *sativa* L.). There are many old
records for this hybrid, but few recent ones, perhaps due to lack of
attention. 44, 49, *56*, 65, *66*, [25, 35, 36, 38, 45–7, 57].

× **arvensis** × **spicata** (*M.* × *smithiana* R. A. Graham; *M. rubra* Sm.,
non Mill.). Recorded between Longstowe and Bourn (CGE) and from
Gamlingay. *35*, [25].

× **spicata** (*M.* × *piperata* L.). There are several records for the Pepper-
mint, including Barrington (CGE). 35, [34, 45, 56, 41].

4. Mentha spicata L.   Spear-Mint   Skrimshire, 1813
M. viridis sensu Bab., et Evans
Garden escape naturalized in a few localities. 38, 39, 45, 58, 41, 50,
[56]. Naturalized in many places; native of C. Europe.

181

LABIATAE

5. Mentha longifolia (L.) Huds. Horse Mint Ray, 1685
M. sylvestris L.
Recorded from a number of localities in waste or grassy places. 35, 45, *54*, 58, [34, 44, 56, 66]. Scattered generally but doubtfully native.

× **rotundifolia** (*M.* × *niliaca* Juss. ex Jacq.; *M. alopecuroides* Hull). Recorded from a number of localities; in the past has been confused with *M. rotundifolia*. 38, 39, 45, 54, 56, *67*, [35, 44, 46, 66].

(It is doubtful if **M. rotundifolia** (L.) Huds. occurs in the county. All specimens so named in CGE are referable to *M. longifolia* × *rotundifolia*.)

Mentha rotundifolia × spicata (*M.* × *cordifolia* Opiz). A specimen from Newnham collected in 1957 was so determined by R. A. Graham. 45.

LYCOPUS L.

1. Lycopus europaeus L. Gipsy-wort Ray, 1660
Common in wet places throughout the county. All squares except 26, 64, 65, 30, 41, [49]. General, but rare in E. Scotland.

ORIGANUM L.

1. Origanum vulgare L. Marjoram Ray, 1660
A local species of grassy and waste places recorded only from a few localities. The virtual absence of this species from the Cambridgeshire chalk is a striking phytogeographical fact. 38, 43, 44, *54*, [23, 25, 34, 45, 46, 55, 65]. General, but local in Scotland and N. Ireland.

THYMUS L. (Thyme)

1. Thymus pulegioides L. Henslow, 1821
T. ovatus Mill.; *T. glaber* Mill.; *T. chamaedrys* Fr.
Occasional in grassland on the chalk and sands, not as common as *T. drucei*. Distinguished from *T. drucei* by the flowering stems which are hairy only on the angles, and by the characteristic oily smell. 24, *25*, 34, 35, 38, 45, 54, 55, 66, [44]. Widespread in S. and E. England, rare in N. England and Wales, almost absent from Scotland and Ireland.

2. Thymus serpyllum L. W. W. Newbould, ?c. 1850
Distinguished from *T. drucei* and *T. pulegioides* by having the flowering stems equally hairy all round. Known only from a specimen collected by Newbould on the Devil's Dyke near Dullingham (BM.) and another

182

collected by J. E. Little at Morden Grange Plantation in 1921 (CGE). [23, 66.] Only in the Breckland region of East Anglia.

3. Thymus drucei Ronn.   ?S. Corbyn, 1656
                              First certain record, L. Jenyns, 1825
T. serpyllum sensu. Bab. pro parte; *T. neglectus* Ronn.; *T. britannicus* Ronn.; *T. froelichianus* sensu Evans; *T. carniolicus* auct.
Distinguished by having the flowering stems hairy on two opposite sides only, or only slightly hairy on the other two. In undisturbed grassy places on chalk and sands. 23, 24, 26, 34, 35, 43–5, 54–6, 64–6, [36, 46, 57]. General.

CALAMINTHA Mill.

1. Calamintha ascendens Jord.   Ray, 1660
C. officinalis sensu Bab.
A local species of dry grassy places. 34, 46, 55, 56, [24, 25, 35, 44, 45, 54, 66, 67]. Throughout most of England, Wales and Ireland.

2. Calamintha nepeta (L.) Savi   Ray, 1663
In a number of localities on roadsides and other grassy places. Differs from *C. ascendens* in its calyx teeth being shortly and more sparsely ciliate, in the obvious (not short or 0) peduncles, and in the leaves of the main axis being mostly 1–2, not 2–4 cm. 44, 47, 54, 55, 65, 66, [23, 34, 45]. FH. A very local plant of S.E. England.

ACINOS Mill.

1. Acinos arvensis (Lam.) Dandy   Basil-thyme   Ray, 1660
Calamintha acinos (L.) Clairv.
Occasional in open habitats in dry places. 23, 24, 33–5, 38, 44, 45, 55, 64–7, 76, [54, 56, 58]. General but rare in north Britain, and in Ireland mostly in the south-east.

CLINOPODIUM L.

1. Clinopodium vulgare L.   Wild Basil   Ray, 1660
Calamintha clinopodium Benth.
Frequent in hedges, wood borders, scrub and grassland throughout most of the county, but rare in the Fens. 23–6, 33–5, 43–6, 54–6, 57, 64–6, 75, [36]. Britain to Inverness and Argyll but rare in the north; rare alien in Ireland.

LABIATAE

Melissa L.
1. Melissa officinalis L.
Garden escape. 45.

Salvia L.
1. Salvia verticillata L.
Casual. [45.]

2. Salvia pratensis L.
Garden escape or casual. 45, [44, 56].

3. Salvia horminoides Pourr.   Wild Clary   Ray, 1660
S. verbenaca sensu Bab.
Occasional on roadsides and in other grassy places, but absent from the northern Fens. 25, 34, 38, 44–6, *47*, 54–6, 65–7, [58, 64]. Widespread in S. England, less so in Wales and N. England, rare in Scotland, and in Ireland mainly in the south-east.

4. Salvia horminum L.
Casual. *46*.

Prunella L.
1. Prunella vulgaris L.   S. Corbyn, 1656
Common throughout the county in grassy places. All squares except 37. General.

× **laciniata.** Recorded from Hardwick between 1913 and 1918. [35.]

2. Prunella laciniata (L.) L.   G. Goode, 1905–6
This species was first found in the county by G. Goode in a small patch at King's Hedges near Chesterton in 1905–06. In 1911 Mr Graveson found the plant in abundance in a meadow at Hardwick, where it remained until the meadow was ploughed up during the Second World War. One plant was found at Trumpington in 1937 (CGE). *35, 45*, [46]. Very local from Somerset and Gloucester to Kent and Lincs. First recorded in 1887 and perhaps not native.

Betonica L.
1. Betonica officinalis L.   Betony   Ray, 1660
Stachys betonica Benth.; *Stachys officinalis* (L.) Trev.
A characteristic species of the margins of boulder-clay woods; very rare elsewhere. 25, 26, 34–6, 43, 45, 54, *64*, 66, [46, 55, 56]. Widespread in England and Wales, rare in Scotland and Ireland.

STACHYS

STACHYS L. (Woundwort)

1. Stachys arvensis (L.) L. Ray, 1685
A weed of arable land, rare except in the Fens, where it is locally not uncommon. 37, 47, 49, 55, 56, [25, 35, 38, 45, 54, 57, 40]. General but commonest in the west.

2. Stachys palustris L. Ray, 1660
Throughout much of the county by ditches and streams and in marshes; also as a weed of arable land. 34, 35, 37–9, 44–9, 55–9, 66, 68, 69, 20, 30, 31, 40, 41, 50, [29, 36, 65]. General.

× **sylvatica** (*S.* × *ambigua* Sm.). Recorded from several localities. *45*, [*55*].

3. Stachys sylvatica L. Ray, 1660
In hedgerows, woods and copses throughout the county. All squares except 37, 59, 68, 30. General, but commonest in the south.

BALLOTA L.

1. Ballota nigra L. Black Horehound Ray, 1660
In hedgebanks, copses and waste places throughout the county. Our plant has short calyx teeth up to 2 mm. long, and is referable to ssp. **foetida** Hayek. All squares except 59, 64, 68. Common in England and Wales, local in Scotland and Ireland.

GALEOBDOLON Adans.

1. Galeobdolon luteum Huds. Yellow Archangel S. Corbyn, 1656
Lamium galeobdolon (L.) L.
A characteristic species of damp places in boulder-clay woods. 25, 35, 36, 43, 45, *64*, 65, [66]. **B.** Common in England and Wales, rare in S. Scotland and E. Ireland.

LAMIUM L.

1. Lamium amplexicaule L. Henbit Ray, 1660
A common weed of arable land throughout the county. 23–5, 33–6, 39, 44–7, 54–8, 65–7, 76, 40, 41, [38, 49]. General; rare in the west.

185

LABIATAE

2. Lamium hybridum Vill.   Relhan, 1802
L. incisum Willd.
A weed of arable land common in the Fens, local elsewhere. 29, 34,
38, 45, 47–9, 54, 58, 59, 67, 68, 30, 31, 40, 41, 50, [66]. General but local.

3. Lamium purpureum L.   Red Dead-nettle   Ray, 1660
A common weed of arable land and waste places throughout the
county. All squares except 24. General.

4. Lamium album L.   White Dead-nettle   Ray, 1660
A common weed of arable land and grassy and waste places throughout
the county. All squares. General but rare in N. Scotland and mainly
in eastern Ireland.

5. Lamium maculatum L.
Garden escape. 25, 45, 49.

LEONURUS L.

1. Leonurus cardiaca L.   Motherwort   Ray, 1685
Formerly recorded from a number of places in the Cambridge and
Wisbech areas. Last recorded in Babington (1860). [45, 40, 41.]
An alien scattered over the British Isles; native of continental Europe.

GALEOPSIS L. (Hemp-Nettle)

1. Galeopsis angustifolia Ehrh. ex Hoffm.   Ray, 1660
G. ladanum sensu Bab.
A weed of arable land, especially on the chalk; almost absent from the
Fens. 24, 25, 34–6, 44, 45, 55, 56, 66, [46, 67]. Local in England
extending to Dunbarton and Moray in Scotland; only in E. Ireland.

2. Galeopsis tetrahit L. sens. lat.   Ray, 1660
A common weed of arable land throughout the county. The common
segregate is **G. tetrahit** L. sens. strict.; it has the middle lobe of the
lower lip of the corolla entire and its network of markings restricted to
the base and never reaching the margin, and with red-tipped glands
grouped in a bunch below the nodes of the stem. **G. bifida** Boenn.,
with the middle lobe of the lower lip of the corolla deeply emarginate,
and its markings reaching the margin or covering the whole lip, and the

glands of the stem distributed all over the internodes, was recorded by C. C. Townsend at Grantchester in 1959. All squares except 24, 26, 37, [54, 66, 41]. General.

3. Galeopsis speciosa Mill.　　　　　　　　　　　T. Martyn, 1763
G. versicolor Curt.
A common weed of arable land in the Fens, rare elsewhere. *25, 34, 36-8, 47-9, 55-9,* 68, 69, 20, 31, [44-6, 40]. General, but rare and often casual in the south of England and Ireland.

NEPETA L.

1. Nepeta cataria L.　　Cat-mint　　　　　　S. Corbyn, 1656
Occasional on roadsides, banks and in hedgerows, mainly on the lighter soils and absent from the open Fens. *23, 25,* 34, 35, 44, 45, 46, 54-7, 66, *67,* [24, 38, 49, 58, 65, 40]. Local in Britain from Westmorland and Northumberland southwards; in Wigtown and Stirling in Scotland; scattered over Ireland except the south-west and not native.

GLECHOMA L.

1. Glechoma hederacea L.　　Ground Ivy　　　　Ray, 1660
Nepeta glechoma Benth.
Common in woods, grassland and waste places throughout the county. All squares. General, but rare in the far north.

MARRUBIUM L.

1. Marrubium vulgare L.　　White Horehound　　　Ray, 1660
Formerly recorded from a number of scattered localities. Last recorded from Wimblington and Doddington by J. S. L. Gilmour in 1930. *38, 49,* [25, 34, 45, 48, 55, 56, 58, 40, 41]. Local from Dunbarton and Moray and from Dublin and Galway southwards, but perhaps only native near the south coast of England.

SCUTELLARIA L.

1. Scutellaria galericulata L.　　Skull-cap　　　Ray, 1660
By ditches and rivers and in marshy places, common in the Fens, rare elsewhere. 29, 35, 37, 45-8, 55-9, 66, 69, 30, [25, 34, 38, 44, 49, 40]. General, but local in Ireland.

LABIATAE

TEUCRIUM L.

(Dent in Ray (1685) under the name *Chamaedrys vulgo vera existimata* mentioned a plant growing along Quy Water which may have been **Teucrium chamaedrys** L., but Babington (1860) thought there must have been a mistake.)

1. Teucrium scordium L.   Water Germander   W. Turner, 1562
Formerly in a number of wet places in the Fens but only seen recently near Stretham and near Wicken. Very intolerant of competition from tall-growing marsh or fen species, and therefore never persisting long in any one locality, unless open ground is being renewed artificially, as by peat or clay digging. 57, [46–8, 56, 58, 30, 40]. A very rare plant now known only in N. Devon, Cambs and W. Central Ireland. Formerly widespread in the Fenlands and also recorded from near Oxford, E. Norfolk, N. Lincs and Yorks.

2†. Teucrium scorodonia L.   Wood Sage   Relhan, 1785
Known with certainty only from Gamlingay. It is not clear when it was last recorded (although Evans (1939) writes as though he had seen it there), but it has not been seen for many years. The record from Ely by Henslow must surely have been an error. [25.] General.

AJUGA L.

1. Ajuga chamaepitys (L.) Schreb.   Ground Pine   Ray, 1660
This rare species is known only as a weed in chalky fields in the Odsey area. In Ray's time it occurred at Thriplow Heath. 23, [44]. Very local in S.E. England extending to Hants, Beds and W. Suffolk.

2. Ajuga reptans L.   Bugle   Ray, 1660
Locally frequent in damp woods, pastures and fens. *24*, 25, 26, 34–6, 43–5, *46*, 48, 54–6, 64–7, [49, 58, 40]. General.

PLANTAGINACEAE

PLANTAGO L. (Plantain)

1. Plantago major L.   Ray, 1660
Common throughout the county in waste places, especially on well-trodden tracks. All squares. General.

PLANTAGINACEAE

2. Plantago media L.   Ray, 1660

Common in dry grassland, especially on the chalk and sands. All squares except 39, 59, 68, 20, 30, [49]. Generally distributed in S. England and the Midlands, becoming rarer northwards; introduced in some places in Scotland and in Ireland.

3. Plantago lanceolata L.   Ribwort   Ray, 1660

Very common in grassy and waste places throughout the county. All squares. General.

4. Plantago maritima L.   Relhan, 1785

Known only from the banks of the River Nene north of Wisbech. 41, [40]. FA. All round the coasts of the British Isles; also inland, mainly on mountains.

5. Plantago coronopus L.   Buck's-horn Plantain   Ray, 1660

Formerly recorded from the 'Hill of Health', Cambridge and Chippenham. Still to be found at Gamlingay and by the banks of the River Nene between Wisbech and Foul Anchor. 25, 41, [45, 66, 40]. G. General near the coast; inland in England.

LITTORELLA Berg.

1†. Littorella uniflora (L.) Aschers.   Shoreweed   Ray, 1670

Extinct; known to Ray on Hinton Moor and to Relhan in Gamlingay bogs; not seen since. Still occurs near Cavenham in Suffolk. [25, 45.] Margins of acid pools and lakes, general, but rare in C. England.

CAMPANULACEAE

CAMPANULA L.

1†. Campanula latifolia L.   Relhan, 1788

Formerly in a few woods on the boulder clay; last seen by Babington at Comberton about 1852. [35, 65.] General in Britain but absent from the extreme south.

2. Campanula trachelium L.   Ray, 1660

Found only in a few boulder-clay woods. 25, 35, 43, 65, 75, [36]. Scattered throughout Great Britain north to Denbigh and Lincs and in S.E. Ireland.

189

CAMPANULACEAE

3. Campanula rapunculoides L. L. Jenyns, 1822
Garden escape now effectively naturalized in grassy places and on roadsides. *34*, 44–6, 55, 56, 67, 76, [25, 36, 47, 48, 58]. **GG.** Widely introduced in the British Isles; native of continental Europe and Asia Minor.

4. Campanula glomerata L. Clustered Bellflower S. Corbyn, 1656
A local plant of grassy places on the chalk and sands. 24, 25, 35, 44, *45*, 54–6, 65, 66, [26, 34, 36]. **D.** From Dorset and Kent to Cumberland and Kincardine, locally common.

5. Campanula rotundifolia L. Harebell Ray, 1660
Rather local in dry grassland; almost absent from the Fens. 23, 24, 33–5, 44–6, 54–7, 65, 66, [25, 48]. General, but rare in west.

6. Campanula patula L.
Casual. [45, 41.]

Legousia Durande

1. Legousia hybrida (L.) Delarb. Venus's Looking-glass Ray, 1660
Specularia hybrida (L.) A.DC.
Frequent weed of arable land mainly in the south of the county. 23, 34–6, 43–5, *46*, 54–6, 65, [38, 39, 47–9, 58, 66, 40, 41]. Throughout S. and E. England.

Jasione L.

1†. Jasione montana L. Sheep's-bit Ray, 1660
Formerly recorded from Newmarket Heath, Furze Hills, Hildersham and Gamlingay. Last seen at the last-mentioned locality about 1909. [25, 54, 66.] Throughout S. and W. Britain, rare or absent in the north and east.

RUBIACEAE

Sherardia L.

1. Sherardia arvensis L. Field Madder Ray, 1660
A frequent, in some areas abundant, weed of arable land and waste places. 23, 24, 26, 34–6, 39, 43–6, 54–6, 64–7, 69, 76, 41, [29, 38, 47–9, 58, 40]. General.

RUBIACEAE

ASPERULA L.

1. Asperula cynanchica L.   Squinancy-Wort   Ray, 1663
A local plant of old chalk grassland. 23, 24, 34, 35, 44, 45, 55, 56, 66, [46, 54, 65, 67]. **GG**. Locally abundant in S. England and Wales extending northwards to S.E. Yorks and Westmorland; W. Ireland.

2. Asperula arvensis L.
Casual. 36.

CRUCIATA Mill.

1. Cruciata chersonensis (Willd.) Ehrend.   Crosswort   Ray, 1660
Galium cruciata (L.) Scop.
Local on roadsides and in grassy places, mainly on boulder clay, but on the chalk at Cherry Hinton. 24, 26, 34–6, 45, 46, *54*, 56, 65, 76, [25, 47, 55, 58]. Throughout Great Britain north to Moray and Inverness, introduced in Fermanagh and Down in Ireland.

GALIUM L. (Bedstraw)

1. Galium odoratum (L.) Scop.   Woodruff   Ray, 1660
Asperula odorata L.
Very local in boulder-clay woods, as in Hildersham Wood. 35, 45, 54, 64, 65, [25, 55]. General.

2. Galium mollugo L.   Ray, 1660
Common in hedgerows and grassy places throughout the county. Two subspecies occur: ssp. **mollugo** with a broad terminal panicle with spreading branches and flowers 3 mm. in diam., and ssp. **erectum** Syme (*G. erectum* sensu Bab., et Evans) with a narrow panicle with ascending branches and flowers 4 mm. in diam. The latter subspecies is probably frequent on the chalk and boulder clay while the former is generally distributed. 23–6, 33–6, 38, 43–6, 48, 54–7, 64–7, 75, 76, 40, [47, 49]. General, but common only in England.

× **verum** (*G.* × *pomeranicum* Retz.; *G. ochroleucum* Wolf ex Schweigg. & Koerte). Recorded from several localities. *35, 44, 65*, [56].

3. Galium verum L.   Lady's Bedstraw   Ray, 1660
In grassy places throughout the county. All squares except 20, 30, [48]. General, coastal in the west.

191

RUBIACEAE

4. Galium saxatile L. S. Corbyn, 1656

G. hercynicum Weigel

Recorded from several places on the sands but only seen recently at Gamlingay and Chippenham. 25, 66, [55, 67, 76]. **G.** General on acid soils.

5. Galium palustre L. Ray, 1660

In marshes, fens and other wet places, especially in the Fenlands. Two subspecies occur: ssp. **palustre** (var. *witheringii* (Sm.) Bab.) with leaves 0·5–1 cm. long and the flowers 3 mm. in diam., and ssp. **elongatum** (C. Presl) Lange (*G. elongatum* C. Presl) with the leaves 1·5–2 cm. and the flowers 4·5 mm. in diam. These correspond at least in part to diploid and octoploid plants. In addition a tetraploid is known with intermediate characters (ssp. **tetraploideum** Clapham ined.). Some Cambs specimens in CGE seem to be referable to this. 25, 29, 35–7, 39, 43–9, 54–9, 65, *66*, 68, 20, 31, 40, 41, [24, 34, 38]. General.

6. Galium uliginosum L. I. Lyons, 1763

More local than *G. palustre* in marshes, fens and other wet places. 25, 29, 34, 37, 44–7, 54–9, 66, 67, 31, 41, [35, 40]. General, but rare in the west and north.

7. Galium tricornutum Dandy Ray, 1685

G. tricorne sensu Bab., et Evans

A cornfield weed found in a number of scattered localities. 34, 35, 45, *46*, 56, [36, 38, 48, 49, 57, 64]. Local in Britain north to Moray, casual in Scotland.

8. Galium aparine L. Goosegrass; Cleavers Ray, 1660

A plentiful and troublesome weed of arable land, waste and grassy places and hedgerows. All squares. General.

9. Galium spurium L. var. **vaillantii** DC. C. E. Raven, 1936

Recorded from Ely by C. E. Raven in 1936 and by N. D. Simpson in 1947. *58*. In a few scattered localities in S. and C. England, doubtfully native.

10. Galium parisiense L. Relhan, 1785

G. anglicum Huds.

Recorded from a few localities on walls. Last seen at Ely and Cambridge by J. Rishbeth in 1946. *45*, *58*, [66, 40]. Rare and local in S.E. England.

192

CAPRIFOLIACEAE

(The plant which Relhan (1820) records from Crabmarsh near Wisbech on the authority of Skrimshire as *Rubia peregrina* L. is probably referable to the formerly much-cultivated **Rubia tinctorum** L.)

CAPRIFOLIACEAE

SAMBUCUS L.

1. Sambucus ebulus L.   Danewort   Ray, 1660
Formerly recorded from a number of localities, but only seen recently at Burwell, near Coton, and planted on railway banks near Odsey. 23, 45, 56, [34–6, 43, 44, 46, 54, 41]. Scattered over much of the British Isles but questionably native.

2. Sambucus nigra L.   Elder   Ray, 1660
In hedgerows, woods and copses throughout the county. The var. **laciniata** L. with divided leaflets is recorded. All squares. General.

VIBURNUM L.

1. Viburnum lantana L.   Wayfaring Tree   S. Corbyn, 1656
Throughout much of the county in hedgerows, woods and copses but absent from the open Fenlands. 23–6, 33–6, 43–5, 54–6, 64, 65, 75, [46, 47, 41]. **CH.** England and Wales north to Yorks.

2. Viburnum opulus L.   Guelder Rose   S. Corbyn, 1656
In hedgerows, woods and copses mainly on the boulder clay, less common on the chalk and sands, rare in the Fenlands (though common at Wicken Fen). 24, 25, 34, 35, 44, 45, 54–7, 64–6, 75, [36, 46, 48, 49, 67, 40]. **CH.** General, but less common in Scotland.

SYMPHORICARPOS Duham.

1. Symphoricarpos rivularis Suksd.   Snowberry   A. J. Crosfield, 1918
S. albus auct.; *S. racemosus* auct.
Commonly planted in hedgerows throughout the county. *25*, 34–6, 44–6, 54–6, 58, 64–7, 75, 20, 30, *40*. Widely planted in the British Isles and often naturalized; native of N. America.

CAPRIFOLIACEAE

LONICERA L. (Honeysuckle)

1. Lonicera xylosteum L.

Garden escape. [58.]

2. Lonicera periclymenum L.   Ray, 1660

In hedges, woods and copses throughout much of the county, but rare in the open Fenlands. 23, *24*, 25, 26, 34–6, 43–6, *49*, 54–7, 64–6, 76, [38, 48, 58, 40, 50]. General.

3. Lonicera caprifolium L.   I. Lyons, 1763

Recorded from a few localities in copses and hedgerows. It has been known at Cherry Hinton since 1763. 26, 35, 44, 45, 65, [55, 58]. CH. In a few places in S., E. and N. England and S. Scotland, introduced. Native of C. and S. Europe.

ADOXACEAE

ADOXA L.

1. Adoxa moschatellina L.   Moschatel; Townhall Clock   Ray, 1660

Recorded from a few localities but only seen recently at White Wood, Gamlingay, and between Weston Colville and Brinkley. 25, *45*, *46*, 65, [44]. G. In Great Britain widespread but local; in Ireland only in Antrim.

VALERIANACEAE

VALERIANELLA Mill. (Corn Salad; Lamb's Lettuce)

1. Valerianella locusta (L.) Betcke   Ray, 1660

V. olitoria (L.) Poll.

A local weed of arable land and railway banks. 24, 25, 29, 33, 35, 36, 38, 45, 46, 30, 40, [48, 54, 41]. General, but rather local.

2. Valerianella rimosa Bast.   W. W. Newbould, 1852

V. auricula DC.

Recorded by Babington (1860) from Eversden, between Caxton and Eltisley, and Wisbech, and by J. S. L. Gilmour from Gamlingay in 1933. None of these records is supported by herbarium sheets in CGE. *25*, [35, 40]. Local in England, Wales and Ireland; in Scotland from Fife only.

VALERIANACEAE

3. Valerianella dentata (L.) Poll.   J. Holme, *c*. 1824

A local weed of arable land. The var. **mixta** (Vahl) Dufr. with hispid fruits is recorded. It should not be mistaken for *V. eriocarpa* Desv., from which it differs in its laxer cymes and unequal calyx teeth. 23, *25*, 34, 35, 44, 45, *49*, 54, 55, [38, 46, 48]. General, but very rare in the north and west.

VALERIANA L.

1. Valeriana officinalis L.   Valerian   S. Corbyn, 1656

Including *V. sambucifolia* sensu Bab., et Evans

Locally frequent by ditches and in damp woods and grassy places. A very variable plant. Both tetraploids and octoploids occur in the county, and differ somewhat in average leaf-shape, but more diagnostically in pollen size. The tetraploid plant occurs on the chalk at Cherry Hinton and on the edges of boulder-clay woods; the octoploid, with typically broader leaflets, is the Wicken Fen plant, *V. sambucifolia* sensu Bab. *24*, 25, 34–6, 43, 45, 55–7, 59, *64*, 65, 66, [44, 47, 48, 54, 58, 40]. General; the octoploid is widespread, the tetraploid rather restricted to calcareous soils in Southern England.

2. Valeriana dioica L.   Ray, 1660

Local in fens and marshy places. 34, 44, 45, 55–7, *64*, 66, [35, 48, 58, 65, 40]. **W.** Throughout England, Wales and S. Scotland.

CENTRANTHUS DC.

1. Centranthus ruber (L.) DC.   Red Valerian   I. Lyons, 1763

Kentranthus ruber (L.) DC.

Frequent garden escape, naturalized on old walls as at Ely. 34, 45, 47, 56, 58, 65, [54, 55]. Frequently cultivated and widely naturalized; native of S. and C. Europe, N. Africa and Asia Minor.

DIPSACACEAE

DIPSACUS L.

1. Dipsacus fullonum L. ssp. **sylvestris** (Huds.) Clapham   Teasel

Ray, 1660

D. sylvestris Huds.

By fields and in waste and grassy places throughout the county, particularly common on the clays. All squares except 39, 67, 30, 41. General, but rare in Scotland and Ireland.

DIPSACACEAE

2. Dipsacus pilosus L. Ray, 1660

Formerly recorded from a number of localities but only seen recently at Alder Carr, Hildersham, by the Roman Road near Worstead Lodge, at Cherry Hinton and at Little Linton. Has been known for many years in the churchyard of Little St Mary's Church, Cambridge, where it is still to be found. 45, 54, 55, [35, 43, 46, 56–8, 66]. Local in England and E. Wales.

CEPHALARIA Schrad.

1. Cephalaria gigantea (Ledeb.) Bobr.

Casual. 58.

KNAUTIA L.

1. Knautia arvensis (L.) Coult.   Field Scabious   Ray, 1660

Scabiosa arvensis L.

Common weed of arable land and waste and grassy places throughout the county. 23–6, 33–6, 38, 43–7, 54–6, *57*, 64–8, 75, 40, 41, [29, 48, 49]. General, but rare in N. Scotland.

SCABIOSA L.

1. Scabiosa columbaria L.   Small Scabious   Ray, 1660

A widespread but local species of old chalk grassland. 23, 24, 33–5, 43–5, 54–7, 66, 67, [26, 36, 46, 65]. **GG.** Great Britain northwards to Stirling and Angus.

SUCCISA Haller

1. Succisa pratensis Moench   Devil's Bit   Ray, 1660

Scabiosa succisa L.

A local plant of fens and marshy places. 25, 35, 44–6, 55–7, 65–7, [24, 34, 36]. **W.** General.

COMPOSITAE

HELIANTHUS L.

1. Helianthus annuus L.

Garden escape. 35, 45, 58, *66*.

2. Helianthus tuberosus L.

Garden escape. 34, 55.

3. Helianthus laetiflorus Pers.

Garden escape. 45, 67.

COMPOSITAE

RUDBECKIA L.
1. Rudbeckia laciniata Ι.
Garden escape. 34.

GUIZOTIA Cass.
1. Guizotia abyssinica (L.f.) Cass.
Casual. 45.

BIDENS L. (Bur-marigold)
1. Bidens cernua L.   Ray, 1660
Formerly recorded from a number of marshy places mostly in the Fenlands. Only seen recently on the Welney Washes. 59, [25, 45–7, 56, 58, 67]. **WW**. General; very rare in Scotland.

2. Bidens tripartita L.   Ray, 1660
Local in marshes and by the margins of lodes, ponds and pits. Both this and *B. cernua* have apparently decreased in the present century. 29, 35–9, 45, 47, 48, 56, 57, 59, [24, 25, 46, 49, 58, 67, 40]. **W**. General, but rare in the north.

GALINSOGA Ruiz. & Pav.
1. Galinsoga ciliata (Raf.) Blake
Rare casual in and around the Botanic Garden; not established. 45.

XANTHIUM L.
1. Xanthium strumarium L.
Casual. 57, [45].

2. Xanthium spinosum L.
Casual. *49*.

SENECIO L. (Ragwort)
1. Senecio jacobaea L.   Ragwort   Ray, 1660
Common in grassland and in waste places throughout the county. All squares. General.

2. Senecio aquaticus Hill   Marsh Ragwort   Ray, 1660
By rivers and lodes and in marshy places through much of the county. 29, 34, 36, 37, 39, 45, 47–9, 55, *56*, 57–9, 30, [46, 66]. General.

197

COMPOSITAE

× **jacobaea.** Recorded by E. M. Rosser from Welney Washes in 1959. 59.

3. Senecio erucifolius L. Ray, 1660
Throughout the county in grassy places, especially roadsides. Best distinguished from *S. jacobaea* by its achenes which are all pubescent while in *S. jacobaea* at least the outer are glabrous. All squares except 20. Common in England and Wales, rare in S. Scotland and Ireland.

4. Senecio squalidus L. Oxford Ragwort E. A. George, 1939
Evans's (1939) note on this species is misleading. The plant was grown in the Cambridge Botanic Garden in the last century, but the supposed hybrids with *S. vulgaris* were almost certainly the radiate form of the latter species. Since E. A. George recorded it at Chesterton Ballast Pits and Coldham's Common in 1939 it has been found in many parts of the county, principally on railway tracks, but is not common in Cambridge city. 25, 26, 29, 34–6, 38, 39, 44–9, 58, 68, 20, 40, 41. Certainly in cultivation in the Oxford Botanic Garden before the end of the seventeenth century, and by the end of the eighteenth 'very plentiful on almost every wall in Oxford' (Smith (1799), p. 600). Now widespread in England and Wales after a recent rapid spread, particularly on railway lines. Details of the spread of this species are given by D. H. Kent (1956, 1960). Native of Sicily and S. Italy.

× **viscosus.** Recorded from Wisbech in 1959. 40.

5. Senecio sylvaticus L. Relhan, 1802
Native at Gamlingay and on the sands in the east of the county, elsewhere probably casual. 25, 35, 44, 54, 57, 66, [38, 39, 45, 47, 49, 56]. **FH.** General.

6. Senecio viscosus L. Ray, 1660
Weed of waste ground and dumps, probably always introduced. 29, 34, 38, 39, *45*, *46*, 49, 58, 40, 50. Scattered throughout lowland Great Britain where it is possibly native; has become more widespread in recent years; rare and local in Ireland.

7. Senecio vulgaris L. Groundsel Ray, 1660
A common weed of arable land and waste places throughout the county. A rayed variety is recorded from Ely and Cambridge; it occurs regularly with the type in the University Botanic Garden (cf. *S. squalidus*). All squares. General.

8†. Senecio paludosus L.   S. Corbyn, 1656

This plant, which is now extinct in the British Isles, formerly occurred in a number of fen ditches. For a full account of the history of the species see A. Bennett (1899). [38, 56–8.] Formerly in Lincs, Cambs, Norfolk and Suffolk.

9†. Senecio palustris (L.) Hook   Ray, 1660

S. congestus (R.Br.) DC. var. *palustris* (L.) Hyl.

Formerly occurred in a few fen ditches but not seen since 1830. [38, 49, 58, 66, 50.] Formerly in East Anglia, Lincoln, Yorks, and Sussex, but now apparently extinct.

10. Senecio integrifolius (L.) Clairv.   Ray, 1660

S. campestris (Retz.) DC.

A very local species of old chalk grassland. 55, 56, 66, [45, 65]. F. Local from Dorset and Kent northwards to Gloucester, Lincs and Westmorland; *S. spathulifolius* auct. brit. in Anglesey.

LIGULARIA Cass.

1. Ligularia clivorum (Maxim.) Maxim.

Effectively naturalized in damp woodland on the site of Sawston Moor in the grounds of Sawston Hall. 44.

DORONICUM L.

1. Doronicum pardalianches L.   Leopard's-bane

Skrimshire, 1818

A garden escape naturalized in a number of places. 34, 35, *36*, 45, 54, 65, [40]. Naturalized in many parts of the British Isles; native of W. Europe. Formerly cultivated as a medicinal drug.

TUSSILAGO L.

1. Tussilago farfara L.   Coltsfoot   Ray, 1660

An abundant weed of arable land and waste places throughout the county. All squares except 24. General.

COMPOSITAE

PETASITES Mill.

1. Petasites hybridus (L.) Gaertn., Mey. & Scherb.   Butterbur
Ray, 1660
P. vulgaris Desf.
Local by streams and rivers mainly in the southern half of the county.
Only the male plant is known in Cambs. 24, 25, 34, 36, 44, 45, 54, 56,
[46, 66, 40]. The female plant, which is easily recognized in summer
by its elongated infructescence, is common in the Midlands. General.

2. Petasites fragrans (Vill.) C. Presl   Winter Heliotrope
S. Hiley, 1863
Naturalized in the Cambridge and Wisbech areas. *36*, 45, 65, 40, 41,
50. Widely naturalized particularly in the south and west of the
British Isles; native of the W. Mediterranean region.

CALENDULA L.

1. Calendula officinalis L.
Garden escape. 34, 54, 58, 67, 41, 50.

INULA L.

1. Inula helenium L.   Elecampane   Ray, 1660
Recorded from a few localities, recently only from Boxworth and
Hardwick. 35, 36, *45*, [41]. Formerly much grown for medicinal
purposes and now widely scattered in the British Isles. Probably native
of C. Asia.

2. Inula conyza DC.   Ploughman's Spikenard   S. Corbyn, 1656
A rare species of roadsides and old pits, mainly on the chalk; occurs in
quantity on a steep chalk slope of the Coldham's Lane chalk-pits in
Cambridge city. 44, 45, 54, *55*, *66*, [65]. Locally common in England
and Wales north to Durham.

PULICARIA Gaertn. (Fleabane)

1. Pulicaria dysenterica (L.) Bernh.   Ray, 1660
Common on damp roadsides and in grassy places throughout the
county. All squares except 37, 39, *58*. England, Wales, Ireland and
S. Scotland.

2†. Pulicaria vulgaris Gaertn.   Ray, 1660

Formerly recorded from a few localities, but not seen since 1833. A note by W. Marshall in 1832 (CGE) gives its habitat as 'wet stiff clayey ground, liable to inundations in wet seasons'. [45–7, 58, 41.] Very rare in S. England and decreasing.

FILAGO L. (Cudweed)

1. Filago germanica (L.) L. sens. lat.   Cudweed   Ray, 1660

Including *F. apiculata* G. E. Sm. and *F. spathulata* C. Presl

A weed of sandy waste places or neglected arable land in a number of scattered localities, perhaps formerly more frequent. The specific distinction of *F. spathulata* and *F. apiculata* is difficult to uphold. Plants occur with a ±prostrate branching habit and with somewhat reflexed tips to the involucral bracts, and are particularly characteristic of seasonally wet places on arable land. All intermediates to the typical erect *F. germanica* can be found. Material from Gamlingay in CGE has reddish tips to the bracts and a generally erect habit; this and similar plants have been called *F. apiculata*. 24, *25*, 26, 44, 45, 54, 55, 65, 66, 76, 40, [23, 34, 35, 38, 46–9, 58, 67, 41]. **FH.** General except N. Scotland.

2. Filago minima (Sm.) Pers.   Ray, 1660

Recorded from a few dry sandy places. Common in the Breckland. 25, 76, [23, 45, 46, 48, 54, 66]. **G.** General and usually calcifuge.

GNAPHALIUM L. (Cudweed)

1†. Gnaphalium sylvaticum L.   Ray, 1685

Found only at Gamlingay where it was apparently still known to Evans (1939). *25.* General, but local.

2. Gnaphalium uliginosum L.   Cudweed   Ray, 1660

A local weed of damp, open ground. 25, 38, 39, 46, 48, 49, *54*, 55, 58, 59, 68, 76, 20, [24, 35, 45, 47, 40, 41]. General.

3†. Gnaphalium luteoalbum L.   Relhan, 1802

The only record is from between Hauxton and Little Shelford by Relhan in 1802. [45.] Perhaps native in a few eastern counties and certainly so in the Channel Islands.

COMPOSITAE

ANTENNARIA Gaertn.

1†. Antennaria dioica (L.) Gaertn.   Cat's-foot   S. Corbyn, 1656
Formerly occurred in several localities. Last seen at Balsham in 1864.
There is material from Cambs in CGE of both sexes. [25, 45, 46, 55, 66, 76.] A northern species very rare in southern England; decreasing.

SOLIDAGO L. (Golden Rod)

1†. Solidago virgaurea L.   Ray, 1663
Known only from Gamlingay where it was last seen in 1916. [25.]
General but rare in the east.

2. Solidago canadensis L.
Garden escape. 44–6, *58*, 67.

3. Solidago gigantea Ait. var. **leiophylla** Fern. (*S. serotina* Ait., non
 Retz.)
Garden escape. 45, *56*.

ASTER L.

1. Aster tripolium L.   Sea Aster   Ray, 1660
Now known only from the banks of the tidal River Nene north of
Wisbech, but formerly on banks of rivers in a few places in the Fen-
lands south to Sutton Gault. The var. **discoideus** Rchb. without ray-
florets occurs. 41, [47, 48, 40]. FA. All round the coasts of the British
Isles; rarely inland.

2. Aster novae-angliae L.
Garden escape. 46, *55*.

3. Aster simplex Wood
Garden escape. *35*.

4. Aster novi-belgii L.
Garden escape, sometimes naturalized. 45, *46*.

 × **laevis** L. (*A.* × *versicolor* Willd.). Garden escape. *45*.

5. Aster lanceolatus Willd.
Naturalized garden escape. 45.

6. Aster salignus Willd. J. Brown, 1864

This species was found on the edge of Wicken Fen in 1864 and has persisted ever since. In 1959 it was found in the middle of an area of fen at Fowlmere Water-cress beds. It is presumably a naturalized garden escape. 44, 57. W.

ERIGERON L.

1. Erigeron acer L. Ray, 1660

Occasional on disturbed ground particularly on light sandy soils, and on walls. 25, 29, 33–6, 44, 45, *46*, 54, 55, *57*, 65, 66, 40, [24, 38, 58, 67, 41]. Throughout England, Wales and C. Ireland, very rare in Scotland.

CONYZA Less.

1. Conyza canadensis (L.) Cronq. Mr Venn, 1878

Erigeron canadensis L.

Since this species was first recorded at Cambridge in 1878, it has become a frequent weed of waste places in all parts of the county. 25, 29, 34–9, 44–6, 48, 49, 55, 58, 65–7, 76, 20, 40, 41. Introduced. Throughout England, Wales and S. Scotland; native of N. America.

BELLIS L.

1. Bellis perennis L. Daisy Ray, 1660

Common in pastures and other grassy places throughout the county. All squares. General.

EUPATORIUM L.

1. Eupatorium cannabinum L. Hemp Agrimony Ray, 1660

Common in wet places throughout most of the county. 25, 29, 34, 38, 39, 44–6, 48, 49, 55–9, 66–8, 20, 30, 40, [24, 35, 47, 54, 65]. General, but rarer in the north.

ANTHEMIS L.

1. Anthemis tinctoria L.

Garden escape. [25.]

2. Anthemis cotula L. Stinking Mayweed Ray, 1660

A weed of arable land occurring throughout the county but rare in the Fens. The unpleasant smell distinguishes this species from the rarer *A. arvensis*, which has the scent of Chamomile; the receptacular scales will confirm the identification—they are linear-acute in *A. cotula* and

COMPOSITAE

broader, lanceolate-cuspidate in *A. arvensis*. 24, 26, 35, 36, 44–6, 55, 58, 64–7, 76, [25, 34, 38, 47–9, 56, 57]. General, but rarer in the north.

3. Anthemis arvensis L.   Corn Chamomile   Ray, 1660
A rather rare weed of arable land almost absent from the Fens, formerly more frequent. 44, 48, *55*, *57*, 65, *66*, 67, [34–6, 38, 45, 47, 49]. General but common in S. and E. England; rare in Ireland.

(**Chamaemelum nobile** (L.) All. (*Anthemis nobilis* L.) is recorded from Crab Marsh near Wisbech by J. Balding (Babington, 1860), but there is no specimen in CGE.)

ACHILLEA L.

1. Achillea millefolium L.   Yarrow   Ray, 1660
Common on roadsides and grassy and waste places throughout the county. All squares. General.

2. Achillea ptarmica L.   Sneezewort   Ray, 1660
Rare in fens and marshes. *37*, 47, 48, *49*, 55, 56, [25, 29, 38, 44, 46, 67]. General, but rare in S. Ireland.

TRIPLEUROSPERMUM Schultz Bip.

1. Tripleurospermum maritimum (L.) Koch ssp. **inodorum** (L.) Hyland.
     ex Vaarama   Scentless Mayweed   Relhan, 1786
Matricaria inodora L.
A common weed of arable land and waste places throughout the county. The commonest 'Mayweed', readily distinguished by the absence of receptacular scales between the disc florets and the very weak smell. The ripe achene, with two conspicuous dark brown oil glands on the outer face, is very different in detail from that of *Matricaria* and *Anthemis* species. All squares. General.

MATRICARIA L.

1. Matricaria recutita L.   Wild Chamomile   T. Martyn, 1763
M. chamomilla sensu Bab., et Evans
An occasional weed of arable land and waste places. Differs from *Tripleurospermum maritimum* ssp. *inodorum* in the strong 'Chamomile' scent, in the hollow (not solid) receptacle, and in the absence of oil glands on the achene. Readily distinguished from *Anthemis* species by

204

the absence of receptacular scales. 25, 26, 34–7, 39, 44–9, 55, 58, 68, 30, [38, 54, 50]. Throughout England and Wales, probably introduced in Scotland and Ireland.

2. Matricaria matricarioides (Less.) Porter   Evans, *c.* 1909
M. suaveolens (Pursh) Buchen., non L.; *M. discoidea* DC.
First recorded in the county in 1909; now a common weed of well-trodden waste land, trackways and farmyards. All squares. General; introduced from N. America towards the end of the nineteenth century.

CHRYSANTHEMUM L.

1. Chrysanthemum segetum L.   Corn Marigold   Ray, 1660
Formerly widespread as an arable weed, now rare and usually casual. 34–6, 46, *56*, 58, 20, [25, 45, 47, 54, 66, 67, 40]. General. Native of the Mediterranean region and W. Asia.

2. Chrysanthemum leucanthemum L.   Ox-eye Daisy   Ray, 1660
In grassy places throughout the county; sometimes abundant in old pasture. All squares except 68, 30, [38]. General.

3. Chrysanthemum parthenium (L.) Bernh.   Feverfew   Ray, 1660
Matricaria parthenium L.
A common garden escape established in hedgerows and waste places throughout the county. 26, 27, 35, *36*, 38, 39, 44–6, 48, 49, 57, 58, 20, 40, 41, 50, [25, 34, 47, 56, 66]. Throughout Great Britain and in a few places in Ireland; probably always introduced. Formerly cultivated as a medicinal herb and used as a febrifuge, hence its common name. Probably native of S.E. Europe.

4. Chrysanthemum serotinum L.
Naturalized near Stretham. 57.

5. Chrysanthemum corymbosum L.
Casual. [45.]

6. Chrysanthemum vulgare (L.) Bernh.   Tansy   Ray, 1660
Tanacetum vulgare L.
Occasional on roadsides and in hedgerows and waste places and in railway yards. 38, 44, 45, *46*, 48, *49*, 55, 68, 20, 30, 50, [25, 35, 36, 39, 56, 40]. General. Formerly cultivated as a medicinal- and pot-herb.

COMPOSITAE

ARTEMISIA L.

1. Artemisia vulgaris L.   Mugwort   Ray, 1660
Common in hedgerows and waste places throughout the county. All
squares. General.

2. Artemisia verlotorum Lamotte
Casual. 29.

3. Artemisia biennis Willd.
Casual. 67.

4. Artemisia absinthium L.   Wormwood   Ray, 1660
In a few localities in waste places, formerly more frequent. 25, 29, 38,
45, 40, [34–6, 44, 46–9, 55, 57, 58, 30, 41]. General, but rare in Scotland
and Ireland. Formerly grown as a medicinal herb.

5. Artemisia maritima L.   Relhan, 1788
Known only from the banks of the tidal River Nene north of Wisbech.
41. FA. Local in many coastal districts of the British Isles.

(**A. campestris** L. is said by Ray (1660) to have been reported from Newmarket Heath by Mr Sare in Howe's *Phytographia Britannica*, but he (Ray)
could not find it there. There is a specimen in the Martyn herbarium (CGE)
labelled: '*Abrotanum campestre* CB., Hort. Cantab., Aug. 23, 1729, Sam.
Dale misit ad Hortum ex agro Cantab.' One of the best known of the continental rarities of the Breckland; still to be found in Norfolk and Suffolk.)

ECHINOPS L.

1. Echinops sphaerocephalus L.
Garden escape. 34.

CARLINA L.

1. Carlina vulgaris L.   Carline Thistle   Ray, 1660
Local on dry grassland mainly on the chalk. 23, 24, 33–5, 45, 55, 56, 64,
66, 41, [36, 44, 47, 54, 58]. GG. General in the lowlands on dry soils.

206

ARCTIUM L. (Burdock)

The three species of *Arctium* which occur in the county have often been misnamed, and their distribution is uncertain. They are best distinguished by the length of the phyllaries which are 15–22 mm. in *A. lappa*, 13–18 mm. in *A. pubens*, and 10–15 mm. in *A. minus*, and by the diameter of the mature head, *c.* 4·2 cm. in *A. lappa*, *c.* 3·3 cm. in *A. pubens* and *c.* 2·5 cm. in *A. minus*.

1. Arctium lappa L.   T. Martyn, 1763

A. majus Bernh.

Throughout much of the county in grassy and waste places, but commonest in the Fenlands. 34–9, 44–7, 49, 54–9, 65, 66, 68, 75, 40, [48]. S. and E. England and S.E. Ireland.

2. Arctium pubens Bab.   First certain record, Babington, 1853

A. vulgare sensu Evans; *A. tomentosum* sensu Bab.

This is probably the common species in the county, although it is replaced to some extent by *A. lappa* in the Fenlands. It is badly underrecorded and many of the records of *A. minus* are probably referable to it. 25, 34, 36, 38, 45, *46*, 47–9, 54, *56*, 58, 20, 30, 40, 50, [24, 29, 35, 44, 55, 57, 66]. Distribution unknown but probably general.

3. Arctium minus Bernh.   First certain record, Henslow, 1824

This species has been recorded for nearly all squares, but many of the records are almost certainly referable to *A. pubens*. It seems to be the characteristic species of the boulder-clay woodlands and is scattered elsewhere. ?All squares except 47, 20, [46]. Probably confined to the southern half of the British Isles.

CARDUUS L.

1. Carduus tenuiflorus Curt.   Skrimshire, *c.* 1800

Recorded from Outwell Bridge by Skrimshire and from Cherry Hinton by C. C. Townsend in 1955. In both places possibly only a casual. 45, [50]. General, locally frequent near the coast.

2. Carduus nutans L.   Musk Thistle   Ray, 1660

On roadsides and in waste and grassy places throughout the county. An attractive plant with its large nodding flower-heads which are deliciously scented. All squares except 25, 64. General, but rare in the north and west.

COMPOSITAE

3. Carduus acanthoides L.   Welted Thistle   Ray, 1663
C. crispus sensu Bab., Evans, et Clapham, Tutin & Warburg
Field margins and waste places throughout much of the county though
not common in the Fenlands. Similar in general appearance to
Cirsium palustre, but certainly distinguished by the simple, not feathery,
pappus hairs, 23–6, 33–7, 43–6, 48, *54*, 55, *56*, 57, 64–6, 75, 40, [29, 38,
39, 49, 50]. General in lowlands, but rare in Scotland and Ireland.

× **nutans**. Recorded from between the Coach and Horses Inn and
Barley by P. D. Sell in 1954 (CGE). 44.

CIRSIUM Mill.

1. Cirsium eriophorum (L.) Scop.   Woolly-headed Thistle
                                           S. Corbyn, 1656
Carduus eriophorus L.
In a few localities in grassy and waste places mainly on the boulder clay.
The most handsome of the native Thistles, unmistakable with its very
large heads with cobwebby spinous bracts. *23, 34*, 35, 36, 44, *54*, 64, 65,
75, [45, 46, 56, 58, 66]. Local in England and Wales.

× **vulgare** (*C.* × *gerhardtii* Schultz Bip.). Recorded with both parents
at Croydon by C. E. Raven in 1945. The identification was confirmed by
J. E. Lousley. *34.*

2. Cirsium vulgare (Savi) Ten.   Spear Thistle   Ray, 1660
Carduus lanceolatus L.; *Cirsium lanceolatum* (L.) Scop., non Hill
A very common weed of grassland, hedgerows and waste places
throughout the county. All squares. General.

3. Cirsium palustre (L.) Scop.   Ray, 1660
Carduus palustris L.
Throughout much of the county in marshy and other damp places,
though apparently absent from the arable Fens. 25, 34–6, 43–7, 54–7,
59, 64–7, [24, 38, 48, 49]. General.

4. Cirsium arvense (L.) Scop.   Field Thistle   Ray, 1660
Carduus arvensis (L.) Hill
An abundant and very troublesome weed of arable land and waste
places throughout the county. Variable in leaf shape and hairiness; the
var. **mite** Wimm. & Grab. (var. *setosum* auct.) with very shallowly

lobed leaves and weak spines, and the var. **incanum** Ledeb. with leaves white-tomentose beneath and strong spines, have both been recorded for waste places in Cambridge, but do not seem to be common. All squares. General.

5. Cirsium acaulon (L.) Scop.   Stemless Thistle   Ray, 1660
Carduus acaulos L.
A local species of old grassland especially on the chalk. 24–6, 33–6, 44, 45, 49, 54–7, 64–7, 20, [38, 46–8, 58, 40]. **GG.** Local in England north to Durham.

6. Cirsium dissectum (L.) Hill   Ray, 1660
Carduus pratensis Huds.; *Cirsium pratense* (Huds.) Druce, non DC.
Formerly in a number of fens but now known only at Wicken, Dernford, and Chippenham. 45, *49*, 56, 66, [44, 46, 54, 55, 58]. **W.** Local in England and Wales north to Yorks, and in Ireland.

× **palustre** (*C.* × *forsteri* Sm.; *Carduus forsteri* (Sm.) Bab.). There is a specimen in CGE collected by Babington in 1840 at Bottisham Fen which has been confirmed by W. A. Sledge. [56.]

7. Cirsium tuberosum (L.) All.   Tuberous Thistle   W. H. Mills, 1919
This very rare species is still by the Mare Way near Eversden where it was first found in 1919. It is the only locality in eastern England. 35. Very rare in Wilts, Glamorgan and Cambs; casual in E. Gloucester.

SILYBUM Adans.

1. Silybum marianum (L.) Gaertn.   Milk-Thistle   Ray, 1660
Casual in a few localities in waste places, rarely persisting. 34, 35, 45, 54, 67, 50, [29, 38, 46, 47, 49, 55, 58, 40]. Widespread as a casual, sometimes naturalized; native of the Mediterranean region.

ONOPORDUM L.

1. Onopordum acanthium L.   Cotton Thistle   Ray, 1660
Scattered over the county in waste places and on roadsides. An impressive plant often several feet high; abundant on the disturbed ground of the excavated Furze Hills at Hildersham. 25, 34, 38, 44–6, 48, *49*, 54–7, 41, [35, 36, 39, 47, 58, 66, 67]. **FH.** Throughout Great Britain but only common in eastern England; doubtfully native.

COMPOSITAE

Centaurea L.

1. Centaurea scabiosa L.   Knapweed   Ray, 1660
On roadsides, in fields and in waste and grassy places throughout the county. All squares except 39, 58, 59, 68, 20, 30, 40. General in lowlands but rare in the north.

2. Centaurea cyanus L.   Cornflower   Ray, 1660
Formerly a frequent cornfield weed, now rarer and occurring mainly as a casual. 35, 36, 44–7, *49*, 55, 20, [23, 25, 38, 54, 57, 66, 67, 40]. Formerly general in the lowlands but now rare owing to cleaning of seed-grain.

3. Centaurea nigra L. sens. lat.   Hardheads   Ray, 1660
Common in grassy and waste places throughout the county. The common segregate is **C. nemoralis** Jord. with the stem not much swollen beneath the heads, the appendages not completely concealing the pale basal part of the bracts, and the teeth of the outer bracts longer than the undivided portion. **C. nigra** L. sens. strict. (*C. obscura* Jord.) also occurs; it has the stem conspicuously swollen beneath the heads, the appendages ± completely concealing the pale basal part of the bracts, and the teeth of the outer bracts ± equalling the undivided portion. All squares except 39, 30. General.

4. Centaurea calcitrapa L.   Star Thistle   Ray, 1660
Formerly known in a number of localities on roadsides or waste places in the Fenlands. Only seen recently on a roadside near Doddington and in a field at Duxford. 38, 44, [45, 47, 49, 56, 58, 66, 40]. Probably introduced. A rare plant of S. and E. England and Glamorgan in Wales; casual elsewhere.

5. Centaurea diluta Ait.
Casual. 45.

6. Centaurea melitensis L.
Casual. *46*.

7. Centaurea solstitialis L.   Henslow, 1828
A casual recorded several times as a weed of arable land and waste places. Last seen on the Gog Magogs in 1949. *45*, *46*, [38, 49, 54, 56, 65]. Native of S. and S.E. Europe and W. Asia.

SERRATULA L.

1. Serratula tinctoria L.   Saw-wort   S. Corbyn, 1656
On the margins of a few woods on the boulder clay, and at Sawston
Moor and Chippenham Fen. Very rare elsewhere in East Anglia. 35,
44, 66, [25, 36, 45, 54, 55]. C.; H. England and Wales; rare in S. Scotland and S.E. Ireland.

CARTHAMUS L.

1. Carthamus tinctorius L.
Casual. 45.

CICHORIUM L.

1. Cichorium intybus L.   Chicory   Ray, 1660
By the edges of fields, in farmyards and waste places, and especially on
farm tracks throughout most of the county. Occasionally grown as a
crop and also included in herbage mixture. 24, 29, 34–6, 38, 44–9, 54,
55, 57–9, 64–8, 20, 31, 40. General but rare in Scotland and Ireland.

LAPSANA L.

1. Lapsana communis L.   Nipplewort   Ray, 1660
Common in hedgerows, copses and waste places throughout the county.
All squares. General.

ARNOSERIS Gaertn.

1†. Arnoseris minima (L.) Schweigg. & Koerte   Ray, 1660
Formerly occurred at Gamlingay where it was last seen in 1914. [25.]
Very rare in eastern Britain.

HYPOCHOERIS L. (Cat's-ear)

1. Hypochoeris radicata L.   Ray, 1660
Throughout most of the county in waste and grassy places, not very
common on the chalk, and rather characteristic of old pasture. All
squares except 24, 57, 58, 64, 68. General.

2. Hypochoeris glabra L.   Relhan, 1793
Formerly occurred at Gamlingay, Hinton and Chippenham. Only seen
recently at Kentford Heath. 76, [25, 45, 66]. K. Great Britain northwards to Moray and Inverness, very local in N. Ireland.

COMPOSITAE

3. Hypochoeris maculata L.   Ray, 1663

Formerly in several localities on the chalk but now known only from the Devil's Dyke. 56, [44, 45, 54]. **D.** A rare species with a scattered relic distribution in England and Wales.

LEONTODON L. (Hawk-bit)

1. Leontodon autumnalis L.   Ray, 1660

Apargia autumnalis (L.) Willd.

In pastures and on roadsides throughout the county. All squares except 64, *66*, 67, 20. General.

2. Leontodon hispidus L.   Ray, 1660

Apargia hispida (L.) Willd.

In grassy places throughout much of the county, but commonest on the chalk and local in the Fenlands. 24–6, 29, 34–6, 38, 43–7, 54–8, 64–7, 31, 40, 41, [39, 48, 49]. General, north to the Clyde especially on calcareous soils.

3. Leontodon taraxacoides (Vill.) Merat   Relhan, 1802

Thrincia hirta Roth; *L. leysseri* G. Beck

Rather local in old pastures and grassland on well-drained soils. 25, 38, 44, 45, 49, 55–7, 66, 41, [34–6, 48]. General, except north Scotland.

PICRIS L. (Ox-tongue)

1. Picris echioides L.   Ray, 1660

Helmintia echioides (L.) Gaertn.

Throughout the county in grassy and waste places, but particularly abundant on disturbed ground (e.g. roadsides and the edges of arable fields) on the boulder clay. All squares except 39, 67, 30, 41. Locally common in S. England and Wales, rare in N. England, S. Scotland and Ireland.

2. Picris hieracioides L.   Ray, 1660

Occasional in grassy places on the chalk and sands, rare elsewhere as on the margin of Hardwick Wood. 23, 24, *25*, 33–5, 38, 44–6, 54–6, 66, [48, 58, 65, 67, 40]. Locally common in England and Wales, very rare in S. Scotland, introduced in Ireland.

TRAGOPOGON L.

1. Tragopogon pratensis L.   Goat's-beard   Ray, 1660
Common in grassy and waste places and arable land throughout the
county. The common plant with the florets only about half as long as
the involucral bracts is referable to ssp. **minor** (Mill.) Wahlenb. (*T.
minor* Mill.). Ssp. **pratensis** with the florets almost or quite equalling the
involucral bracts occurs rather frequently in the southern half of the
county, as in the Royston area. All squares except [57]. General, but
absent from the north and west.

2. Tragopogon porrifolius L.
Garden escape or casual. 34, *46*, 65.

× **pratensis**. Recorded from Chesterton by A. Wallis. There are no
specimens to support the record. [46.]

LACTUCA L. (Lettuce)

1. Lactuca serriola L.   Prickly Lettuce   Ray, 1660
L. scariola L.

Formerly rare but now a frequent weed of roadsides and disturbed
ground throughout the county. The rapid spread of this plant in Cambs
(as in Beds, cf. Dony, 1953, p. 385) seems to have taken place between
1930 and 1945. In recent years it may have declined again. In Babington's time it was as rare as *L. virosa*. The oldest specimen in CGE is from
Ely in 1853. All squares except 54, 64, 65, 41, [66]. S. England and Wales.

2†. Lactuca virosa L.   Ray, 1660
Recorded before 1850 from several localities; specimens exist in CGE
from Elm near Wisbech (1844) and from Cherry Hinton chalk-pit
where it was known from 1825 to 1855. There are no recent records for
the county. This species is abundant on sand dunes on the East Anglian
coast, and is readily distinguished from *L. serriola* by the ripe fruit,
which is black, with a conspicuous narrow wing or flange; that of
L. serriola is greyish and lacks the marginal wing. The general habit and
particularly the more nearly simple leaves of *L. virosa* are usually
sufficient to distinguish it. [45, 46, 56, 57.] Scattered throughout
Great Britain; only common in S.E. England.

3. Lactuca saligna L.   Ray, 1660
A rare species, seen recently only at Earith (just within the county) and
Aldreth. Babington's habitat indication, 'chalky places', is misleading;

COMPOSITAE

most of the old records agree with the recent ones in being open habitats on the banks of rivers or ditches. 47, *58*, [45, 46, 48, 56]. S.E. England, mainly on the coast.

MYCELIS Cass.

1. Mycelis muralis (L.) Dumort.   Wall Lettuce   Relhan, 1785
Lactuca muralis (L.) Gaertn.

Recorded from a few localities on walls and under trees; persistent in Cambridge city in and near the Botanic Garden, and locally common in a beech wood on the Wandlebury Estate. Doubtfully native. 45, 55, [44, 66]. **Cam.** General in England and Wales, rare in Scotland and Ireland.

SONCHUS L. (Sowthistle)

1. Sonchus palustris L.   I. Lyons, 1763

Formerly recorded from a few marshy places in the Fens. Last seen at Bottisham in 1843. It has recently spread to just within our county boundaries near Doddington, presumably from Woodwalton Fen, Hunts, where it was introduced. 38, [56–8]. Rare in E. England (cf. Bennett, 1905).

2. Sonchus arvensis L.   Corn Sowthistle   Ray, 1660

A common and serious perennial weed of arable land and waste places throughout the county. This plant is not infrequently recorded as *S. palustris* especially when it is growing on ditch-banks in the Fens. The auricles of the upper cauline leaves afford a quick and safe method of distinguishing the two: in *S. arvensis* they are rounded, in *S. palustris* long-pointed. All squares. General in the lowlands.

3. Sonchus oleraceus L.   Ray, 1660

A common and variable weed of arable land and waste places throughout the county. Lewin (1948) describes variation in leaf-shape, glandular hairiness of peduncles and involucre, achene colour, etc. Some of his Cambs material in CGE illustrates this variation. All squares. General in the lowlands.

4. Sonchus asper (L.) Hill   Ray, 1660

A common and variable weed of arable land and waste places throughout the county. The var. **inermis** Bisch., with simple obovate leaves and small soft marginal spines, was recorded by Lewin from Cambridge in 1941 (CGE), and presumably occurs elsewhere. (The type has acute

214

pinnatifid leaves with strong marginal spines.) *S. asper* differs from *S. oleraceus* in the upper auricles which are rounded, not pointed, and more diagnostically in the smooth, not rugose, achenes. All squares. General in the lowlands.

(A putative hybrid, **S. asper × oleraceus**, was recorded by Lewin from Cambridge (CGE). The parents differ in chromosome number (*S. oleraceus* $2n = 18$, *S. asper* $2n = 36$) and the triploid hybrid is sterile and rare.)

HIERACIUM L. (Hawkweed)

The account of this difficult genus has been prepared by P. D. Sell.

1. Hieracium grandidens Dahlst.   P. D. Sell, 1950

Abundant on a roadside bank just out of Royston where it was probably introduced. A closely allied species as yet unnamed with the branches of the inflorescence ± straight, not curved, and the styles pure yellow, not dark, grows with it. 34. Scattered over Britain.

2. Hieracium sublepistoides (Zahn) Druce   C. D. Pigott, 1948

Common in the beech wood on Wort's Causeway and amongst bushes by the Gog Magogs golf course. Also in a beech plantation at Babraham. 45, 55. **GG.** Widespread in England and Wales.

3. Hieracium vulgatum Fries

Casual. *25.*

4. Hieracium lepidulum (Stenstr.) Omang   P. D. Sell, 1950

On a roadside bank just out of Royston where it was probably introduced. 34. Widespread in S. England, rare elsewhere.

5. Hieracium maculatum Sm.   R. H. Compton, 1913

Sparingly in the beech wood on Wort's Causeway. 45. Widespread in Great Britain but most frequent in the south.

6. Hieracium strumosum (W. R. Linton) A. Ley

C. D. Pigott, S. M. Walters & C. West, 1951

Rare on the Devil's Dyke, and abundant in a railway cutting between Royston and Meldreth. 34, 66. **D.** Common in England and Wales.

7†. Hieracium umbellatum L.   P. Dent in Ray, 1685

Recorded by P. Dent and H. Dixon (1882) from Gamlingay and by J. Martyn from Hildersham Furze Hills. No specimens support these records, but both localities are suitable for the species. [25, 54.] General in sandy places.

8. Hieracium perpropinquum (Zahn) Druce   Ray, 1685
H. boreale sensu Bab., et Evans

The only extant locality as a native plant is Gamlingay, but Relhan's Linton record of *H. boreale* is almost certainly referable to this species. It has recently occurred at Cambridge as a casual. 25, 45, [?54]. G. Very common in England and Wales, rare in Scotland and E. Ireland.

9. Hieracium salticola (Sudre) Sell & West
Casual. *46.*

10. Hieracium vagum Jord.
Casual. *45.*

11. Hieracium pilosella L.   Mouse-ear Hawkweed   Ray, 1660

Throughout most of the county in grassy places, though commonest on the chalk and sands. The var. **pilosella**, with the phyllaries clothed with simple and glandular hairs, and var. **concinnatum** F. J. Hanbury, with glandular hairs only, are both common. Var. **tricholepium** (Naeg. & Peter) Pugsl., with the phyllaries clothed with long, white simple hairs and almost eglandular, occurs at Cherry Hinton. 23–6, 29, 33–9, 43–6, 49, 54–6, 58, 64–7, 76, 41, [48, 40]. General.

12. Hieracium brunneocroceum Pugsl.
Garden escape. *45, 40.*

CREPIS L. (Hawk's Beard)
1. Crepis foetida L.   Ray, 1660

Recorded by Ray and other authors in several localities and by G. S. Gibson just before 1860 (cf. Babington, 1860). Evans (1939) suggests that these records are probably erroneous, as the difference between this species and *C. vesicaria* ssp. *taraxacifolia* was then not fully appreciated. This seems likely, but no herbarium specimens are available to support this suggestion. [45, 54, 66.] Rare in S.E. England.

2. Crepis vesicaria L. ssp. **taraxacifolia** (Thuill.) Thell.

                    C. E. Moss, 1909
C. taraxacifolia Thuill.

This species was not recorded until the present century (but see *C. foetida*). It is now a common weed of grassland, waste places and arable land throughout the county. All squares except 54, 59. An introduction first recorded in 1713, since when it has spread throughout the south of the British Isles. Native of C. Europe.

3. Crepis setosa Haller f.

Introduced with lawn seed to St John's College Backs in 1961. 45.

4. Crepis biennis L. Relhan, 1785

Recorded from a few localities in grassy places. 35, 45, 46, [25, 38, 43, 54, 56]. **DF.** Local in lowlands; very rare in Scotland and Wales.

5. Crepis capillaris (L.) Wallr. Ray, 1660

C. virens L.

A common and variable weed of grassland, waste places and arable land. This yellow-flowered Composite is easily recognized by its small obconical fruiting heads borne on branched leafy stems. It flowers several weeks later than *C. vesicaria* ssp. *taraxacifolia* and has a long flowering season, so that a *Crepis* in flower after July, in Cambs, is almost certainly this species. All squares. General.

TARAXACUM Weber (Dandelion)

In continental Europe, especially Scandinavia, the genus has been divided into numerous apomictic microspecies, but very little serious work has ever been carried out on the taxonomy of the British plants. In Cambs there seems to be only one microspecies of *T. palustre*, and the segregates of *T. laevigatum* and *T. spectabile* could easily be worked out as they are generally in native habitats. *T. officinale*, however, contains many as yet unidentified microspecies, mostly introduced.

1. Taraxacum officinale Weber sens. lat. Ray, 1660

Leontodon taraxacum L.

Common throughout the county on roadsides and in waste and grassy places. All squares. General.

2. Taraxacum palustre (Lyons) DC. sens. strict. I. Lyons, 1763

Leontodon taraxacum var. *palustre* (Lyons) Bab.

Found only in a few marshy places and fens. First described from Hinton Moor by Lyons. This plant with its almost simple leaves and tightly

COMPOSITAE

appressed phyllaries is very distinct from any other species found in the county. 34, 44, *49*, 56, 57, [36, 45]. **W.** *T. palustre* sens. lat. is widespread, but in the strict sense is probably more local.

3. Taraxacum spectabile Dahlst. sens. lat.   Henslow, 1826

In fens and marshes and damp woodland rides. Often to be found growing with, and sometimes confused with, *T. palustre*, as at Wicken Fen. 25, 34, 36, 44, 49, 54, *55*, 56, 65, 66. **W.** General.

4. Taraxacum laevigatum (Willd.) DC. sens. lat.   Ray, 1670

Leontodon taraxacum var. *erythrospermum* (Andrz. ex Bess.) Bab.

The common species of relict grassland and heath on the chalk and sands, occasionally elsewhere. 25, 36, 45, 54–6, 58, 64, 66, 68, 76, 40, [47]. General in lowlands.

MONOCOTYLEDONES

ALISMATACEAE

BALDELLIA Parl.

1. Baldellia ranunculoides (L.) Parl.   Ray, 1660

Alisma ranunculoides L.

In a few localities by the margins of ponds, rivers and lodes; formerly more frequent. 29, 44, *47*, 48, 56, 57, 20, [25, 38, 45, 46, 49, 55, 58, 65, 66]. **W.** Local, but rarer in the north.

ALISMA L. (Water Plantain)

1. Alisma plantago-aquatica L.   Ray, 1660

A. plantago Bab.

In streams, ditches and lodes and by rivers and pits throughout the county: very common and often abundant in fen lodes. All squares except 26, 54. General, but rarer in the north.

2. Alisma lanceolatum With.   Henslow, 1823

Closely resembling *A. plantago-aquatica* but differing in its narrow, lanceolate leaves, cuneate (not cordate) at the base, and achenes with style near the top (not arising from about the middle). The flowers last only about one day in both species, but open in the late morning in

218

ALISMATACEAE

A. lanceolatum whilst in *A. plantago-aquatica* they open in the afternoon. This difference is surprisingly constant and can be appreciated at Wicken Fen, for example, where both species grow together in the Lode. Throughout the county but not as common as *A. plantago-aquatica*. 24, 29, 34, 36, 37, *39*, 43, 45, 46, 48, 49, 55–9, 20, 30, 40. General distribution is not fully known but appears to be mainly in south and east England, a few localities elsewhere.

SAGITTARIA L.

1. Sagittaria sagittifolia L.   Arrow-head   Ray, 1660
Perennial herb with linear, translucent submerged leaves, lanceolate to ovate floating leaves and sagittate aerial leaves. Throughout the county in lodes, rivers and pits; most frequent in the Fens. 29, 34, *35*, 37–9, 44–8, 55–7, 59, 67, 68, 20, 30, 31, 41, [36, 49, 40]. Scattered over England, Wales and east central Ireland; not native in Scotland.

BUTOMACEAE

BUTOMUS L.

1. Butomus umbellatus L.   Flowering Rush   Ray, 1660
A rather frequent species of ditches, lodes, riversides and pits, almost entirely confined to the Fens. 29, 35, 37–9, 45, 48, 49, 56, 57, 59, 20, 30, [36, 46, 47, 58, 66, 40]. **Cam.** Local in England and Ireland, rare in Wales, doubtfully native in Scotland.

HYDROCHARITACEAE

HYDROCHARIS L.

1. Hydrocharis morsus-ranae L.   Frog-bit   Ray, 1660
Rather common in lodes, ditches, ponds and pits throughout the Fens, formerly south to Cambridge; absent from the rest of the county. The smallest of the 'water-lily' aquatics with round floating leaves; the largest *Hydrocharis* leaves rarely exceed 4 cm., whilst *Nymphoides* (see page 163) has leaves 3–10 cm. in diameter. The white flowers are unisexual and the plants dioecious. Rarely, if ever, fruiting in this country. 29, 37–9, *46*, 47–9, 57–9, 30, 40, 50, [45, 56, 67, 68]. Scattered throughout England, Wales and Ireland usually in calcareous waters, absent from Scotland.

219

HYDROCHARITACEAE

STRATIOTES L.

1. Stratiotes aloides L.   Water Soldier   Ray, 1660
Formerly widespread in pits, lodes and ditches in the Fens, but seen recently only in the Stretham area. A stoloniferous herb with large rosettes which are completely submerged in the winter, rising to the surface in spring and remaining there until autumn, when they again submerge. This floating and submerging is said to be due to changes in the amount of calcium carbonate deposited on the leaves (cf. Arber (1920), pp. 50–1). 57, [29, 37, 38, 45, 47–9, 56, 58, 59]. Scattered throughout England, mainly in the east; probably introduced elsewhere. Only female plants normally occur in Britain though plants with hermaphrodite flowers have been recorded.

ELODEA Michx

1. Elodea canadensis Michx   Canadian Pondweed   Babington, 1848
Anacharis alsinastrum Bab.

This water weed was grown in the Botanic Garden, Cambridge, in 1847 and escaped into the River Cam the following year, where it soon spread and became abundant, choking up the fen waterways and causing much trouble to river traffic. It is now widespread in rivers, lodes and pits throughout the county, but is not as abundant as in the period immediately after its introduction. For a full account of its introduction and spread in the British Isles see W. Marshall (1852). 29, 34, *35*, 36–9, 44–9, 55–9, 67, 68, 20, 30, 31, 50. In fresh waters throughout most of the British Isles. First introduced about 1836 in Ireland and 1842 in Britain, it spread rapidly and attained great abundance, then diminished. Native of N. America.

JUNCAGINACEAE

TRIGLOCHIN L. (Arrow-grass)

1. Triglochin palustris L.   Ray, 1660
A local plant of marshy places in the Fens. 44, 45, 48, 55–9, 66, 20, 30, 41, [29, 38, 46, 49, 54, 40]. W. In marshes throughout the British Isles, but local in most southern and midland counties.

JUNCAGINACEAE

2. Triglochin maritima L.   Skrimshire in Relhan, 1802
Known only from the banks of the tidal River Nene north of Wisbech
where it was last seen in 1930. *41.* In salt marshes and grassy places
on rocky shores around the coasts of the British Isles, very rare inland.

(A species of **Zostera** was recorded by C. E. Moss and Evans (1939) from
the Nene near Wisbech, but in the absence of specimens its identity cannot be
ascertained.)

POTAMOGETONACEAE

POTAMOGETON L. (Pondweed)

The following account of the genus, especially the critical species and
hybrids, is based mainly on the determinations of herbarium specimens by
J. E. Dandy and G. Taylor. Anyone wishing to make a critical study of the
group should consult the series of papers by these authors in the *J. Bot., Lond.*,
between 1938 and 1942; and also the monograph of *Potamogetons in the
British Isles* by A. Fryer and A. Bennett, who did much collecting in the Fens
in the last quarter of the nineteenth century.

Key to the species including *Groenlandia densa.*

1	Floating leaves present.	2
	Floating leaves 0.	6
2	Submerged leaves (phyllodes) all linear with no expanded blade; floating leaves with the margins decurrent for a short distance down the stalk, making a flexible joint. **1. P. natans**	
	Submerged leaves with a ± expanded translucent blade; floating leaves not jointed.	3
3	All leaves thin, translucent, distinctly and finely net-veined, often brownish; floating and most of the submerged leaves ± broadly elliptical, all short-stalked. **3. P. coloratus**	
	Floating leaves ± coriaceous, opaque, not translucent.	4
4	All leaves distinctly stalked; usually in very shallow water and then most or all leaves commonly of the floating type, broadly elliptical to elliptical-ovate. **2. P. polygonifolius**	
	Some or all leaves sessile.	5
5	Submerged leaves blunt, with quite entire margins, commonly 6–12 × 1–2 cm., often reddish; stipules 2–5 cm., broad, blunt, robust; peduncles not thickened upwards. **6. P. alpinus**	
	Submerged leaves acute, cuspidate or mucronate, with microscopically denticulate margins; peduncles thickened upwards. **5. P. gramineus**	

POTAMOGETONACEAE

6   Leaves all in opposite pairs or rarely in threes.   **Groenlandia densa**
     Some or all leaves alternate.   7

7   Leaves usually more than 6 mm. in width, or if narrower then
     linear-lanceolate, not parallel-sided.   8
     Grass-leaved; leaves narrowly linear or filiform, not over 6 mm. in
     width, parallel-sided.   13

8   At least the lower leaves ± amplexicaul.   **8. P. perfoliatus**
     Leaves not amplexicaul.   9

9   Stem compressed; leaves commonly 4–9 × 1–1·5 cm., lanceolate,
     margins distinctly serrate and often strongly undulate; beak
     equalling the rest of the fruit.   **14. P. crispus**
     Stem ± terete; leaf-margin not distinctly serrate to the naked eye;
     beak shorter than the rest of the fruit.   10

10   Leaves blunt, margins entire; peduncles not or hardly thickened
     upwards.   11
     Leaves not blunt, margins minutely serrate; peduncles distinctly
     thickened upwards.   12

11   Leaves hooded at tip and rounded at base, commonly 10–18 × 2–
     4·5 cm.   **7. P. praelongus**
     Leaves not hooded at tip and narrowed at the base, commonly
     6–12 × 1–2 cm.   **6. P. alpinus**

12   Plant much branched at base, leaves commonly 2·5–10 × 0·3–1 cm.,
     acute or acuminate; stipules 1–2 cm.   **5. P. gramineus**
     Plant not much branched at base; leaves 8–20 × 2–5 cm.; stipules
     2–6 cm.   **4. P. lucens**

13   Leaves with two large air-filled longitudinal canals, one on each
     side of the midrib, occupying the greater part of their volume;
     stipules adnate below to the leaf-base, forming a stipular sheath
     with a free ligule.   **15. P. pectinatus**
     Leaves not as above; stipules free from the leaf-base throughout
     their length.   14

14   Leaves with 3(–5) principal and many faint intermediate longi-
     tudinal veins; stems strongly compressed or even winged.
         13. P. compressus
     Leaves usually with only 3–5 longitudinal veins or apparently
     1-veined especially when very narrow, but never with many
     faint intermediate longitudinal veins; stems not or slightly
     compressed.   15

15   Leaves mostly 5-veined, the laterals closer to the margin and to
     each other than to the midrib.   **9. P. friesii**
     Leaves mostly 3-veined or apparently 1-veined, rarely 5-veined
     and then veins with equal spacing.   16

16 Stipules tubular below.   **10. P. pusillus**
     Stipules open, convolute.   *17*

17 Leaves rarely exceeding 1 mm. wide, subsetaceous, narrowing to
     the base and tapering to a long fine point; usually no air-filled
     lacunae bordering the midrib; fruit often toothed below and
     tubercled on the dorsal margin, usually one per flower.
                                           12. P. trichoides
     Leaves usually more than 1 mm. wide, subacute to rounded and
     mucronate, with air-filled lacunae bordering the midrib, rarely
     narrower and acute; fruit smooth, usually 4 per flower.
                                           11. P. berchtoldii

1. Potamogeton natans L.   Ray, 1660
Throughout the county in pits, rivers and lodes, common in the Fens.
29, 34–8, 43–8, 55, 56, 59, 65, *66*, 20, 31, 40, [25, 49, 58, 76]. **W.** General.

2. Potamogeton polygonifolius Pourr.   Relhan, 1805
Recorded from Burwell Fen (1805), Bottisham Fen (1829), Foulmire
(1838), Wicken Poor's Fen (1908), and Gamlingay where it has been
seen a number of times and was last recorded by C. D. Pigott in 1948.
It is usually to be found in rather shallow, acid water. *25*, [44, 56, 57].
General, but rare in central and eastern England, being absent from the
chalk areas.

3. Potamogeton coloratus Hornem.   L. Jenyns, 1827
P. plantagineus Du Croz. ex Roem. & Schult.
Frequent in the Fens but rare elsewhere; usually in rather shallow peaty
water. 29, 38, 44, 45, 48, 49, 55, *56*, 57, 66, [24, 34, 35, 39, 47, 58]. **W.**
A species with a very scattered distribution, locally frequent in the
central plain in Ireland and fens in eastern England, otherwise very rare
though as far north as the Outer Hebrides on the west side of Scotland.
    × **gramineus** (*P.* × *billupsii* Fryer). Recorded from Benwick in 1892 by
A. Fryer and Burwell Fen in 1909 by C. E. Moss and by Evans (1939).
No specimen has been seen from Burwell. *56*, [39].

4. Potamogeton lucens L.   Ray, 1660
Common in rivers, lodes and pits in the Fens, rare elsewhere. 25, 29,
35, 37, 39, *45*, *46*, 47–9, 55–7, 59, *66*, 20, 30, 40, [34, 36, 38, 58]. **W.**
Throughout most of the British Isles but frequent only in S. and E.
England and the northern half of Ireland, and very local elsewhere.

POTAMOGETONACEAE

× **natans** (*P.* × *fluitans* Roth). Recorded from the Doddington, Chatteris, Mepal and Whittlesea areas between 1880 and 1890 but not seen since. [29, 48, 49.]

× **perfoliatus** (*P.* × *salicifolius* Wolfg.; *P. decipiens* Nolte ex Koch). Formerly recorded from a number of localities in the Fens where both parents are common. Last seen near Sutton in 1938. *48*, [29, 38, 39, 49, 56].

5. Potamogeton gramineus L.   Relhan, 1802

P. heterophyllus Schreb.

Formerly recorded from a number of localities in the Fens, now rare. Still to be found in the Lt. Wilbraham River. Also recorded from Gamlingay and questionably from West Wratting. 29, 48, 55, *56*, [25, 38, 45, 47, 57, ?65]. Essentially a northern and eastern species in Britain, extremely rare south of the Thames and in Wales. Throughout Ireland but very rare in the south.

× **lucens** (*P.* × *zizii* Koch ex Roth; *P. coriaceus* (Mert. & Koch) A. Benn.). Formerly recorded from a few localities in the Fens. Last seen near Mepal in 1933. *48*, [29, 38, 39, 47, 58].

× **perfoliatus** (*P.* × *nitens* Weber; *P. involutus* (Fryer) H. & J. Groves). Recorded from Blackbush Drain, Whittlesea, by A. Fryer in 1894 and 1895. [29.]

6. Potamogeton alpinus Balb.   ?Henslow, 1835
                                   First certain record, R. C. L. Howitt, 1959

P. rufescens Schrad.

Occurs at Manea where it was first found in 1959. There are old records unsubstantiated by herbarium specimens from Bottisham and Burwell. 48, [?56]. Throughout most of the British Isles but rare in S. and W. England and almost absent from Wales. Rare in southern Ireland, frequent in the north.

7. Potamogeton praelongus Wulf.   Babington, 1849

Frequent in rivers, lodes and pits in the Fens. 37, *38*, *39*, *45*, 46–8, 56, 20, 30, 40, [29, 57–9]. W. Scattered throughout the British Isles northwards from the Thames. Rare in Wales and absent from S.E. Ireland.

8. Potamogeton perfoliatus L.   Ray, 1660
Common in rivers, lodes and pits throughout the Fens, occasional
elsewhere. 29, 36–9, *44*, 45, *46*, 47–9, 56–9, *66*, 68, 20, 30, 31, 40, [34,
35, 55, 67]. **W.** Throughout the lowlands of the British Isles.

9. Potamogeton friesii Rupr.   Ray, 1660
P. compressus sensu Bab.
Now known only in a few localities in the Fens where it was formerly
rather more frequent. *46*, 48, 49, 56, 59, [29, 38, 39, 45, 47, 57, 58, 40].
Mainly in central and eastern England, west to Shropshire and Somer-
set, north to Yorkshire. Otherwise in a few very scattered localities in
Scotland and central Ireland.

10†. Potamogeton pusillus L.   ?Ray, 1660
          First certain records by A. Fryer and H. & J. Groves in 1884

P. panormitanus Biv.
Long confused with *P. berchtoldii* Fieb. to which most of the old
records apply. There is authentic material in CGE from a few localities
in the Fens collected between 1884 and 1888. [38, 39, 47, 48, 57.]
Throughout the lowlands of the British Isles; rare in Ireland.

11. Potamogeton berchtoldii Fieb.   ?Ray, 1660
                          First certain record L. Jenyns, 1823
P. pusillus sensu Bab. pro maxima parte
Formerly in a number of localities, mainly in the Fens; now very
local. *46*, 56, 66, [25, 38, 44, 57, 58, 67]. **W.** Throughout the British
Isles, except for the main mountain areas.

× **coloratus** (*P.* × *lanceolatus* Sm.; *P. perpygameus* Hagstr. ex Druce).
Recorded from Burwell Fen by A. Bennett in 1880. [56.]

12†. Potamogeton trichoides Cham. & Schlecht.
                              H. & J. Groves, 1884
A rare species recorded from Ely (1884), Mepal (1897, 1909) and Cambridge (1915). The last-mentioned record is unsupported by a herbarium
specimen and is possibly incorrect. [?45, 48, 58.] S. and E. England,
Anglesey and Stirling, local or rare.

POTAMOGETONACEAE

13†. Potamogeton compressus L.   Relhan, 1802

P. zosteraefolius Schumach.

Formerly in a number of localities in the Fens; last seen at Roswell Pits, Ely, in 1887. [45, 46, 56–8.] Local in C. and E. England and also Montgomery and Angus.

(**P. acutifolius** Link. Evans (1939) says of this species: 'The only evidence of this as a Cambridgeshire species rests on a specimen in Buddle's herbarium at the British Museum, which, along with a specimen of *P. friesii*, was described by Ray (*Hist. Pl.* 1, 189 (1686)) under the name *Potamogiton caule compresso, folio Graminis canini*, and stated to occur copiously in the Cam near Cambridge and in many other rivers.')

14. Potamogeton crispus L.   Ray, 1660

The commonest species of the genus, occurring in rivers, lodes, streams and pits throughout the county. *25*, 29, 34, *35*, 36–9, 44–9, 55–9, 68, 69, 20, 30, 31, 40, 41, [54, 65–7]. Throughout lowland British Isles, but much rarer in the west.

× **perfoliatus** (*P.* × *cooperi* (Fryer) Fryer). Found by C. R. Billups at Parsonware Drove, Benwick, in 1892 and confirmed by A. Fryer. [39.]

15. Potamogeton pectinatus L.   Ray, 1660

P. flabellatus Bab.

Throughout much of the county in rivers and lodes, though commonest in the Fens and rare in the south. 29, 34, 36, 37, 39, 45–9, 55–8, 67, 20, 30, 31, 40, [38, 41]. W. Throughout the lowlands of the British Isles, but rarer in the north and west.

GROENLANDIA Gay

1. Groenlandia densa (L.) Fourr.   Ray, 1660

Potamogeton densus L.

Now very local in a few localities in the Fens, formerly rather more widespread. 45, 48, 55–7, [29, 35, 37, 38, 44, 46, 47]. W. Frequent in the lowlands of England, except Devon and Cornwall, and in south-central Ireland; extremely rare elsewhere.

RUPPIACEAE

RUPPIA L.

1†. Ruppia maritima L.　　　　　　Skrimshire in Relhan, 1802
R. rostellata Koch

Submerged aquatic herb; found by Skrimshire in ditches below Wisbech but not seen since. [41.] Round most of the coasts of the British Isles, but less frequent in the north.

ZANNICHELLIACEAE

ZANNICHELLIA L.

1. Zannichellia palustris L.　　　　　　　　　　　　Ray, 1660

Submerged aquatic herb; frequent in ditches and lodes in the Fens, rare elsewhere. 35, 36, 38, 39, 43, 45–8, 55, 57, 59, 68, 69, 31, 41, 50, [44, 49, 56, 30]. General, but rare in the north and west.

LILIACEAE

NARTHECIUM Huds.

1†. Narthecium ossifragum (L.) Huds.　　Bog Asphodel　Ray, 1685

Formerly occurred in bogs at Gamlingay. Not given as extinct by Babington in 1860, but there are no later records. [25.] Common in bogs, wet heaths, moors and wet acid places over most of the British Isles, but generally rare in the east.

ASPHODELUS L.

1. Asphodelus fistulosus L.
Garden escape. [58.]

CONVALLARIA L.

1. Convallaria majalis L.　　Lily of the Valley
　　　　　　　　　　　　　J. Martyn in I. Lyons, 1763

Still occurs in its original locality at White Wood, Gamlingay. It has been recorded as a garden escape from Cambridge (1940) and Devil's Dyke (1955). 25, *45*, 66. G. Widespread in England, local in Wales and Scotland. Commonly cultivated and sometimes escaping.

LILIACEAE

ASPARAGUS L.

1. Asparagus officinalis L. ssp. **officinalis**   Asparagus

I. Lyons, 1763

An escape from cultivation frequently established in waste places. *35, 36,* 38, 45–8, 56, *57,* 58, 67, 41. Commonly cultivated as a vegetable; native distribution obscure.

RUSCUS L.

1. Ruscus aculeatus L.   Butcher's Broom   Ray, 1663

Formerly recorded from a number of localities, perhaps always as an escape from cultivation. Now only known in the grounds of Wandlebury on the Gog Magogs where it was presumably originally planted, and at Hinxton Grange. This species was once cultivated at Whittlesford for its use in preparing chamois leather and parchment. It was known by the workers as 'knee-um' (knee-holm or knee-holly). They found that its sharp leaf branches sprinkled water on to the leather better than anything else. 44, 45, *46,* [36, 56]. **GG.** Widespread in southern England but rather local; extending north to Caernarvon, Leicester and Norfolk; commonly cultivated and sometimes escaping in north England and Scotland.

FRITILLARIA L.

1†. Fritillaria meleagris L.   Snake's-head   Relhan, 1785

Recorded by Relhan from some fields at Westhoe Park near Linton but not seen since. It was formerly abundant just over the border in Essex at Bumpstead (cf. Gibson, 1862, p. 316). [54.] Very local from Somerset and Kent to Staffs and Lincs, and in many cases of garden origin.

TULIPA L.

1. Tulipa sylvestris L.   Wild Tulip   G. M. Rhodes, 1952

In long grass between the moat and mound of the old house, Kirtling Towers. It is said to have been there 'for hundreds of years' and was formerly abundant, but the population was greatly reduced by ploughing-up of the grassland in the Second World War. It has since increased again. 65. Naturalized in eastern England and south-eastern Scotland north to Fife, and extending west to Devon, Gloucester, Worcester and Lancs. Native of southern Europe.

228

ORNITHOGALUM L.

1. Ornithogalum umbellatum L.   Star of Bethlehem   Relhan, 1788
Occasional in grassy places; doubtfully native and often obviously a garden escape. 24, 34, 35, *37*, 44, 45, *47*, 55, 56, 76, 41, [38, 46, 58, 66, 67, 40]. Probably native in eastern England, elsewhere naturalized.

2. Ornithogalum nutans L.   Henslow, 1821
An infrequent garden escape which sometimes persists. 35, 45, *54*, 65, [55]. Naturalized in eastern, central and northern England; native of S. Europe and Asia Minor.

3†. Ornithogalum pyrenaicum L.   Bath Asparagus   Ray, 1685
Known to the seventeenth- and eighteenth-century botanists from a 'bushy close' near the church at Little Eversden where it was apparently last seen in 1774. [35.] Native in Somerset, W. Gloucester, Wilts, Berks, Sussex, Norfolk, Beds, Hunts, and Cambs; a rare casual elsewhere.

ENDYMION Dumort.

1. Endymion non-scriptus (L.) Garcke   Bluebell   Ray, 1660
Scilla non-scripta (L.) Hoffmanns. & Link; *E. nutans* (Sm.) Dumort.
Common and often dominant over large areas in boulder-clay woods and in woods on the sands at Gamlingay. 23, *24*, 25, 26, 34, 35, 36, 43–7, 54–6, 64, 65, [38, 48]. **B.** General.

2. Endymion hispanicus (Mill.) Chouard
Garden escape. 44, 45.

MUSCARI Mill.

1. Muscari atlanticum Boiss. & Reut.   Henslow, 1828
M. racemosum sensu Bab., et Evans
A rare species of hedgebanks on the chalk. It is still to be found in its original locality between Cherry Hinton and the Gog Magog Hills. 44, 45, 55, 56. Native in dry places in Norfolk, Suffolk, and Cambs.

COLCHICUM L.

1. Colchicum autumnale L.   Meadow Saffron   Relhan, 1785
Recorded by Relhan from Wood Ditton, and well known to the Cambridge botanists in a wood at Weston Colville, in the first decade of the twentieth century, where it was last seen by Shrubbs in 1912. In 1909 Evans collected it at West Wratting, and F. H. Perring and

LILIACEAE

S. M. Walters recorded it in the grounds of Hildersham Hall in 1957 where it is presumably planted. 54, [65]. Local from Devon and Kent to Durham and Cumberland, also in S.E. Ireland. Occasionally naturalized in Scotland.

Paris L.

1. Paris quadrifolia L.   Herb Paris   Ray, 1660

A characteristic species of boulder-clay woods. 25, 35, 43, 54, 55, *64*, 65, *67*, [45]. **B.** In Britain on basic soils from Somerset and Kent to Caithness, but local and with an eastern tendency and absent from areas of the west.

JUNCACEAE

Juncus L. (Rush)

1 Stems leafless, with brownish sheaths at the base, and with an apparently lateral inflorescence exceeded by a cylindrical bract.   2

Stems leafy, leaves with flat or sub-cylindrical lamina; inflorescence obviously terminal.   4

2 Stem glaucous, with less than 20 prominent ridges, pith interrupted.
6. J. inflexus

Stem not glaucous, with at least 30 less prominent ridges, pith continuous.   3

3 Stems easily cracked, rather obviously ridged especially near the inflorescence, not shining; capsule shortly mucronate.
8. J. conglomeratus

Stems more pliable without cracking, smooth and shining when fresh; capsule not mucronate.   **7. J. effusus**

4 Leaves channelled or flattened, solid, not septate.   5

Leaves subcylindrical or laterally compressed, hollow, obviously septate or setaceous.   9

5 Leaves all radical, very stiff and reflexed; plant densely caespitose.
1. J. squarrosus

Cauline leaves present (sometimes only near base).   6

6 Perianth-segments pale greenish or straw-coloured, acute; capsule obviously longer than broad.   7

Perianth-segments brownish to blackish, obtuse; capsule sub-globose.   8

7 Annual; capsule oblong, blunt.   **5. J. bufonius**

Perennial; capsule ovoid, obtuse-mucronate.   **2. J. tenuis**

8 Rhizome short; capsule obtuse, *c.* 1·5 times perianth-segments.
                                                    3. J. compressus
   Rhizome long; capsule acuminate, ± equalling perianth-segments
   (salt-marshes).   **4. J. gerardii**

9 Leaves setaceous (plants usually less than 20 cm.).   *10*
   Leaves thicker, subcylindrical or laterally compressed (more
   robust plants, often more than 30 cm.).   *11*

10 Leaves deeply channelled; perianth-segments acute; annual.
                                                   5. J. bufonius
   Leaves showing a double tube in cross-section; perianth-segments
   obtuse; perennial.   **12. J. bulbosus**

11 Basal leaves and stems very similar, cylindrical, with both longitudinal and transverse septa; perianth-segments pale, obtuse.
                                                9. J. subnodulosus
   All leaves cauline with transverse septa only; perianth segments
   darker brown, acute.   *12*

12 Erect with few (generally only 3) straight subcylindrical leaves;
   inflorescence richly branched; capsule gradually acuminate.
                                                  10. J. acutiflorus
   Usually ascending or decumbent with many (up to 7) somewhat
   curved compressed leaves; inflorescence less richly branched;
   capsule rather suddenly contracted to a point. **11. J. articulatus**

(The hybrid **J. acutiflorus × articulatus** can be recognized by its intermediate
characters and small sterile capsules.)

1†. Juncus squarrosus L.   Ray, 1685
Recorded by the seventeenth- and eighteenth-century botanists at
Hinton and Teversham Moors; and also at Gamlingay where it was
still known by Babington in 1860, but has not been recorded since.
Still occurs on Cavenham Heath, W. Suffolk. [25, 45.] General on acid
soils.

2. Juncus tenuis Willd.   T. J. Foggitt, 1919
Recorded only from Gamlingay in 1919, Twenty Pence Ferry on the
River Ouse in 1940, Coton in 1954 and Conington in 1959. 36, 45, *47*,
[25]. Naturalized in much of the British Isles, but particularly in the
south and west. Native of N. and S. America.

3. Juncus compressus Jacq.   Ray, 1660
Recorded in a number of localities in winter-wet pasture, but never
abundant. 34, 36, 38, 44, 45, 55–7, *58*, 59, 40, *41*, 50, [25, 46–8, 66].
DF. Almost confined to England, not common.

231

JUNCACEAE

4. Juncus gerardii Lois. ?Babington, 1860

First certain record A. Fryer, 1881

?*J. compressus* sensu Bab. pro parte

Known only with certainty from the tidal River Nene at Foul Anchor and also near Parsons Drove. Old records from the Fens almost certainly refer to *J. compressus*. 31, 41. **FA.** All round the coasts of the British Isles, rarely inland.

5. Juncus bufonius L.   Toad-Rush   Ray, 1660

Throughout the county in seasonally wet open ground. A small annual intolerant of competition. All squares except 24, 37, 54, 64, [26, 36, 58]. Abundant throughout the British Isles.

6. Juncus inflexus L.   Hard Rush   Ray, 1660

J. glaucus Sibth.

Throughout the county in wet places, often abundant. All squares. General, but avoiding acid soils and local in Scotland and parts of Wales.

7. Juncus effusus L.   Soft Rush   Ray, 1660

In wet places throughout the county but generally less common than *J. inflexus*. The var. **compactus** Hoppe with the inflorescence condensed into a single rounded head is recorded. This variety should not be confused with *J. conglomeratus*. All squares except 24, [41]. General.

8. J. conglomeratus L.   Ray, 1660

Rather rare, mainly in boulder-clay woods. The inflorescence is typically compact but can be somewhat effuse in shady conditions. The species behaves as a calcifuge in East Anglia, and its rare occurrence in boulder-clay woods is reminiscent of *Holcus mollis*. 25, 29, 35, 43, 45, *49*, 54, *64*, 65, [34, 38, 46–8, 55, 56, 58, 41]. Throughout the British Isles.

9. Juncus subnodulosus Schrank   Relhan, 1820

J. obtusiflorus Ehrh. ex Hoffm.

The characteristic rush of peat fens, locally abundant at Wicken. 25, 34, *36*, 37, 38, 44, 45, *46*, 48, 49, 55–7, 59, 66, 67, 20, 30, 40, 50, [24, 29, 35, 39, 47, 58]. **W.** England, Wales, Ireland and S. Scotland, locally abundant; usually in base-rich fens or coastal marshes.

232

10. Juncus acutiflorus Ehrh. ex Hoffm.   Ray, 1660
Rare, but formerly more widespread. The only recent records backed
by herbarium material are from Gamlingay Cinques, where it is still
abundant, and Hauxton by S. M. Walters in 1949. This species seems to
be somewhat calcifuge, and has declined with the nineteenth-century
drainage of acid or neutral bog or marsh. It is no longer found at
Wicken or Chippenham Fens, though there are old records for both of
these places. The late flowering and suberect habit of growth distinguish this species readily from at least the common forms of *J. articulatus*, but the hybrid *J. acutiflorus* × *articulatus* can easily be confused
with it (see below). 25, *45, 49*, [26, 35, 47, 56, 57, 65, 66]. General,
especially on acid soils.

× **articulatus** (*J.* × *surrejanus* Druce). The sterile hybrid has been recorded from four localities in the Cambridge area, and may well be more
common than these suggest. It seems to be characteristic of wet
permanent pasture which is roughly grazed; in three localities it was
accompanied by *J. articulatus*, whilst in one (Hauxton) both *J. articulatus*
and *J. acutiflorus* were present. For a detailed description of this
hybrid see Timm & Clapham (1940). 44, 45.

11. Juncus articulatus L.   Ray, 1660
J. lampocarpus Ehrh. ex Hoffm.
In wet ground throughout the county. All squares except 24, 39, 54, 64,
41. W. General.

12. Juncus bulbosus L.   Ray, 1685
J. supinus Moench; includ. *J. kochii* F. W. Schultz
Very variable in habit and stature. Now only known at Gamlingay and
Blackbush Drove near Whittlesey. Formerly occurred in a number of
other localities in the Fenland; presumably these were mainly local
acid 'Moors'. 25, 29, [26, 38, 46, 49, 55, 56, 58, 66]. G. General,
mostly on acid soils.

LUZULA DC. (Woodrush)

1. Luzula pilosa Willd.   I. Lyons, 1763
Local in a few woods on the boulder clay. 25, 35, 54, 64, 65, [36]. B.
General.

2†. Luzula sylvatica (Huds.) Gaudin   J. Downes, 1832
The only record for the county is from Wood Ditton by J. Downes in
1832 (CGE). The Gamlingay plant recorded by C. E. Moss is *L. pilosa*.

JUNCACEAE

Large-leaved forms of *L. pilosa* should not be mistaken for this species. They can be distinguished by the leaf apices, which in *L. pilosa* end in an obtuse callus, but in *L. sylvatica* are very acute. [65.] General, but most abundant in the west and north.

3. Luzula campestris (L.) DC.   Ray, 1660

In dry grassland throughout most of the county. Flowers in April, a month earlier than the following species. 25, 29, 34–9, 44–6, 49, 54–6, 65–7, 76, 41, 50, [23, 47, 48, 64, 40]. General.

4. Luzula multiflora (Retz.) Lejeune   Ray, 1660

In a few scattered localities in marshy ground. The var. **congesta** (Thuill.) Koch, with subsessile, rounded or lobed, inflorescence heads, is recorded. 25, 29, 45, *49*, 56, [48, 55, 66]. G. General, chiefly on acid or peaty soils.

AMARYLLIDACEAE

ALLIUM L.

(**Allium ampeloprasum** L.   Wild Leek

Recorded as being 'quite naturalized' by the Madingley Road by Miss G. Bacon in 1923. There are no specimens in CGE.)

1. Allium vineale L.   Crow Garlic   Ray, 1660

Fairly common on roadsides, in waste places and as a weed of arable land and permanent pasture on the boulder clay, rather scarce elsewhere. In Cambs plants the inflorescence consists of bulbils only. 25, 26, 34–7, 44–6, 48, 55, 65, 66, [47, 54]. Common in England and Wales, local in Scotland and Ireland.

2†. Allium oleraceum L.   Relhan, 1820

Recorded from Hinton by Relhan and between Coton and Barton by S. W. Wanton and W. W. Newbould, but Babington could only find *A. vineale* in the latter locality. There are no specimens to support these records. There are specimens in CGE from Hildersham Furze Hills collected by F. R. Tennant in 1901 and R. H. Lock in 1905. [?45, 54.] Dry grassy places from Devon and Kent to Wigtown and Moray with a distinct eastern tendency; in Ireland very local and only in the eastern coastal counties.

3. Allium paradoxum (Bieb.) G. Don   A. J. Crosfield, 1928

Well established by the Madingley Road and recorded from several other localities in the vicinity of Cambridge. 45, 46, 55. Naturalized

AMARYLLIDACEAE

in a number of places in the British Isles and still spreading; native of the Caucasus, N. Persia and Mountain Turkmenia.

4. Allium ursinum L.   Ramsons   Ray, 1660
A very local plant of boulder-clay woods. 35, 44, 45, 56, 64, 65, [25, 54, 40]. **DPW.** General.

GALANTHUS L.

1. Galanthus nivalis L.   Snowdrop   I. H. Burkill, 1890
Recorded as a garden escape or casual in a number of localities. 34, 44–6, 54, 56, 65, [38]. Probably native in Britain north to Dunbarton and Moray, but commonly planted and often naturalized.

NARCISSUS L.

1. Narcissus pseudonarcissus L.   Daffodil   Ray, 1660
In a few scattered localities in woods and plantations perhaps originally planted, but known to Relhan at Whittlesford where it is still to be found. 300 acres of cultivated varieties are grown in the Wisbech and Manea areas. 34, 44, 45, 54, [41]. Locally native in England and Wales; naturalized in Scotland and Ireland.

(**N. poeticus** L. sens. lat. is recorded by Ray (1685) as *Narcissus medioluteus*, by J. Martyn as *N. sylvestris*, and T. Martyn (1763) as *N. poeticus*; but was excluded by Relhan. The localities given are Barnwell Abbey and other places. There are no herbarium specimens.)

2. Narcissus × biflorus Curt.
Garden escape. [56, 58.]

IRIDACEAE

SISYRINCHIUM L.

1. Sisyrinchium bermudiana L.
Casual. *25.*

IRIS L.

1. Iris foetidissima L.   Ray, 1660
Occasional in woods and copses, mainly on the boulder clay and chalk. 23, 25, 26, 35, 36, 45, *65*, [44, 46, 49, 55, 56]. **B.** S. England and Wales, naturalized in Scotland and Ireland.

235

IRIDACEAE

2. Iris pseudacorus L.   Yellow Flag   Ray, 1660
In ditches and marshes, and by pits and rivers throughout the county.
All squares except 26, 64, 20. General.

CROCUS L.

1. Crocus purpureus Weston
Garden escape. 45.

2. Crocus sativus L.   Saffron   Gerard, 1599
Formerly cultivated, especially about Hinton, and listed by J. Ray and
T. Martyn and Relhan, but it had disappeared by the time Babington
wrote his *Flora* in 1860. The stamens and stigmas yielded an important
orange-red dye used in medieval times for cloth dyeing, paints and
food colouring. It was first cultivated in England as early as 1350 but
the first record for Cambridgeshire is given by Gerard in 1599. The
centre of cultivation in East Anglia lay between Cherry Hinton and
Saffron Walden. The industry continued until the last quarter of the
eighteenth century when saffron was replaced by chemical dyes. [45.]

DIOSCOREACEAE

TAMUS L.

1. Tamus communis L.   Black Bryony   Ray, 1660
In hedgerows, woods and copses throughout most of the county.
23–6, 33–6, 43–6, 49, 54–8, 64–7, 75, 20, 50, [38, 47, 48, 40]. Native
in England and Wales, a rare introduction in Ireland. The absence of
this Mediterranean–Atlantic species from the Irish native flora is a
remarkable fact.

ORCHIDACEAE

CEPHALANTHERA Rich.

1. Cephalanthera damasonium (Mill.) Druce   Henslow, 1825
C. grandiflora Gray; *C. latifolia* Janchen
In a few localities in dry chalky woods or copses, usually under beech.
Unknown to the early botanists and presumably not present until the
planting of the beech woods (cf. p. 26). 24, 34, *44*, 45, 55, [54, 56]. GG.
Local from Kent to Somerset, Wilts, and Dorset north to Cambs.
Once reported from E. Yorks. (See map, page 14, and plate 7.)

ORCHIDACEAE

EPIPACTIS Sw.

1. Epipactis palustris (L.) Crantz   Ray, 1660
Formerly found in a number of fens but now only known at Chippenham and Dernford. *44, 45, 56,* 66, [25, 58, 67]. C. In fens and dune-
slacks throughout Great Britain and Ireland northwards to Perth.

2. Epipactis helleborine (L.) Crantz   Ray, 1660
E. media sensu Bab.
In a few localities in woods, mainly on the boulder clay. 24, 25, 33, 35,
44, 54, [36, 65, 66]. H. General except the extreme north.

3. Epipactis purpurata Sm.   J. C. Faulkner, 1962
Found only in a wood at Hildersham. 54. S.E. England and the
Midlands.

4. Epipactis phyllanthes G. E. Sm.   Shrubbs, 1896
A rare species found by Shrubbs on Robinson Crusoe's Island, Cambridge, in 1886, and seen there again by J. Drayner, *c.* 1946; also found
by D. P. Young at Stetchworth in 1950, and by Mrs G. Crompton at
Thriplow in 1958. Our plants have the labellum perfectly formed, the
anthers sessile or subsessile, and the epichile longer than the hypochile;
they are referable to var. **vectensis** (T. & T. A. Stephenson) D. P. Young.
44, *45,* 66. T. Very local from Devon and Kent north to Cambs in the
east, and Lancs in the west.

SPIRANTHES Rich.

1. Spiranthes spiralis (L.) Chevall.   Autumn Lady's Tresses
                                                                    Ray, 1660
S. autumnalis Rich.
Formerly in a number of localities in rather wet grassy places. Now
known only at Tydd Gote. 41, [44–6, 54, 55, 66, 40]. In England,
Wales and Ireland north to a line between Mayo and Yorkshire but
common only in the south.

LISTERA R.Br.

1. Listera ovata (L.) R.Br.   Twayblade   S. Corbyn, 1656
Widespread in woods and grassy places, though rare in the Fens. 25,
33–6, 43–6, 54–6, 64–6, 75, 30, [48, 40]. General.

ORCHIDACEAE

1. Neottia nidus-avis (L.) Rich.   Bird's-nest Orchid   Relhan, 1785
A rather rare plant of boulder-clay woods. Saprophytic herb with short
creeping rhizomes concealed in a mass of short, thick, fleshy, blunt
roots, shaped like a bird's nest, from which the plant gets its name. 25,
35, *44*, 54, 64, *65*, [36]. **H.** Throughout Britain northwards to Banff
and Inverness, and in Ireland.

HAMMARBYA Kuntze

1†. Hammarbya paludosa (L.) Kuntze   S. Dale & P. Dent, 1684
Malaxis paludosa (L.) Sw.
Formerly occurred in bogs on Hinton Moor and at Gamlingay, and was
last seen at the latter by Babington in 1855. [25, 45.] Local; very rare
in S. and C. England, and usually in wet *Sphagnum*.

LIPARIS Rich.

1†. Liparis loeselii (L.) Rich.   Ray, 1660
Sturmia loeselii (L.) Reichb.
Locally abundant in the early nineteenth century on Burwell Fen, and
recorded from Hinton, Teversham, Fulbourn and Sawston 'Moors'
in the eighteenth century. This rare fen orchid persisted at Chippenham
and Wicken into the present century, but is now apparently extinct, the
last certain Wicken record being made in 1945. The very precise
habitat requirement of *Liparis* probably explains its decline and
extinction. It grows exclusively in moss communities (dominated by
Acrocladium cuspidatum and spp. of *Campylium*) in very wet fen, and
there is evidence that such communities do not persist permanently in
the Fens, but need some type of interference such as the peat-digging
operations provided in the last century. *56*, [44, 45, 55, 66]. A rare
plant now confined to a few fens in Norfolk and Suffolk and to dune
slacks in Glamorgan and Carmarthen. Extinct in Cambs, Hunts, and
E. Kent.

HERMINIUM R.Br.

1†. Herminium monorchis (L.) R.Br.   Musk Orchid   Ray, 1663
Formerly recorded from grassy chalk-pits at Hinton and on the Gog
Magog Hills; and near Linton until at least 1825. [45, 54.] A rare and

local plant of chalk-downs and limestone pastures in S. and E. England from Dorset and Kent north to Glamorgan, Monmouth, Gloucester, Oxford, Bucks, Beds, and Norfolk.

COELOGLOSSUM Hartm.

1. Coeloglossum viride (L.) Hartm.   Frog Orchid   Ray, 1660
Habenaria viridis (L.) R.Br.
Formerly in a number of scattered localities. The only recent record is from a grazed pasture at Soham where it was still to be found in 1960. 67, [25, 35, 45, 46, 54–8, 65, 66]. Pastures throughout the British Isles but especially in the north and usually on calcareous soils.

GYMNADENIA R.Br.

1. Gymnadenia conopsea (L.) R.Br.   Fragrant Orchid   Ray, 1660
A local species formerly slightly more frequent. Ssp. **conopsea** is a short to medium plant with a short, laxly-flowered spike (6–10 cm. long), and is to be found on chalk grassland flowering in June. Ssp. **densiflora** (Wahlenb.) G. Camus, Bergon & A. Camus is a tall plant with long dense-flowered spikes (8–16 cm. long), and flowers in fens in July. Both can be found where chalk-grassland and fen merge. 45, 56, 66, [35, 36, 44, 46, 55, 65]. C.; D. General on base-rich soils.

PLATANTHERA Rich.

1. Platanthera chlorantha (Custer) Reichb.   Greater Butterfly Orchid
Ray, 1660
Habenaria chlorantha (Custer) Bab., non Spreng.
A rather local plant of boulder-clay woods. 25, 26, 35, 45, *54*, 64, 65, [34, 36, 43, 46, 49, 55]. B. General, except the extreme north.

(The records for **Platanthera bifolia** (L.) Rich. by Evans (1939) from Hildersham and Moor Barns, Madingley, and by R. Ross from Croydon Wilds, are unsupported by herbarium specimens and must be regarded as doubtful.)

OPHRYS L.

1. Ophrys apifera Huds.   Bee Orchid   Ray, 1660
A local plant in grassland, amongst scrub and by the edges of copses mainly on the chalk, but also on the boulder clay and sands. 23, 24, *25*,

ORCHIDACEAE

34, 35, *36*, 44–6, 55, 56, 59, 65–7, [54, 58]. Throughout England, Wales and Ireland.

2†. Ophrys sphegodes Mill.   Spider Orchid   Ray, 1663
O. aranifera Huds.
Known to the early botanists in several places on the chalk. Last seen at Abington in 1837 by Babington. [45, 54.] S. and E. England from Dorset and Wilts to Kent and northwards to Cambs and Suffolk; also in Denbigh, Northants and Jersey. Not now known north of the Thames.

3. Ophrys insectifera L.   Fly Orchid   Ray, 1660
Formerly recorded from a number of localities on the chalk, boulder clay and sands. The only recent records are Ditton Park Wood (1952 and 1955) and Buff Wood (1956). 25, 65, [24, 35, 45, 46, 54–6, 66]. In suitable habitats in lowland England, Glamorgan, Anglesey and Denbigh in Wales, Perth in Scotland, and central Ireland.

HIMANTOGLOSSUM Spreng.

1. Himantoglossum hircinum (L.) Spreng.   Lizard Orchid
Shrubbs, 1920
This rare species was first found at Hinxton in 1920 but did not flower until 1921. In 1922 it flowered again but was burnt over by a tramp's fire and was not seen there again. In 1923 a specimen was found on the Gog Magog Hills near Fulbourn and in 1931 another was recorded on the Devil's Dyke. In 1946 it was refound on the Devil's Dyke and has been recorded in most years since, with a maximum of 22 plants in 1948. In 1959 a solitary plant was found on an entirely different part of the Dyke, and by 1961 a small colony was established here. Twelve plants were found near Harston in 1954 but it has not been seen there since. 45, 56, 66, [44, 55]. D. Very rare and local in S. and E. England. After a rapid increase during the 1930's it is now apparently very rare again.

ORCHIS L.

1†. Orchis ustulata L.   Burnt Orchid   Ray, 1660
Formerly recorded in a few places on the chalk. Last seen on the Fleam Dyke in 1941. *55*, [44, 45, 54, 66]. A very local plant of chalk-downs and limestone pastures throughout England.

240

2. Orchis morio L.   Green-winged Orchid   Ray, 1660
A rare plant of old pastures, formerly more frequent. *25, 34, 35*, 36,
45, 46, 56, 57, *58*, *67*, [23, 38, 48, 54, 55, 64–6, 40]. Throughout England
and Wales, but becoming rare in the north; C. Ireland.

3. Orchis mascula (L.) L.   Early Purple Orchid   Ray, 1660
A characteristic spring-flowering species in boulder-clay woods. 25, 26,
35, 36, 43, 44, *45*, *46*, 54, 55, 64, 65, [49, 66, 40]. **B.** General.

DACTYLORCHIS (Klinge) Vermeul.

All the Cambridgeshire material of *Dactylorchis* in CGE has been revised
by J. Heslop Harrison.

1 Stem solid throughout most of its length; leaves spotted; upper
   non-sheathing bract-like leaves two or more in number; lower
   bracts relatively narrow (not over 3 mm.); flower spur slender
   (not over 2 mm.).   **1. D. fuchsii**
   Stem usually hollow throughout most of its length; leaves usually
   unspotted; bract-like leaves on upper part of stem, 0, 1 or 2,
   non-sheathing; lower bracts relatively wide (over 3 mm.); spur
   usually stout (over 2 mm.).   2

2 Stem cavity wide (usually exceeding ½ total diam.); leaves yellow-
   green, erect, narrowing almost from the base, apex markedly
   hooded; flowers relatively small; labellum generally less than
   8 mm. in width (*D. incarnata*).   3
   Stem cavity medium or small (often less than ½ total diam.); leaves
   dark green often spreading, broadest above the base, ovate
   (linear in *D. traunsteineri*), apex not or only slightly hooded;
   flowers mostly large; labellum generally considerably greater than
   8 mm. in width.   6

3 Anthocyanin pigmentation totally absent from the plant; flowers
   creamy or yellowish-white.   **2. D. incarnata** ssp. **ochroleuca**
   Anthocyanin pigment present in the flowers.   4

4 Flower colour purple, magenta, ruby or crimson-red (to maroon on
   drying).   **2. D. incarnata** ssp. **pulchella**
   Flower colour flesh-pink.   5

5 Labellum small (not over 8 mm. in width, not over 7 mm. in length);
   leaves usually 4 or 5 in number; plant up to 40 cm.
                          2. D. incarnata ssp. **incarnata**
   Labellum large (more than 9 mm. in width, more than 7 mm. in
   length); leaves 6 or more; plant often greatly exceeding 40 cm.
                          2. D. incarnata ssp. **gemmana**

ORCHIDACEAE

6 Leaves very narrow (mostly not over 1·5 cm. in width) lanceolate or linear-lanceolate, few in number (usually 3 or 4, more rarely 5), often marked with small transverse purple or brownish bars, never with rings or blotches; inflorescence often lax, few-flowered; labellum deltoid or obcordate. **4. D. traunsteineri**
Lower leaves usually broad (more than 1·5 cm.) ovate or lanceolate, generally 6 or more in number, usually (in Cambs) unspotted; inflorescence usually many-flowered and dense; labellum elliptical, obscurely 3-lobed with angular incised lateral lobes.
3. D. praetermissa

1. Dactylorchis fuchsii (Druce) Vermeul. ssp. **fuchsii**   Ray, 1660
Orchis maculata sensu Bab.; *Orchis fuchsii* Druce
In damp woods and pastures mainly on the chalk and clays. 24, 25, 34, 35, 43–5, 54–6, 64–6. DF. The following have been recorded as *Orchis maculata* and presumbably refer to *D. fuchsii*, as *D. maculata* ssp. *ericetorum* (E. F. Lint.) Vermeul. has not been recorded from the county. *36, 46*, [23, 38, 48, 49, 58, 67, 40]. Throughout most of Britain on base-rich soils.

× **praetermissa** (*Orchis mortonii* Druce). This hybrid has been recorded with certainty only from Madingley in 1826 and Thriplow in 1954. 44, [36].

2. Dactylorchis incarnata (L.) Vermeul.   ?Ray, 1660
First certain record, Babington, 1852
Orchis latifolia L.; *O. incarnata* L.; *O. strictifolia* Opiz
Locally frequent in fens and wet meadows. The Wicken Fen population consists mainly of plants referable to ssp. **incarnata** and ssp. **pulchella** (Druce) H.-Harrison f. Ssp. **ochroleuca** (Böll.) H.-Harrison f. also occurs there, and a plant in CGE (1900) has been queried as ssp. **gemmana** (Pugsl.) H.-Harrison f. Specimens in CGE from Quy (1825) and Devil's Dyke (1952) have been queried as ssp. **gemmana** and ssp. **ochroleuca** respectively. 44, 45, 55–7, 66, 30, [25, 49, 54]. W. General.

3. Dactylorchis praetermissa (Druce) Vermeul.   ?Ray, 1660
First certain record, H. C. Gilson, 1941
Orchis latifolia sensu Bab. pro parte; *O. praetermissa* Druce
Locally frequent in a few localities in marshy places. Flowers 2–3 weeks later than *D. incarnata*. A white variety occurs at Thriplow. 24, 34, 35, 44, 45, 55, 66. T. Chiefly in fens in S., C. and E. England and Wales.

242

4. Dactylorchis traunsteineri (Sauter) Vermeul.　　C. E. Moss, 1913

Specimens in CGE from Dernford Fen (1913) and Chippenham Fen (H. Godwin, 1951) have been referred to this species by Heslop Harrison. Another specimen collected by Godwin in 1951, at Chippenham, is either this species or *D. praetermissa* crossed with *D. fuchsii*. The population at Chippenham in 1961 consisted of numerous individuals of *D. traunsteineri* and its hybrids with *D. praetermissa*. In 1962 *D. praetermissa* was also seen. 66, [45]. Local in England and Ireland but distribution imperfectly known.

ACERAS R.Br.

1. Aceras anthropophorum (L.) Ait. f.　　Man Orchid　　Relhan, 1785

A very rare plant of chalk grassland. Known to Babington (1860) in several localities but only seen recently at Haslingfield and Abington (see plate 6). 45, 54, [35, 55, 65, 66]. Rare on chalk in S., E. and C. England. Now very rare north of the Thames.

ANACAMPTIS Rich.

1. Anacamptis pyramidalis (L.) Rich.　　　　　　Ray, 1660

Orchis pyramidalis L.

A local species of old grassland and banks mainly on the chalk. 23, 24, 26, 35, 44, 45, 54–6, 66, [36, 58, 65, 40]. D. In suitable habitats throughout Great Britain north to Fife, and Ireland.

ARACEAE

ACORUS L.

1. Acorus calamus L.　　Sweet Flag　　　　Ray, 1690

In a few scattered localities by rivers and lodes in the Fens. The asymmetrical midrib of the long linear leaves, and the unmistakable sweet scent when bruised, characterize this plant. 37, 47, 57, 58, 68, 69, [45, 46, 56, 59]. R. Introduced into Europe by 1557 and recorded as naturalized in England by 1660. Now scattered throughout the British Isles. Native of S. Asia and central and western N. America. (See plate 8.)

ARACEAE

ARUM L.

1. Arum maculatum L.　Lords and Ladies; Cuckoo Pint　Ray, 1660
Throughout the county in woods, copses and hedgerows. Plants occur
with spotted leaves, var. **maculatum**, and with unspotted leaves, var.
immaculatum Mutel, the latter being probably the more frequent. Plants
with a yellow, not purple, spadix have been recorded from Cherry
Hinton and Wandlebury. 24–6, 34–6, 38, 39, 43–7, 49, 54–8, 64–6, 75,
76, 20, [48, 40]. Throughout England, Wales and Ireland, rather local
in Scotland and perhaps only native in the south and east.

LEMNACEAE

LEMNA L. (Duckweed)

1. Lemna polyrhiza L.　　　　　　　　　　　　　　I. Lyons, 1763
Occasional in the Fens in still waters, rare elsewhere. Readily distinguished from *L. minor* by its larger size (5–8 mm.) and numerous
rootlets. 29, 37, 36, 39, 45, 46, 48, *56*, 57–9, 20, 30, [25, 35, 38, 47, 49].
England except the extreme north and south-west; local in Wales; rare
in S. Scotland, and in Ireland from Wexford to Down and west to
Limerick and Clare.

2. Lemna trisulca L.　　　　　　　　　　　　　　　Ray, 1660
Throughout the county in still waters, but most frequent in the Fens.
The only submerged species. 29, 35–9, 43–9, 55–9, 65, 68, 76, 20, 30,
40, 41, 50, [25, 64, 66]. Britain north to Caithness; Ireland except in
the extreme north-west and south-west.

3. Lemna minor L.　　　　　　　　　　　　　　　　Ray, 1660
The commonest species, occurring in still waters throughout the county.
All squares except 66. General except for the extreme north, but
generally absent from acid waters.

4. Lemna gibba L.　　　　　　　　　　　　　　　　Relhan, 1785
Scattered throughout the county in still waters but much less frequent
than *L. minor*; resembling the latter in size but with spongy swollen
tissue beneath. 29, 35–9, 45, 46, 48, 49, 55–9, 20, 30, 40, 41, 50, [25, 44,
47]. C. and S. England, rare in Wales; in Ireland scattered and mainly
in the east.

SPARGANIACEAE

SPARGANIACEAE

SPARGANIUM L. (Bur-reed)

1. Sparganium erectum L.   Ray, 1660

S. ramosum Huds.

Throughout the county in ditches, rivers, lodes and pits. Four varieties are to be found in Cambs, and are distinguished on fruit characters. Var. **erectum** has the fruits cuneate-obpyramidal, dark brown to black above with a distinct shoulder 6–8(–10) mm. long (excluding style), 4–6 mm. wide at shoulder and upper part flattened. Var. **microcarpum** (Neuman) Glück is similar but has the fruits obpyramidal with a rounded apex, smaller, 6–7(–8) mm. long, 2·5–4·5 mm. wide, upper part domed and wrinkled below the style. Var. **neglectum** (Beeby) Fiori & Paoletti has the fruits with an indistinct shoulder, upper and lower parts uniform, shiny, light brown, ellipsoidal, 7–9 mm. long and 2–3·5 mm. wide. Var. **oocarpum** Čelak. is similar but has the fruits ± spherical, 5–8 mm. long and 4–7 mm. wide. All squares except 24, 26, 64, [54]. General.

2. Sparganium emersum Rehm.   Ray, 1660

S. simplex sensu Bab., et Evans

Throughout the county in ditches, rivers, lodes and pits, most frequent in the Fens. 29, 34, 37, *44*, 45, 46, 48, 49, 55–9, 20, 30, [35, 38, 47, 66, 67]. General.

3. Sparganium minimum Wallr.   Relhan, 1788

Formerly in a number of localities in the Fens, but recently recorded only from a ditch by Wicken Fen. 57, [29, 44, 47, 55, 56, 58, 66]. **W.** Scattered throughout the British Isles in peaty waters.

TYPHACEAE

TYPHA L. (Reed-mace)

1. Typha latifolia L.   Ray, 1660

Throughout the county in ditches, marshes and lodes and by the margins of pits and rivers. All squares except 24, 26, 64, 67, [54]. **W.** General but less frequent in the north.

TYPHACEAE

2. Typha angustifolia L.   Relhan, 1785

Occasional in pits, fens and lodes, not so frequent as *T. latifolia*. 29, 34, 36–8, *45*, 46–9, 56, 58, 40, 41, [25, 35, 44, 57, 66]. **W.** Locally common in parts of Britain, rare and local in Scotland and Ireland.

(There is a specimen of **T. minima** Hoppe in CGE labelled as collected by Shrubbs at Wicken Fen in 1893. This species has never been recorded wild in Britain; it is very probable that the sheet is labelled incorrectly, and the plant could have originated from the University Botanic Garden.)

CYPERACEAE

ERIOPHORUM L. (Cotton-grass)

1. Eriophorum angustifolium Honck.   S. Corbyn, 1656

Formerly in several scattered localities but now reduced by drainage and cultivation to relic populations at Sawston and Teversham. A large stand was found on the wet floor of a chalk-pit at Barrington in 1960. 35, 44, 55, *56*, [25, 45, 49, 66]. General in bogs and fens, now rather rare in lowland England.

2†. Eriophorum latifolium Hoppe   J. Holme in J. E. Smith, 1824

Formerly in a few localities in fens and marshy places. Last seen at Chippenham Fen in 1887. [44, 45, 66.] Scattered throughout the British Isles but local; generally calcicolous.

SCIRPUS L.

1†. Scirpus cespitosus L.   I. Lyons, 1763

Trichophorum cespitosum (L.) Hartm.

Recorded by Lyons and Relhan from near Stourbridge, by Relhan from Hinton Moor, and from Wicken Fen by Babington and H. Dixon. The Wicken record may be a misidentification of *Eleocharis quinque-flora*, q.v. [45, ?56.] Scattered throughout the British Isles but absent from base-rich soils.

2. Scirpus maritimus L.   Ray, 1660

Frequent by ditches and rivers around Wisbech, and by the Washes near Earith. Formerly in a number of localities by rivers in the Fens as far south as Stretham. 37, 30, 31, 40, 41, 50, [29, 38, 39, 47–9, 57, 58]. Round the coasts of the British Isles except the extreme north, in ditches and ponds and by the margins of tidal rivers; rarely inland.

3. Scirpus lacustris L. Ray, 1660
Schoenoplectus lacustris (L.) Palla
Throughout most of the county in pits, rivers, ditches and lodes in fairly deep water. 25, 34–7, 44–8, 55–8, 20, 30, [29, 38, 54, 66, 40]. General.

4. Scirpus tabernaemontani C. C. Gmel. T. Martyn, 1763
Schoenoplectus tabernaemontani (C. C. Gmel.) Palla
In a few scattered localities in ditches, rivers, pits and marshy places, mostly in the Fens, but south to Thriplow. Differs from *S. lacustris* in having glumes papillose on the back, and usually 2 stigmas, whereas *S. lacustris* has smooth glumes and usually 3 stigmas. 34, 39, 44, 55, 56, 58, 40, *41*, [29, 35, 38, 45, 48, 49]. Scattered throughout the British Isles especially near the sea.

5. Scirpus setaceus L. Ray, 1685
Isolepis setacea (L.) R.Br.
In a few places in short, damp (winter-wet) turf. Usually associated with *Juncus bufonius*. 25, 44, 45, 66, [38, 48, 56, 57]. G. Recorded for every county in the British Isles except Huntingdon.

6. Scirpus fluitans L. Henslow, 1835
Eleogiton fluitans (L.) Link
A rare species of drains and ditches mostly in the Fens. Still found at Blackbush Drove near Whittlesey where it was recorded by Babington (1860). *29, 49, 56*, 20, [25, 57]. Throughout most of the British Isles but local.

ELEOCHARIS R.Br.

Key to *Eleocharis* and the smaller species of *Scirpus* (i.e. species with a single terminal spike).

1 Floating or submerged aquatics with slender filiform stems. 2
  Marsh or reed-swamp plants; stems often stout. 3
2 Stems branched, rooting at nodes, with linear leaves; spikes produced above water surface. **S. fluitans**
  Stems unbranched, leafless, from a slender rhizome; spikes rarely produced when plants growing in water. **E. acicularis**
3 Stems slender with 1 or 2 well-developed filiform leaves; plant usually annual. **S. setaceus**
  Stems leafless or with a tiny lamina on uppermost sheath; plant perennial. 4

CYPERACEAE

4 Stems filiform; spike not more than 4 mm. **E. acicularis** (land form)
   Stems not filiform, slender or rather stout; spike 5 mm. or more.   5

5 Plant with thick creeping rhizome; stigmas 2; nut ± biconvex.   6
   Plant densely caespitose or spreading by offsets on delicate underground stolons; stigmas 3; nut ± triangular in section.   8

6 Single sterile glume practically encircling base of spikelet.
                                                              E. uniglumis
   Two subequal sterile glumes at base of spikelet.   (*E. palustris*) 7

7 Fruit (excluding style-base), *c.* 1·5 mm., middle glumes more than
   3·75 mm.; stomatal length 0·07–0·08μ.   **E. palustris** ssp. **vulgaris**
   Fruit (excluding style-base), *c.* 1·25 mm.; middle glumes less than
   3·75 mm.; stomatal length 0·055–0·06μ.
                                        E. palustris ssp. **microcarpa**

8 Uppermost sheath with a small (2–3 mm.) lamina (acid bog and
   heath, extinct in Cambs).   **S. cespitosus**
   Uppermost sheath with no trace of lamina.   9

9 Plant somewhat caespitose, producing offsets on very slender
   stolons; basal glume more than half as long as spikelet.
                                                          E. quinqueflora
   Plant densely caespitose, offsets 0; basal glume less than ¼ length of
   spikelet (acid bog and heath, extinct in Cambs).   **E. multicaulis**

1. Eleocharis acicularis (L.) Roem. & Schult.

J. Holme in Relhan, 1785

Occasional in lodes, ditches and rivers throughout the Fens. It is most
often seen completely submerged in the vegetative state, sometimes
covering the bottom. It fruits freely when exposed on mud dredged up
from the ditch during summer cleaning, or when the water-level falls
sufficiently to expose the plant in late summer. 38, 47, 48, 56, 68, 20,
[29, 37, 57, 58, 67]. **W.** Scattered throughout the British Isles.

2. Eleocharis quinqueflora (F. X. Hartmann) Schwarz   Relhan, 1785
E. pauciflora (Lightf.) Link; *Scirpus pauciflorus* Lightf.

Formerly known in a few localities in fens and marshy places, now
only on the Main Drove at Wicken Sedge Fen where it was first recorded
by S. M. Walters in 1945. 56, [25, 44, 45]. **W.** Scattered throughout the
British Isles but commoner in the north; usually in open base-rich
marshy ground.

3†. **Eleocharis multicaulis** (Sm.) Sm.   Relhan, 1785
Recorded from Ickleton by G. S. Gibson (in Babington, 1860), from
Gamlingay (where it was known until 1860) and from Sheep's Green,
Cambridge, by H. Dixon in 1881. Only the Gamlingay locality, from
which we have seen a specimen, can be regarded as certain. [25, ?44,
?45.] Scattered throughout the British Isles but mainly in the south
and west; strictly calcifuge in wet heath and bog communities.

4. **Eleocharis palustris** (L.) Roem. & Schult.   Ray, 1660
Throughout the county in wet and marshy places. The most frequent
plant is ssp. **vulgaris** S. M. Walters, but ssp. **microcarpa** S. M. Walters
occurs. 25, 29, 35–7, 39, 44–8, 55–9, 66, 67, 30, 40, [38, 49]. General.

5. **Eleocharis uniglumis** (Link) Schult.   A. Fryer, 1880
Rare, in wet peaty ground. Occurs on the Main Drove at Wicken
Sedge Fen, together with *E. palustris* and *E. quinqueflora*. The Wicken
population is variable and cytologically heterogeneous; plants with
haploid chromosome numbers of 23, 31–7 and about 46 have been
found there, those with about 46 having significantly large fruits. *45,
55, 56, 59,* [48]. **W.** Locally abundant in dune slacks and brackish
marshes near the sea; rather rare inland in calcareous marshes and fens.

BLYSMUS Panz.

1. **Blysmus compressus** (L.) Panz. ex Link   I. Lyons, 1763
A rare plant of boggy meadows which has been lost in its old localities
but found in three new ones. There is an old record from near Wisbech
which may be in the county. Easily confused with small plants of
Carex disticha. Blysmus has spikelets strictly in two rows (i.e. di-
stichous), whereas this is not true of the *Carex* despite its name. The
bristles surrounding the fruit of *Blysmus* are easily seen with a lens;
of course they are not present in *Carex.* 44, 45, 56, [?40]. **T.** Scattered
throughout England, N. Wales and S. Scotland, mainly on base-rich
soils.

SCHOENUS L.

1. **Schoenus nigricans** L.   Bog-rush   Ray, 1660
A rare plant of fens and peaty places, formerly more frequent. 44, 56,
66, [25, 34, 45, 46, 55, 58]. **C.** Throughout the British Isles in peaty
places, especially near the sea, and sometimes in salt-marshes; abundant
in W. Scotland and Ireland.

249

CYPERACEAE

RHYNCHOSPORA Vahl

1†. Rhynchospora alba (L.) Vahl   Ray, 1685

Used to occur in the bogs at Gamlingay Heath where it was last seen by Babington before 1860. [25.] Scattered throughout the British Isles in acid bogs; rare in the east.

CLADIUM Browne

1. Cladium mariscus (L.) Pohl   Sedge   Ray, 1660

In a few fens and peaty places, locally abundant or dominant; the plant was probably widespread and abundant in the ancient fens. It was considered to be a valuable natural crop and used for the thatching of houses and barns, fire lighting, and, according to Evans (1939), was cut up as chaff for horse-fodder. It is still abundant at Wicken Fen, where the poor of the village still have the right to cut 'sedge' and 'litter' on the Poor's Fen, on the third Monday in July. In its few other extant localities, it is much reduced. 44, 45, 55, 56, 66, [34, 38, 48, 49, 58]. W. Thinly scattered over the British Isles, but frequent only in Norfolk and W. and C. Ireland.

CAREX L.

'Fruit' as used in this key means the nut and the investing perigynium i.e. utricle.

1   Spike solitary, terminal.   *2*
    Spikes more than one.   *3*

2   Dioecious (male spikes sometimes with one female flower at base
     (f. *isogyna*)).   **41†. C. dioica**
    Monoecious, spikes with 3–10 female flowers at base.
                           40. C. pulicaris

3   Spikes obviously of two kinds; the upper all or predominantly
     male, the lower all or predominantly female.   *4*
    Spikes more or less similar in appearance, most or all of them con-
     taining both male and female flowers.   *30*

4   Fruits hairy.   *5*
    Fruits glabrous.   *10*

5   Fruits more than 5 mm. long; leaf sheaths always hairy at top.
                           18. C. hirta
    Fruits not over 5 mm. long; leaf sheaths always glabrous.   *6*

6   Leaves glaucous at least beneath; female spikes medium- or long-
     peduncled.   **17. C. flacca**
    Leaves not glaucous beneath; female spikes sessile, or, if peduncled,
     then peduncles entirely included in sheathing part of bract.   *7*

7   Fen plant, more than 40 cm. tall; leaves very narrow, channelled.
                                          19. C. lasiocarpa
    Plant of dry grassland and heath, not over 35 cm. tall; leaves flat,
    or folded.   *8*

8   Female glumes dark purple-brown, rounded at the apex; female
    bracts dark purple-brown, glumaceous or with a setaceous
    point.   **21. C. ericetorum**
    Female glumes brown, acute to acuminate at the apex; female
    bracts light brown or greenish, long-pointed.   *9*

9   Lowest bract shortly sheathing.   **22. C. caryophyllea**
    Lowest bract not sheathing.   **20. C. pilulifera**

10 Ripe fruits flattened, stigmas usually two.   *11*
    Ripe fruits trigonous or inflated, stigmas usually three.   *13*

11 Plant forming large dense tufts; bracts usually setaceous; lower
    sheaths leafless, the margins becoming filamentous in decay.
                                              23. C. elata
    Plants shortly creeping; bracts leaflike; lower sheaths not as above. *12*

12 Rather slender plant, leaves usually not over 3 mm. wide; male
    spike usually solitary.   **25. C. nigra**
    Robust plant; leaves usually more than 3 mm. wide; male spikes
    usually 2–4.   **24. C. acuta**

13 Fruits beakless, or with a short entire beak not over 0·5 mm.   *14*
    Fruit with obvious beak.   *16*

14 Leaves hairy beneath, green.   **15. C. pallescens**
    Leaves glabrous, usually glaucous.   *15*

15 Male spikes usually more than one; fruits many, closely packed
    together; lowest bract with little or no sheath.
                                              17. C. flacca
    Male spike one; fruits not over 20 in a spike, loosely arranged;
    lowest bract with obvious sheath.   **16. C. panicea**

16 Female spikes not more than 20 mm. long.   *17*
    Female spikes more than 20 mm. long.   *22*

17 Fruit not more than 3 mm. long; female spikes usually crowded
    below male spike (part of *C. flava* agg.).   **6. C. serotina**
    Fruit more than 3 mm. long; female spikes usually some distance
    below male spike.   *18*

18 Female spikes ovoid, subsessile or the lowest long-stalked; bracts
    longer than inflorescence (part of *C. flava* agg.).   *19*
    Female spikes cylindrical, with fairly long stalks; bracts seldom
    exceeding inflorescence.   *20*

251

CYPERACEAE

19 Fruit half-elliptic, ± curved, beak 1·5–2 mm. long.
                                                    4. C. lepidocarpa
      Fruit broadly elliptic, straight, beak *c*. 1 mm. long.   **5. C. demissa**

20 Male and female glumes with broad, silvery hyaline margins.
                                                    2. C. hostiana
      Male and female glumes without hyaline margins or with, very
          narrow ones.   *21*

21 Plant shortly creeping; female glumes dark purple-brown; male
          spike ± clavate.   **3. C. binervis**
      Plant tufted; female glumes brownish; male spike cylindrical.
                                                    1. C. distans

22 Male spike one.   *23*
      Male spikes two or more.   *27*

23 Female spikes not more than 4 mm. in diameter.   *24*
      Female spikes more than 5 mm. in diameter.   *25*

24 Leaves 6–10 mm. wide; female spikes 1·5–2 mm. wide; fruit
          3·5 mm. long; beak 0·2 mm. long, emarginate. **14†. C. strigosa**
      Leaves 3–6 mm. wide; female spikes 3–4 mm. wide; fruit 4–5 mm.
          long; beak 1–1·5 mm. long, bifid.   **7. C. sylvatica**

25 Leaves 2–5 mm. wide.   **3. C. binervis**
      Leaves more than 5 mm. wide.   *26*

26 Female spikes 30–50 × *c*. 10 mm.   **8. C. pseudocyperus**
      Female spikes 70–160 × 5–7 mm.   **13. C. pendula**

27 Male spikes usually 4–6; fruit *c*. 8 mm. long. **11. C. riparia**
      Male spikes usually 2–3; fruit 6 mm. long or less.   *28*

28 Leaves more than 7 mm. wide; beak of fruit emarginate, not bifid.
                                                    12. C. acutiformis
      Leaves not more than 7 mm. wide; beak of fruit bifid.   *29*

29 Fruit 4–5 mm. long, gradually narrowed into the beak; female
          glumes long-acuminate.   **10. C. vesicaria**
      Fruit 5–6 mm. long, abruptly contracted into the beak; female
          glumes shortly acute.   **9. C. rostrata**

30 Stems not tufted, rhizome creeping.   *31*
      Stems tufted, not creeping.   *33*

31 Uppermost spike usually entirely male; fruit broadly winged in the
          upper part; plant of open sandy ground.   **31. C. arenaria**
      Uppermost spike female or mixed; fruit narrowly winged or wing-
          less; plant of grassy, peaty or muddy places.   *32*

32 Fruit usually narrowly winged, 4 mm. or more in length.
                                        30. C. disticha
    Fruit not winged, not over 3·5 mm. in length.   **28†. C. diandra**

33 Slender plant with widely spaced, small spikes in the axils of long
       bracts.   **37. C. remota**
    Not as above.   *34*

34 Plant forming large tussocks (young growth on 'pedestal' of dead
       remains).   *35*
    Plant not forming large tussocks.   *36*

35 Leaves 3–7 mm. wide.   **26. C. paniculata**
    Leaves 1–2 mm. wide (Wicken Fen).   **27. C. appropinquata**

36 At least some spikes with male flowers at top; common species.   *37*
    At least the upper spikes with male flowers at base; rare calcifuge
       species.   *40*

37 Leaves more than 4 mm. wide; stem stout; plant of wet places.
                                        29. C. otrubae
    Leaves 4 mm. wide or less; stem slender; plant of dry places
       (*C. muricata* agg.).   *38*

38 Lower leaf-sheaths, and often the glumes and bracts, tinged
       reddish-vinaceous; ligule about 5 mm. long; fruits with a broad
       swollen base.   **34. C. spicata**
    Lower leaf-sheaths, glumes and bracts sometimes pinkish but
       never reddish-vinaceous; ligule not more than 2 mm. long;
       fruits narrower at base.   *39*

39 Inflorescence 40–140 mm. long; lowest bract longer, often very
       much longer, than its spike; lowest nodes of inflorescence
       15–50 mm. apart; beak of fruit 1–1·5 mm. long.   **32. C. divulsa**
    Inflorescence 30–70 mm. long; lowest bract shorter than or some-
       times longer than its spike; lowest nodes of inflorescence not
       over 25 mm. apart; beak of fruit *c.* 2 mm. long.
                                       33. C. polyphylla
    Inflorescence 10–35 mm. long; lowest bract shorter than or longer
       than its spike; lowest nodes of inflorescence not over 8 mm.
       apart; beak of fruit 1–2 mm. long.   **35. C. muricata** (*C. pairaei*)

40 Fruits spreading stellately, not over 10 in a spike.   **36. C. echinata**
    Fruits erect or erectopatent, more than 10 in a spike.   *41*

41 Spikes pale greenish white; fruit 2–3 mm. long.   **38†. C. curta**
    Spikes brownish; fruit 3·5–5 mm. long.   **39. C. ovalis**

1. Carex distans L.   I. Lyons, 1763
Local in wet meadows and fens, usually on peat. 34, 44, 45, *46*, 55–7,
66, [25, 35, 36, 54, ?40, 41]. W. General, but local; confined to basic
habitats inland, and to coastal marshes.

CYPERACEAE

2. Carex hostiana DC.   Relhan, 1802
C. fulva sensu Bab.

In a few marshy localities. This plant has in the past been confused
with *C. distans* which seems to be the commoner species in Cambs.
34, 44, *45*, 55, 56, 66, [35, 46]. C. General.

× **lepidocarpa** (*C.* × *fulva* Gooden). This hybrid formerly grew at
Shelford Common, and still occurs at Chippenham Fen where it grows
with the parents. 66, [45]. C.

3. Carex binervis Sm.   First certain record, Henslow, 1825

Three sheets in CGE from Shelford Common (1825), Little Eversden
(1854) and Chesterton Ballast Pits (1930) are referable to this species.
Recently found in wet meadows on part of the old 'Whittlesford
Moor'. All other records must be regarded as questionable. 44, *46*,
[?25, ?34, 35, 45, ?54, ?56]. General but usually on acid heath or
moorland.

4. Carex lepidocarpa Tausch   ?Ray, 1660.
                                                 First certain record, Henslow, 1825
C. flava sensu Bab.

A local plant of base-rich marshes and fens. 44, 45, 55–7, 66, [25, 38,
46, 48, 49]. W. General.

5. Carex demissa Hornem.   C. E. Moss, 1909
?*C. oederi* sensu Bab. pro parte; ?*C. oederi* var. *oedocarpa* sensu Evans

There are specimens in CGE from Adventurers' Fen (1909), and also
from Gamlingay Cinques (1913) where it still occurs. 25, [56]. G. A
calcifuge species found throughout the British Isles.

6. Carex serotina Mérat   Relhan, 1802
C. oederi sensu Bab., et Evans pro parte

Formerly in a few localities in fens; now known only at Wicken Fen,
where it is characteristic of disturbed peaty ground where there is no
competition from taller-growing species. The Gamlingay plant recorded
as *C. oederi* was probably *C. demissa*. 55, 56, [?25, 38, 44, 45, ?46, 47,
?48, ?57, 66]. W. Scattered throughout the British Isles, generally in
base-rich habitats and often on bare, wet limestone.

7. Carex sylvatica Huds.   Ray, 1660
Frequent in woods, mainly on the boulder clay. 25, 26, 34–6, 43–6, 54,
55, 64, 65, 75, [49]. B. General, but rare in the north-east.

8. Carex pseudocyperus L.   Ray, 1670

By rivers, ditches, pits and in marshy places; formerly in a number of
scattered localities, mostly in the Fens, now known only at Whittlesford,
Burwell and several places near Earith. 37, 44, *46*, 47, 56, [25, 35, 45,
48, 54, *55*, 57, 58, 65, 66, 40]. Scattered throughout England, Wales
and Ireland; in Scotland recorded from Moray.

9. Carex rostrata Stokes   Relhan, 1793

C. ampullacea Gooden.

Formerly in a number of widely scattered localities mostly in the Fens.
Last seen at Quy Fen in 1949. *25, 56*, [44, 45, 48, 58, 66, 67, 40].
General, but not recorded from Middlesex, Huntingdon and the
Channel Islands.

10. Carex vesicaria L.   Mr Newton in Relhan, 1793

Formerly in a number of scattered localities. Specimens in CGE from
Stretham Ferry (1833), Chittering (1932) and Cottenham (1933) are
certainly correct, but some of the field records may well be misidentifica-
tions of *C. rostrata* . It was last seen by C. D. Pigott at Aldreth in 1948.
46, 47, 57, [29, 35, 45, 48, 56, 58, 40]. General.

11. Carex riparia Curt.   Ray, 1660

Common by ditches, rivers, ponds, etc., throughout the county. The most
abundant sedge in the Fenlands. All squares except 26, [54, 66].
General, but rare or absent in the north and west.

12. Carex acutiformis Ehrh.   J. Martyn, 1763

C. paludosa Gooden.

Common by ditches, lodes, rivers and pits, and in marshy places, but
generally less abundant than *C. riparia*. 25, 29, 34, *35*, 37, 39, 44, 45, 47,
48, 54–7, 59, 66, 20, [38, 46, 49, 58, 40]. General, but rare in the north
and west.

13. Carex pendula Huds.   Ray, 1663

A local plant of damp woodland rides and the sides of ditches. The
tallest British *Carex*, easily distinguished by its extremely broad leaves
even in the vegetative state. 34, *35*, 36, 43, 44, 48, 54, 64, 65, 75, [55].
DPW. Scattered throughout the British Isles but absent from
N. Scotland.

CYPERACEAE

14†. Carex strigosa Huds.   Relhan, 1802

The only record is from Hall Wood, Wood Ditton, by Relhan. It occurs just outside the county in Pondbottom Wood, Barley, Herts. [65.] Scattered through England, Wales and Ireland.

15. Carex pallescens L.   Relhan, 1793

A rare plant of woods mainly on the boulder clay. 35, 43, 54, *64*, 65, [25, 47, 56]. **DPW**. General, but often local.

16. Carex panicea L.   J. Martyn in I. Lyons, 1763

A local plant of damp grassy places. *25*, 34, 44–6, *49*, 55–7, 66, [35, 38, 47, 48]. **W**. General.

17. Carex flacca Schreb.   Relhan, 1785

C. glauca Scop.

Throughout the county in all types of grassland, scrub, marshes and fens. Very variable in size and colour of inflorescence. Readily distinguished from the only other sedge with obviously glaucous leaves, *C. panicea*, by the sheathless lower bract, the smaller more tightly packed fruits, and the presence of more than one male spike. In the vegetative state, the reddish colour of the basal sheaths of *C. flacca* contrasts with the pale brown or buff of *C. panicea* and usually provides a safe means of identification. 24–6, 29, 34–6, 38, 39, 43–6, *47*, 49, 54–8, 64–7, 20, 41, [48]. General.

18. Carex hirta L.   I. Lyons, 1763

Roadsides, waste places, grassland and marshes throughout the county. Easily distinguished from any other common sedge by the usually obvious hairiness of the sheathing part of the leaf. All squares except 26, [24]. General, but absent from the far north.

19. Carex lasiocarpa Ehrh.   Relhan, 1785

C. filiformis sensu Bab.

Formerly known in a number of places in marshes and fens, now only at Wicken and until recently at Quy. 56, [25, 38, 39, 45, 47, 49, 55, 66, 67]. **W**. General; local in the north, very rare in the south.

20. Carex pilulifera L.   I. Lyons, 1763

Only known with certainty from Gamlingay, where it still occurs. An old field record from Hildersham Furze Hills (1883) may be correct, but records from the chalk, unsupported by herbarium sheets, are unlikely. The species appears to be rather strictly calcifuge, at least in eastern England. 25, [?45, ?54, ?55, ?66]. **G.** General.

21. Carex ericetorum Poll.   J. Ball, 1838

A very rare plant of old chalk grassland. First recorded as a British plant by Babington (1863) but collected on the Roman Road as early as 1838 by J. Ball, and identified by him tentatively as *C. ericetorum*. Letters from Ball to Babington in 1860–1, preserved in the University Herbarium, show that it was the stimulus of Babington's new *Flora of Cambridgeshire* which reminded Ball of the neglected sedge, new to Britain, lying in his herbarium. *65*, 66, [45]. **D.** A rare plant occurring in East Anglia, Lincs, and scattered localities in northern England, a distribution which resembles that of *Linum anglicum*.

22. Carex caryophyllea Latourr.   Ray, 1660
C. praecox sensu Bab.

A local plant of dry grassland mainly on the chalk and sands. This and *C. flacca* are the only common sedges of the chalk. 44, 45, 54–6, 66, [23, 25, 34, 35, 46, 49, 58, 76, 40]. **GG.** General, but mainly on dry grassland.

23. Carex elata All.   Ray, 1660
C. stricta Gooden., non Lam.

By ditches, lodes, rivers and pits mainly in the south of the county. 34, 44, 45, 55–8, 66, [36, 38, 46, 48, 49]. **W.** Scattered throughout England and Ireland; very rare elsewhere.

24. Carex acuta L.   Ray, 1670
C. gracilis Curt.

Widespread but local, by ditches, lodes, pits and rivers in the Fens, rare elsewhere. 34, 35, 37, 44, 45, *47*, 56, *57*, 58, 67, [29, 36, 38, 48, 55, 66, 40]. **W.** General, but local.

25. Carex nigra (L.) Reichard   Relhan, 1802
C. vulgaris Fries; *C. goodenowii* Gay

A rare plant of marshy places, formerly more widespread. 25, 44, 45, 56, 59, 67, [29, 34, 48, 55, 65, 66]. **G.** General. Cambs, Hunts and Northants are the counties where this sedge can be considered rare.

(A specimen in CGE, collected in a pond near Wicken in 1932 by J. S. L. Gilmour, was named **C. trinervis** Degl. by E. Nelmes in 1956. This specimen

CYPERACEAE

does not seem to match Norfolk material very well and both it and the Norfolk plant differ considerably from the continental plant. It is closely allied to the very variable *C. nigra*.)

26. Carex paniculata L. J. Martyn in I. Lyons, 1763

In a few scattered localities in marshy places and by ditches, absent from Wicken, where *C. appropinquata* replaces it. *44, 55*, 56, 67, 68, [25, 34, 45, 58, 66, 40]. Scattered throughout the British Isles; rare in Scotland.

27. Carex appropinquata Schumach. A. Fryer, 1885

C. paradoxa Willd., non J. F. Gmel.

Scattered throughout Wicken Fen but not known elsewhere. Both this species and *C. paniculata* occur together in several fens on the Breckland margin in W. Suffolk, as at Icklingham and Wangford. 56. W. Rare plant of mid- and east England, and in two counties in both Scotland and Ireland.

28†. Carex diandra Schrank. Relhan, 1802

C. teretiuscula Gooden.

Recorded from Grantchester Meadows and Fulbourn by Relhan, and known from Thriplow Peat Holes in the first half of the nineteenth century, where it was last recorded in 1860. [44, 45, 55.] Scattered throughout the British Isles.

29. Carex otrubae Podp. Ray, 1660

C. vulpina sensu Bab., et Evans

One of our most widespread sedges by ditches and in marshy places. This species has long been included under *C. vulpina* sens. lat. *C. vulpina* L. sens. strict. has not yet been recorded from the county. All squares except 26, 54, 67, [24]. General, but in the west essentially a coastal species.

× remota (*C. pseudoaxillaris* K. Richt.; *C. axillaris* Gooden., non L.). A hybrid of characteristic appearance recorded from several localities but recently only from near Dullingham. *25*, ?*64*, *65*.

30. Carex disticha Huds. Lyons, 1763

Throughout the county in marshy places. 25, 29, 34–9, 44–9, 55–9, 66, 67, [30, 40]. General, but local in Scotland.

31. Carex arenaria L. Evans, 1939

Found only on the sands at Kennett Heath and Chippenham. Abundant on the Breckland just over the Suffolk border. 67, 76. All round the coasts of the British Isles; inland in Hants, Surrey, W. Suffolk, W. Norfolk, Cambs, Lincs, and S.E. Yorks.

The taxonomy of the following four species is very confused and awaits clarification. In addition, plants from Fordham, Sawston and the Gog Magog Hills are intermediate between *C. divulsa* and *C. polyphylla* and have been referred to a fifth species **C. chabertii** F. Schultz. The most up-to-date revision of the group is E. Nelmes (1947 for 1945).

32. Carex divulsa Stokes   I. Lyons, 1763
A local plant of roadsides and woods. *34, 36, 43, 45, 54, 55, 64, 65*, [*25, 26, 35, 38, 46, 49, 57*]. Scattered throughout England, Wales and S. Ireland.

33. Carex polyphylla Kar. & Kir.   A. Fryer, 1879
?*C. divulsa* sensu Bab. pro parte
In a few grassy localities, mostly roadsides. 34, 36, 45, 54, 55, [38]. Scattered over southern Britain; possibly elsewhere, as distribution is not well known.

34. Carex spicata Huds.   Henslow, 1825
C. muricata sensu Bab. pro parte; *C. contigua* Hoppe
Frequent on roadsides and in other grassy places. This seems to be much the commonest of the sedges in the *muricata* group in the county. 25, 34–6, 43–6, *47*, 54–8, 66, [38, 48, 49, 65, 76, 41]. General, but rare in the north and in Ireland.

35. Carex muricata L.   E. F. Warburg, 1941
C. pairaei F. W. Schultz; ?*C. muricata* sensu Bab. pro parte
Grassy places in several localities on the sands. 25, 54, *66*, 67. FH. General. This is the commonest species of the *muricata* aggregate in S.W. Britain.

(**C. elongata** L. is recorded from the Washes of the Nene by H. Dixon in 1883. There is no specimen to support the record, which needs confirmation.)

·36. Carex echinata Murr.   Relhan, 1785
C. stellulata Gooden.
Previously recorded from several localities in the south of the county but now known only at Gamlingay. 25, [36, 45, 66]. G. General on wet acid heath and in bogs.

CYPERACEAE

37. Carex remota L. Ray, 1663

A local plant mostly of woods on the boulder clay. 25, 26, *35*, 43, 45, 54, 64, 65, [34, 36, 44, 49]. **DPW.** General, but rarer in Scotland and absent from the extreme north.

38†. Carex curta Gooden. Relhan, 1793
C. canescens sensu Evans

Formerly occurred at Gamlingay where it was last seen in 1853. [25.] Scattered throughout most of the British Isles on acid soils; commoner in the north and absent from the extreme south-west.

39. Carex ovalis Gooden. Relhan, 1786
C. leporina sensu Evans

A rare plant now only found at Hildersham Wood, and also at Gamlingay where it is abundant in the Cinques. There is an old record (1895) from Hildersham Furze Hills that is probably correct, but old records from the Fens are unsupported by herbarium sheets, and must be considered doubtful. 25, 54, [?44, ?46, ?56, ?40]. **G.** General on acid mineral soils.

40. Carex pulicaris L. Ray, 1670

Formerly found in several places in the south-eastern fens and still occurring at Chippenham and Quy. 56, 66, [44–6]. **C.** General.

41†. Carex dioica L. Ray, 1696

Formerly found in a few boggy places; last seen at Gamlingay in 1841. Some specimens from Gamlingay in CGE are monoecious, and are forma **isogyna** (Ångström ex Fr.) Kükenth. [25, 45, 66.] Scattered throughout the British Isles; frequent only in the north.

GRAMINEAE

PHRAGMITES Adans.

1. Phragmites communis Trin.   Reed Ray, 1660

Common in ditches, rivers, lodes, pits and marshes and in fens where it is sometimes dominant over large areas. It is similar in vegetative appearance to *Phalaris arundinacea*, with which it often grows, but can be distinguished by having a ring of hairs instead of a ligule. It is still cut and sold for thatching at Wicken Fen. All squares except 24, 26, 64. General, but often coastal in the north and west.

GRAMINEAE

MOLINIA Schrank

1. Molinia caerulea (L.) Moench   I. Lyons, 1763

In a few localities in fens and marshy places. Still locally abundant at Chippenham and Wicken Fens. Easily recognized by its tussocky habit and absence of ligule. 25, 44, 45, 55, 56, 66, [35, 47, 49]. W. Throughout the British Isles, often occurring abundantly, and frequently dominating large areas, especially in acid moorland in the north and west.

SIEGLINGIA Bernh.

1. Sieglingia decumbens (L.) Bernh.   I. Lyons, 1763

Triodia decumbens (L.) Beauv.

Rare in old grassland on the sands, the chalk and in the Fens; frequent in the Cinques at Gamlingay. The only other British grass without a ligule. 25, 44, *49*, 55, 56, 66, [35, 45–8, 76]. C. General.

GLYCERIA R.Br.

The following three closely allied species are all perennial herbs of marshy ground or shallow water, and are eagerly grazed by cattle.

1. Glyceria fluitans (L.) R.Br.   Ray, 1660

Fairly common by ditches, rivers and ponds throughout most of the county. *24*, 25, 29, *35*, 36–9, 43–9, 55–7, 59, 64, 65, 67–9, 20, 30, 40, *41*, 50. General.

× **plicata** (*G.* × *pedicellata* Townsend). This male-sterile hybrid, widely distributed in the British Isles, is frequent by streams and ponds, but seems to be absent from the northern Fens. Because of its powers of vegetative spread, this hybrid can occur in the absence of one or both parents. 34, 36–8, 44, 45, *47*, 48, 55–8, 64, [65].

2. Glyceria plicata Fr.   Babington, 1845

By ditches and ponds in a number of localities, mainly in the southern half of the county. 29, 35, 36, 44–6, 55–7, 59, 64, [38, 48, 49]. Generally distributed through England and extending into S. and E. Scotland; also in Wales and in widely scattered localities in Ireland; probably restricted to base-rich water.

GRAMINEAE

3. Glyceria declinata Bréb.   S. M. Walters, 1945

This small species of *Glyceria*, with a characteristic 3-toothed lemma and arcuate ascending habit of growth, was not distinguished by the earlier botanists. It occurs by several ponds and streams in the Cambridge area, particularly on muddy cattle-trampled ground. 35, 36, 44–6. CC. Scattered throughout the British Isles, but detailed distribution still not worked out, though apparently commoner than *G. plicata* in the north and west.

4. Glyceria maxima (Hartm.) Holmberg   Ray, 1660
G. aquatica (L.) Wahlb., non J. & C. Presl

In ditches, rivers and pits throughout the county. This species, *Phragmites* and *Phalaris arundinacea* are the three common 'Reeds' of the Cambridgeshire countryside. It is readily distinguished by its bright green, parallel-sided leaves. All squares except 24, 26, 64, 65. Scattered throughout most of the British Isles, becoming rare in the north and west, and in Ireland mainly in the southern part of the central plain.

FESTUCA L.

1. Festuca pratensis Huds.   Meadow Fescue   I. Lyons, 1763
A local plant scattered through the county in damp meadows. 25, 34–6, 38, 43–6, 48, *55, 56*, 57, 59, 64–6, 41, [24, 47, 49]. General.

2. Festuca arundinacea Schreb.   I. Lyons, 1763
On roadsides and in old pastures throughout the county. Much commoner than the preceding species and distinguished by its coarser habit, the whitish persistent leaf bases and more branched inflorescence. 23, 25, 26, 29, 34–6, 43–7, *49*, 55–8, 64–6, 20, 30, 31, 41, 50, [48, 40]. General.

3. Festuca gigantea (L.) Vill.   T. Martyn, 1763
In woods and shady places mainly on the boulder clay. Superficially similar to the other large woodland grass, *Bromus ramosus*, with which it often grows, but easily distinguished by the glabrous sheaths, the conspicuous purplish nodes and untidy, twisted awns. In *B. ramosus* the sheaths have abundant deflexed hairs, the nodes are not conspicuous and the awns are straight and neat. *25*, 26, 34, 35, 43–6, 54, 64–6, 75, [24, 36, 38, 48, 49, 56]. General, but rare in N. Scotland.

4. Festuca heterophylla Lam.   J. L. Gilbert, 1955

A specimen from a roadway near Madingley Park, collected by J. L. Gilbert in 1955, was determined as this species by W. O. Howarth. 36. It was probably introduced into the British Isles as a fodder plant in the early nineteenth century and planted in woodlands. It now occurs naturalized in scattered localities throughout Britain. Native in C. and S. Europe and S.W. Asia.

5. Festuca rubra L.   Red Fescue   Babington, 1855

Throughout the county in a great variety of habitats from old walls to all types of grassland. The ssp. **commutata** Gaudin (*F. fallax* Thuill.), a densely tufted perennial without rhizomes, was collected on Clare College cricket ground, Cambridge, in 1930 by J. S. L. Gilmour and determined by T. G. Tutin. This subspecies is often sown for lawns and may occur elsewhere in the county. All squares except 30. A very variable species, occurring throughout the British Isles.

6. Festuca ovina L.   Sheep's Fescue   I. Lyons, 1763

Throughout much of the county in dry grassland. The var. **hispidula** (Hack.) Hack., with shortly hairy lemmas, and the ssp. **tenuifolia** (Sibth.) Peterm., with mucronate, not awned, lemmas are recorded. *F. rubra* and *F. ovina* can best be distinguished vegetatively as follows. *F. rubra* is usually rhizomatous, often purplish at the base, has leaf-sheaths entire when young, and culm leaves often flat. *F. ovina* is densely tufted with setaceous leaves, whitish at the base and with split leaf-sheaths. 23, 24, 26, 29, 33–5, *36*, 37, 44–6, *47*, 48, 54–6, 64–7, 76, 40, [38, 49, 57, 41]. General, but not as common as *F. rubra* and probably sometimes recorded incorrectly for that species.

FESTUCA × LOLIUM = × FESTULOLIUM Aschers. & Graebn.

F. pratensis × Lolium perenne

× *Festulolium loliaceum* (Huds.) P. Fourn.; *Festuca pratensis* var. *loliacea* (Huds.) Bab.

This sterile hybrid is recorded from a number of scattered localities in the county, and can usually be found where both parents occur. 44, 55, *56*, 30, 50, [34, 38, 45, 49, 54, 58]. Recorded from most of lowland England, and also in S. Wales and Ireland.

GRAMINEAE

LOLIUM L.

1. Lolium perenne L.   Rye-grass   Ray, 1660
Throughout the county in a great variety of habitats. All squares.
Throughout the British Isles. A valuable grazing and hay grass,
cultivated in England for nearly 300 years.

2. Lolium multiflorum Lam.   Italian Rye-grass   Babington, 1860
L. italicum A. Braun
Throughout most of the county in pastures, or as a casual. *24*, 25, 26,
34–7, 39, 44–6, 54, 55, 58, 59, 66, 68, 69, 20, 30, 40, 41, 50, [38]. In most
parts of the British Isles, as an escape from cultivation, often becoming
well naturalized. A valuable fodder plant introduced into Britain about
1830, and now much sown for hay or grazing. Native of C. and S.
Europe, N.W. Africa, and S.W. Asia.

3. Lolium temulentum L.   Darnel   Ray, 1660
Formerly known in a number of localities as a cornfield weed; only
recent record as a casual. 45, [35, 38, 47, 48, 56, 65]. Scattered over the
British Isles as a casual in waste places, rare and inconstant. Formerly a
not uncommon weed. Native of the Mediterranean region.

VULPIA C. C. Gmel.

1. Vulpia bromoides (L.) Gray   W. Vernon, before 1700
Festuca sciuroides Roth
Frequent in sandy and gravelly places, by railway tracks and on walls.
25, 29, 33, *35*, 38, 39, 45, *47*, 48, 49, 54, 65, 76, 40, [34]. G. General.

2. Vulpia myuros (L.) C. C. Gmel.   I. Lyons, 1763
Festuca myuros L.
In a number of scattered localities in waste or cultivated ground on dry
sandy or gravelly soils, also by railway tracks and on walls. 25, 39, 45,
46, 55, 56, 65, 67, 50, [47, 48, 54]. G. Local in S. England, Wales and
S. Ireland; casual in N. England and Scotland.

3. Vulpia ambigua (Le Gall) More   C. D. Pigott, 1951
Known only from Kennett Heath, Isleham gravel pit, a sand pit at
Chippenham, and by the railway at Dullingham. It is locally abundant

on the Breckland just over the Suffolk border. 65–7, 76. **K.** On coastal sands in S. England and the Channel Isles, and on sandy heaths and warrens in Norfolk, Suffolk and Cambs, generally rather rare but sometimes locally abundant.

PUCCINELLIA Parl.

1. Puccinellia maritima (Huds.) Parl.   I. Lyons, 1763

Sclerochloa maritima (Huds.) Lindl. ex Bab.; *Glyceria maritima* (Huds.) Wahlb.

Only to be found in marshy areas by the tidal River Nene north of Wisbech. 31, 41. FA. All round the shores of the British Isles, being most abundant in the east and south; the main constituent of grassy salt-marshes or saltings, covering extensive areas of mud flats; inland in brackish areas in Worcestershire and Staffordshire.

2. Puccinellia distans (L.) Parl.   ?Relhan, 1786
First certain record, Henslow, 1831

Sclerochloa distans (L.) Bab.; *Glyceria distans* (L.) Wahl.

Locally abundant on brackish marshes near the tidal River Nene and its tributaries. The records by Relhan from 'By the sides of dunghills, common near Cambridge, and Cottenham', are unsupported by herbarium sheets, but may be correct as the species occasionally occurs as a casual on waste ground inland. 31, 41, [?45, ?46]. FA. Widely distributed round the shores of the British Isles, mostly on mud in the higher parts of salt-marshes; occasionally in river-meadows, on waste land and rubbish tips both near the sea and inland.

CATAPODIUM Link

1. Catapodium rigidum (L.) C. E. Hubbard   Ray, 1670

Sclerochloa rigida (L.) Link; *Scleropoa rigida* (L.) Griseb.; *Desmazeria rigida* (L.) Tutin

Throughout the county in dry places, especially on walls, round chalk-pits and by railway lines. 23, 24, *25*, 29, 34, 35, 38, 43–6, 55, 56, *57*, *58*, 65–7, 20, 40, 41, 50, [48, 49, 54]. General, but rare in the north and west.

2†. Catapodium marinum (L.) C. E. Hubbard   Relhan, 1785

Sclerochloa loliacea Woods ex Bab.; *Desmazeria marina* (L.) Druce

Recorded by Relhan from Wisbech but not seen there since. There are

GRAMINEAE

specimens in CGE from the Court of Clare College, Cambridge, collected by A. J. Crosfield in 1913; it was presumably introduced with gravel from the sea-coast. [45, 40.] Round the coasts of the British Isles but local in the north-east.

POA L.

1. Poa annua L.   Ray, 1660

Common throughout the county in a great variety of habitats, especially common on dry, rather bare, well-trodden ground, arable land and walls. All squares except 59. Although not an ideal lawn grass, it is often abundant in closely mown turf where it supplies a fine green sward except under dry conditions. General.

2. Poa nemoralis L.   Relhan, 1785

Local, mainly in shady places in woods or copses, but recorded also on walls. A delicate grass with a characteristic appearance. The inflorescence superficially resembles an *Agrostis*; the culm leaves are borne at right angles and have a very short ligule. 25, 36, 43–5, 55, *64*, 65, 20, 41, [35, 66]. GG. General, but rare in W. Ireland.

3. Poa compressa L.   I. Lyons, 1763

Occasional on walls, also on disturbed chalky ground. A late-flowering *Poa* with a small inflorescence and flattened culm. 25, 26, 35, 36, 44, 45, *46*, 55, 56, 58, 65, [24, 34, 38, 47, 48, 41]. **Cam.** General, except in N. Scotland and Ireland.

4. Poa pratensis L. sens. lat.   Meadow Grass   Ray, 1660

A very variable species, common throughout the county in a great variety of habitats. All squares except 68, [26]. General.

The three following species of this complex, partly apomictic aggregate occur in Cambridgeshire.

Poa pratensis L. sens. strict.   ?Ray, 1660
                                             First certain record, Henslow, 1821

Recorded from a few localities. 34, 39, 45, 46, 66, 40, 50. General.

Poa angustifolia L.   I. Lyons, 1763

P. pratensis ssp. *angustifolia* (L.) Gaudin

This is the most frequent segregate in Cambs, especially on the chalk, flowering in April and early May. Easily distinguished by its very long narrow leaves; superficially resembling *Festuca rubra*, but awnless. 24,

266

25, 34–6, 38, 45, 47, 54–6, 64, 65, 67, 76, 20. Distribution not yet known in detail but perhaps mainly in S. and E. England.

Poa subcaerulea Sm.   J. Ball, 1838

P. pratensis var. *subcaerulea* (Sm.) Hook.; *P. pratensis* ssp. *subcaerulea* (Sm.) Tutin

Recorded from Cambridge by J. Ball in 1838, from Linton by C. E. Moss in 1909 and from Hildersham by A. Bennett (see Evans, 1939). [45, 54.] Frequent in the northern part of the British Isles, less so in the south.

5. Poa trivialis L.   I. Lyons, 1763

Common throughout the county in a great variety of habitats from all types of grassland to woods, and as a weed of arable land. This is the Rough-stalked Meadow-grass readily distinguished from *P. pratensis* by its long pointed ligule. After flowering the dead inflorescence remains erect (not nodding as in *P. pratensis*). All squares. General.

6. Poa palustris L.   R. A. Lewin, 1941

Known only from Wicken Fen where it occurs in small quantity. Although first collected in 1941, it was not recognized until 1952. 56. W. Generally assumed to be introduced in Britain, and certainly occurring not infrequently as a casual on waste ground. Its riverside marsh habits, and its more recently discovered fen habitats, do however correspond with native habitats on the Continent, and its relatively late recognition in British Floras probably represents the failure of most British botanists to recognize the grass rather than evidence of a late introduction. Scattered throughout Britain; very rare in Ireland. Native in continental Europe, temperate Asia and N. America.

CATABROSA Beauv.

1. Catabrosa aquatica (L.) Beauv.   Ray, 1670

Formerly in a number of localities by the muddy margins of ponds and ditches especially in areas trodden by cattle. Recently seen only at Coe Fen and Coldham's Common, Cambridge, and at Swaffham Prior. 45, 56, [35, 44, 46, 49, 54, 55, 58]. CC. Irregularly distributed through the British Isles and in most districts rare.

GRAMINEAE

DACTYLIS L.

1. Dactylis glomerata L.   Cock's-foot   Ray, 1660
Abundant throughout the county in grassy places. All squares. A very important pasture and hay grass, frequent and often abundant throughout the British Isles.

CYNOSURUS L.

1. Cynosurus cristatus L.   Dog's-tail   Ray, 1660
Found throughout the county especially in old grassland. All squares except 24, 37. General.

2. Cynosurus echinatus L.
Casual. *45, 46, 54.*

BRIZA L.

1. Briza media L.   Quaking Grass   Ray, 1660
Throughout much of the county in old grassland, particularly chalk and fen. 24, *25*, 26, 34–8, 43–6, 49, 54–7, 59, 64, 66, 67, 41, [47, 48, 40]. General, except N. Scotland, in a great variety of habitats on neutral or calcareous, wet or dry, soils.

MELICA L.
1. Melica uniflora Retz.   Relhan, 1786
Known only in a few woods on the boulder clay in the east of the county. 64, 65. **DPW**. General, but rare in the north.

BROMUS L.
1. Bromus erectus Huds.   Relhan, 1785
Zerna erecta (Huds.) Gray
Locally abundant on old grassland, roadsides, and in waste places, on the chalk and sands, very rare or absent elsewhere. On the Gog Magog Hills, and the Devil's and Fleam Dykes it is the dominant grass over large areas. The forma **erectus** with minutely rough spikelets is the common form, but forma **villosus** (Mert. & Koch) Kunth with softly hairy spikelets occurs. 23, 24, 33–6, 44, 45, 54–6, 64, 66, [46, 58, 65]. **GG**. Chalk and limestone in England and Wales but rare in the west. Probably introduced in Scotland and Ireland.

2. Bromus ramosus Huds.   I. Lyons, 1763

B. asper Murr.; *Zerna ramosa* (Huds.) Lindm.

A characteristic species of open woodland, wood-margins and hedge-rows on the boulder clay and sands; almost absent from the Fenlands. 24–6, 34–6, 43–7, *49*, 54–6, *57*, 58, 64–6, 20, [38, 48]. General.

3. Bromus inermis Leyss.

Zerna inermis (Leyss.) Lindm.

Casual. 45.

4. Bromus sterilis L.   Ray, 1660

Anisantha sterilis (L.) Nevski

Throughout the county by roadsides, hedgerows, field-margins and in waste places. The most conspicuous annual grass in spring, flowering abundantly in May. All squares. Almost throughout the British Isles; abundant in the lowlands, absent from some parts of the north and west.

5. Bromus diandrus Roth   S. M. Walters, 1952

Anisantha gussonii (Parl.) Nevski

Recently recorded in several localities and locally abundant on road-sides and waste ground around Fordham and Chippenham, apparently increasing. Distinguished from the preceding species by the very large spikelets (7–9 cm. including the awn) and the obviously hairy inflorescence branches. 39, 45, 48, 56, 66, 67. A rare weed of cultivated land or waste places in England; native of the Mediterranean region.

6. Bromus tectorum L.

Casual. 45.

The following eight species are difficult taxonomically and some of the field records may be inaccurate. There are, however, over a hundred specimens in CGE, all of which have been checked by T. G. Tutin, and which form a sound basis for this treatment.

7. Bromus mollis L.   Ray, 1660

Serrafalcus mollis (L.) Parl.; *B. hordeaceus* sensu Evans

Throughout the county in meadows, on roadsides and in waste places. Largely self-pollinated and very variable, it shows the differentiation of local populations characteristic of many self-pollinated species. All squares except [49]. General.

269

GRAMINEAE

8. Bromus thominii Hardouin   T. G. Tutin, 1931

Scattered throughout the county on roadsides and in waste places; apparently absent from the Fens. 23, 25, 34, 35, *36*, 38, 43–6, 55, 59, 76. **Cam.** Widespread in lowland British Isles.

9. Bromus lepidus Holmberg   J. S. L. Gilmour, 1926
B. britannicus I. A. Williams

In a few localities on roadsides and in waste places and in sown leys; apparently absent from the Fens. 35, *36*, 44, 45, *46*, 54, 55, 67, [65]. Probably widely scattered throughout the British Isles.

10. Bromus racemosus L.   Relhan, 1820
Serrafalcus racemosus (L.) Parl.

In a number of localities by tracks and field margins and on roadsides. *35*, 37, 45, *47*, 54, *55*, *57*, [34, 56, 65]. Scattered throughout the British Isles but uncommon.

11. Bromus commutatus Schrad.   ?Relhan, 1802
        First certain record, Henslow, 1833
Serrafalcus commutatus (Schrad.) Bab.

Locally frequent by the margins of fields, on tracks and by roadsides. 25, 34, 35, 39, 45, 46, 54–6, 64, [23, 36, 38, 47–9, 57, 65, 40]. Scattered throughout the British Isles; local.

12. Bromus interruptus (Hack.) Druce   A. M. Barnard, 1849

Recorded from a few localities by the edges of fields but only seen recently in two places near Little Abington. The Odsey locality is the first record for Britain and those near Little Abington are the only ones in the country where the plant has been recorded since 1950. 54, *65*, [23, 45]. Scattered localities in S. and E. England.

13. Bromus arvensis L.   W. W. Newbould, 1856
Serrafalcus arvensis (L.) Godr.

Recorded from a few localities by tracks and in waste places. Only seen recently at Eversden and Cambridge. 35, 45, [34, 36, 54, 65]. A rare introduced weed of waste places; native of continental Europe, Siberia and W. Asia.

270

14. Bromus secalinus L. J. Martyn, 1727

Serrafalcus secalinus (L.) Bab.

Recorded from a number of localities by tracks and the edges of fields but only seen recently at Croydon (1951), Hardwick (1951) and Eversden (1954). Two forms occur: forma **secalinus**, with glabrous spikelets, and forma **hirtus** (F. Schultz.) A. & G., with softly and shortly hairy spikelets. 34, 35, *45, 46, 55,* [38, 47, 54, 56, 57, 66, 41]. A widespread casual, native of continental Europe, N. Africa and W. Asia.

15. Bromus carinatus Hook. & Arn.

Ceratochloa carinata (Hook. & Arn.) Tutin

Casual. *45.*

16. Bromus unioloides Kunth

Ceratochloa unioloides (Willd.) Beauv.

Casual. *45.*

BRACHYPODIUM Beauv.

1. Brachypodium sylvaticum (Huds.) Beauv. Ray, 1670

Frequent throughout most of the county in woods and by hedgerows and ditches, but rather local in the Fens. 23–6, 33–6, 43–7, 54–6, *58,* 64–7, 75, 20, [38, 49]. General.

2. Brachypodium pinnatum (L.) Beauv. Relhan, 1785

A rather local species of ungrazed grassland, and wood margins and rides on the boulder clay, rarely elsewhere. Forms loose to compact tufts and often invades grassland vegetatively in circles like fairy rings, as on the Fleam Dyke. Plants with hairy or glabrous spikelets occur in the county. *25,* 26, 34–6, 44, 45, 55, 56, 66, [24, 54, 40, 41]. H. Frequent in south England, rare elsewhere. Not known further east than the Devil's Dyke.

AGROPYRON Gaertn.

1. Agropyron caninum (L.) Beauv. T. Martyn, 1763

Triticum caninum L.; *Roegneria canina* (L.) Nevski

Local in woods and hedgerows. 24, *25,* 34–6, 43, 45, 64, 65, 75, [38, 46, 49, 66]. H. General, but local in N. Scotland and Ireland.

2. Agropyron repens (L.) Beauv. Couch or Twitch Ray, 1660

Common throughout the county as a bad weed of cultivated land; also by roadsides and in waste places. Forming tufts or large patches and

271

GRAMINEAE

spreading extensively by creeping wiry rhizomes. Very variable in awn length. The long-awned forms can be distinguished from *A. caninum* by the rhizomatous habit and by the absence of hairs just below the node. All squares. General in the lowlands.

3. Agropyron pungens (Pers.) Roem. & Schult.   I. Lyons, 1763

Triticum pungens Pers.; *Elytrigia pungens* (Pers.) Tutin

Locally abundant by the tidal River Nene. 41, [40]. Along the coasts of England, from Yorks and Cumberland to Cornwall, S. Wales and the southern half of Ireland.

(**Triticum aestivum** L., Wheat, is a cereal grown throughout the county (64,000 acres in Isle of Ely, 48,700 in S. Cambs) while 'Rivet Wheat', **T. turgidum** L., is seen very rarely. **Secale cereale** L., Rye, is still cultivated fairly frequently (500 acres in Isle of Ely, 80 in S. Cambs) and is extensively grown on the light Breckland soils of Suffolk.)

HORDEUM L.

1. Hordeum secalinum Schreb.   T. Martyn, 1763

H. pratense Huds.; *H. nodosum* sensu Evans

Throughout the county in undisturbed grassland particularly on the clays. 24–6, 34–8, 44–8, 54–9, 64, 65, 68, 20, 31, [49, 40, 41]. Frequent and often locally abundant in the southern part of the British Isles.

2. Hordeum murinum L.   Ray, 1660

Common throughout the county in grassy and waste places, on walls and by the margins of cultivated land. All squares except *66*. General, but absent from the north and west.

3. Hordeum marinum Huds.   Henslow, 1829

H. maritimum Stokes

Formerly found by the River Nene north of Wisbech, where it was last recorded in 1881. [40, 41.] South and east coasts of Britain from the Severn to the Forth.

4. Hordeum jubatum L.

Casual. *46*.

(**Hordeum vulgare** L., Barley, is cultivated throughout the county (20,000 acres in the Isle of Ely and 77,500 in Cambs).)

KOELERIA Pers.

1. Koeleria cristata (L.) Pers. Relhan, 1785
K. gracilis Pers.; *K. britannica* Domin
A characteristic plant of old grassland on the chalk and sands, rare elsewhere. 23–5, 33–6, 43–5, 54–7, 66, 67, 76, *41*, [46, 64]. A common plant of dry grassland throughout the British Isles.

TRISETUM Pers.

1. Trisetum flavescens (L.) Beauv. Ray, 1670
Throughout the county in dry grassy places and waysides; less common in the Fens. 23–6, 29, 34–8, 43–8, 54–7, 64–7, 75, 76, 41, [49, 40]. General, though less frequent or rare in Scotland and Ireland.

AVENA L.

1. Avena fatua L. Wild Oat Ray, 1660
A serious weed of arable land throughout the county, which has increased in recent years. All squares except 66, 67, 30, [24, 25, 29, 34]. General in cultivated regions.

2. Avena ludoviciana Durieu J. M. Thurston, 1951
A weed of arable land as yet recorded from a few scattered localities. Easily distinguished from *A. fatua* when ripe by the fact that the two or three florets are shed as a single unit. 25, 35, 36, 45, 54. Widespread locally common weed in South England introduced with wheat-seed in the First World War. Probably native of C. Asia.

3. Avena sativa L. Cultivated Oat
This species, which is grown as a crop throughout the county (7000 acres in the Isle of Ely, 9900 in S. Cambs), is sometimes found as a relic of cultivation. Such plants can be distinguished from *A. fatua* and *A. ludoviciana* by the glabrous lemma; if an awn is present it is only on the lowest floret. 34–6, 44, 45, 48, 59.

HELICTOTRICHON Bess.

1. Helictotrichon pratense (L.) Pilg. Ray, 1688
Avena pratensis L.
A characteristic species of chalk grassland. 24, 34, 35, 44–6, 54–7, 66, [23, 36, 43]. GG. In dry grassland throughout Great Britain, but with

GRAMINEAE

an eastern tendency, mainly on chalk and limestone, but on quite acid rocks in Scotland.

2. Helictotrichon pubescens (Huds.) Pilg. I. Lyons, 1763
Avena pubescens Huds.
Scattered throughout the county, but most frequent on the chalk. 25, 35–7, 43–6, 54–7, 64, 66, 67, 76, 41, [23, 34, 38, 47, 48, 58, 65, 40]. General.
A species of damper and deeper soils than *H. pratense*, flowering about a month earlier.

ARRHENATHERUM Beauv.

1. Arrhenatherum elatius (L.) Beauv. ex J. & C. Presl Oat-grass
I. Lyons, 1763
A. avenaceum Beauv.
Common throughout the county in rough grasslands, hedgerows, roadsides and waste ground. The commonest flowering grass on Cambridgeshire roadsides in June. The var. **bulbosum** (Willd.) Spenner is recorded. This variety has the short basal internodes of the culms bulbous or pear-shaped, and is sometimes a troublesome weed of arable land. All squares. General.

HOLCUS L.

1. Holcus lanatus L. Yorkshire Fog Ray, 1660
In a great variety of habitats, especially meadows and pastures, abundant throughout the county. All squares. General. Probably, with *Plantago lanceolata*, one of the very few ubiquitous species in the British Isles.

2. Holcus mollis L. Relhan, 1785
A very local plant of open copses and woodland on the boulder clay and sands. Readily distinguished from the preceding species by its rhizomatous habit, the pubescence restricted to the nodes and by its obviously awned inflorescence. 25, 43, 54, 65, [35, 36, 38, 45, 49]. DPW. General, usually calcifuge.

DESCHAMPSIA Beauv.

1. Deschampsia cespitosa (L.) Beauv. Ray, 1660
Aira cespitosa L.
In rough grassland, marshy places and ditch-sides throughout the county. A coarse tussocky grass with a delicate inflorescence and

leaves so rough that it is difficult to draw the fingers over the surface towards the base. All squares. Wet and badly drained soils throughout the British Isles.

2. Deschampsia flexuosa (L.) Trin. I. Lyons, 1763

Aira flexuosa L.

Known only from the dry sandy soils at Gamlingay. Vegetatively similar to *Festuca ovina*, but easily distinguished by the rather long acute ligule. 25. G. General on acid, sandy and peaty soils.

AIRA L.

1. Aira praecox L. I. Lyons, 1763

In a few localities in dry sandy or gravelly places, abundant in the Breckland. 25, *37*, 39, 54, 65-7, 76, [23, 38, 47]. G. General, but preferring acid sandy soils.

2. Aira caryophyllea L. Ray, 1660

A rare species of dry sandy or gravelly places only seen recently at Hildersham Furze Hills and Isleham plantation. 54, 67, [25, 37-9, 47, 58, 66, 76]. General in suitable habitats, locally abundant.

CALAMAGROSTIS Adans.

1. Calamagrostis epigejos (L.) Roth I. Lyons, 1763

Occasional in damp woods, by streams and pits, and in fens, but apparently completely absent from the open Fenland of the extreme north. 25, 29, 34-6, 44-6, 54-7, 65, 66, [37, 48, 58]. W. Widely distributed in England, sparse in Wales and Scotland, and very rare in Ireland. A coastal plant in the west.

2. Calamagrostis canescens (Weber) Roth Relhan, 1820

C. lanceolata Roth

Covers a similar area in Cambs to *C. epigejos* but is very local, and usually in fens or damp woodlands. Flowers in May and early June, six weeks earlier than the preceding species. It can be distinguished vegetatively by its narrower leaves which are always hairy above. In *C. epigejos* the leaves are twice as broad (5-10 mm.) and usually glabrous above. 25, 35, *37*, 45, 54, 56, 57, 65, 66, [34, 38, 47-9, 58]. W. Scattered throughout England but very local, and in a few places in Wales and E. Scotland.

GRAMINEAE

AGROSTIS L.

1. Agrostis canina L.   Brown Bent   Relhan, 1785

A rare plant locally frequent within the county only at Gamlingay, but abundant in the Breckland. It is a very variable species and two subspecies are recorded: ssp. **canina** (*A. canina* var. *fascicularis* Sinclair), the Gamlingay plant, which is stoloniferous with no rhizomes and has the panicle little contracted after flowering, and ssp. **montana** (Hartm.) Hartm. (*A. canina* var. *arida* Schlechtst.) which is rhizomatous with no stolons and usually strongly contracted and spike-like after flowering. 25, *49*, 56, 68, [34, 37, 45, 46, 66]. General. The ssp. *canina* is usually in damp acid places and sometimes floating in ditches, while ssp. *montana* is on heath and usually dry acid grassland.

× **tenuis** Sibth. Recorded from Chatteris in 1957. 38.

2. Agrostis tenuis Sibth.   Common Bent   Ray, 1670

A. vulgaris With.

Evans's (1939) statement that this is 'probably our most plentiful grass and most universal both in its larger and smaller forms' is quite wrong. He must have confused it with *A. stolonifera*. It is frequent on the sands at Gamlingay and in the east of the county but rare elsewhere. 25, 38, 39, 45, 46, *49*, 54, 56, 65–7, 76, 20, 40, 41, 50, [34, 37, 47, 48, 55]. FH. General, especially prevalent on poor, dry, acid soils.

3. Agrostis gigantea Roth   Babington, 1855

Scattered over the county as a weed of arable land and waste places. Usually easily distinguished from *A. stolonifera* by its obvious rhizome, broader leaves and longer panicle. The var. **dispar** (Michx.) Philipson is recorded; this has the culms erect or geniculate, the sterile shoots erect or shortly procumbent as stolons, and the panicle frequently with few branches bearing rather scattered spikelets. 23, 24, 34, 35, 38, 39, 44, 45, *46*, 55, 67, 20, 40, [48, 49, 56, 66]. Widely distributed in lowland districts of the British Isles.

4. Agrostis stolonifera L.   Creeping Bent   I. Lyons, 1763

A. alba sensu Bab.

Abundant throughout the county in a great variety of habitats. The var. **palustris** (Huds.) Farw. occurs; it has extensively creeping stolons which mat loosely together and do not form a turf as var. **stolonifera**. All squares. General.

276

APERA Adans.

1. Apera spica-venti (L.) Beauv.   Ray, 1670

There are several records from the sands in the east of the county and at Gamlingay, but it has only recently been seen at Kennett Heath in 1951. *54, 76*, [25, 64]. K. Probably native in dry sandy fields of E. England where it is locally abundant; elsewhere only a casual.

2. Apera interrupta (L.) Beauv.   Babington, 1852

Now known only from the sands in the east of the county where it is very rare. It was also found by J. Stratton in the ancient ditch near Pampisford Hall in 1855. 67, [54, 76]. Probably native in the Breckland of East Anglia; elsewhere an introduced weed of arable land, or casual.

PHLEUM L.

1. Phleum bertolinii DC.   Ray, 1660

P. pratense sensu Bab. pro parte; *P. nodosum* sensu Evans

Common throughout the county in dry places. It was formerly not distinguished from *P. pratense* from which it differs in being more slender and smaller in all its parts. It has only 14 chromosomes while *P. pratense* has 42, and should perhaps only be regarded as a subspecies of *P. pratense*. All squares except 48, 58, 30. General on a wide range of soils. An important constituent of old pastures; some strains of particularly leafy growth are of great palatability to cattle.

2. Phleum pratense L.   Timothy   Ray, 1660

Throughout the county in meadows, roadsides and waste places. All squares except 25. General, but rare in the north. Probably only native on the moist soils of water-meadows and other low-lying grasslands. A very important fodder plant and often sown for hay grass.

3. Phleum phleoides (L.) Karst.   I. Lyons, *c.* 1780

P. boehmeri Wibel

Formerly in a few localities on dry soils in the east of the county. Now known only at Hildersham Furze Hills and Isleham Plantation. 54, 67, [56, 66]. FH. Dry soils in W. Norfolk, W. Suffolk, Cambs, Beds, Herts and Essex.

4. Phleum arenarium L.   J. Hempsted in J. E. Smith, 1795

Recorded from Newmarket Heath by J. Hempsted in 1795, Chippenham by Babington in 1838, between Chippenham and Fordham by

277

GRAMINEAE

T. G. Tutin in 1936, and still at Isleham gravel pit in 1961. 67, [66]. On dunes and in sandy fields all round the coasts of the British Isles; occurs inland on sandy soils in the Breckland.

5. Phleum paniculatum Huds.

P. asperum Jacq.

Casual, recorded from the Gog Magog Hills by Mr Woodward (Withering, *Bot. Arrang. Brit. Pl.* ed. 2, 62, 1787), from Bourn Bridge by Mr Crowe (Relhan, 1786), and from Newmarket Heath by Mr Miller (Huds. *Fl. Angl.* ed. 2, 26, 1778). There is a specimen from Bourn Bridge collected by J. Gibson around 1850, in the Saffron Walden Museum. The species is an occasional casual in the British Isles and is a native of the Mediterranean region. [45, 54, 66.]

ALOPECURUS L.

1. Alopecurus myosuroides Huds.   Black Grass or Black Twitch
                                        Ray, 1660

A. agrestis L.

A common weed of arable land and waste places through much of the county, but less frequent in the Fens. 24–6, 34–7, 39, 43–9, 54–6, 58, 64, 65, 69, 75, 31, [38, 41]. Scattered throughout the British Isles, but most frequent in cultivated areas in southern England.

2. Alopecurus pratensis L.   Meadow Foxtail   T. Martyn, 1763

In meadows, roadsides and other grassy places throughout most of the county. Similar in general appearance to *Phleum pratense* but flowering at least a month earlier (May–June); it can immediately be distinguished from *Phleum* by the presence of long awns on the lemmas. 24–6, 29, 34–9, 43–8, 56–9, 64–8, 30, 40, [23, 49, 54, 41]. General.

3. Alopecurus geniculatus L.   Ray, 1660

Fairly frequent in wet meadows in the Fenlands especially by the muddy margins of pools and in damp depressions; rare elsewhere. 34, 36–9, 43–9, 56–9, 67, 31, 40, 41, [25]. General.

4. Alopecurus aequalis Sobol.   Relhan, 1820

A. fulvus Sm.

Rare on pond margins and water-meadows in the Fens. Distinguished from *A. geniculatus* by the shorter anthers which are orange not violet at maturity, and the short awns of the lemmas. *37, 47, 48,* 54, 56, 40, [35, 36, 46, 57]. Widely distributed but local in England, rare in Wales.

278

MILIUM L.

1. Milium effusum L.   Relhan, 1820

A rather local plant of boulder-clay woods. 25, 35, 36, 43, 54, 64, 65, [49]. **B.** General, but often local.

ANTHOXANTHUM L.

1. Anthoxanthum odoratum L.   Sweet Vernal Grass   I. Lyons, 1763

In damp places in old pastures on neutral and acid soils, local. 25, 36, 38, 39, 43–8, *49*, 54–8, 66, 67, 41, [34, 35, 40]. **G.** General.

2. Anthoxanthum puelii Lecoq & Lamotte   J. E. Little, 1911

Known only at Gamlingay where it was last seen in 1930. *25.* Introduced from France in the latter half of the nineteenth century with fodder plants; for some time widely scattered in England and very rare in Wales and Scotland; now very rare everywhere; native of S.W. Europe.

PHALARIS L.

1. Phalaris arundinacea L.   Reed-grass   Ray, 1660

Common in ditches, rivers, lodes, pits and marshy places throughout the county. It is rather similar in vegetative appearance to *Phragmites communis* with which it often grows, and from which it can be distinguished by the presence of a long ligule. All squares except [26, 54]. General.

2. Phalaris canariensis L.   Canary Grass   Skrimshire, 1828

Frequent casual of rubbish dumps and waste places. 34, 36, 45, *46*, 55, 66, 40, 41, 50, [38, 48]. A widely distributed casual especially in E. England; native of the western Mediterranean region. Widely cultivated in warm-temperate regions for its seeds which are used as food for cage-birds.

3. Phalaris minor Retz.

Casual. *66, 76.*

PARAPHOLIS C. E. Hubbard

1. Parapholis strigosa (Dumort.) C. E. Hubbard   Relhan, 1788

Lepturus incurvatus sensu Bab.; *Pholiurus filiformis* sensu Evans; *Lepturus filiformis* auct.

Known only from brackish areas by the tidal River Nene north of Wisbech, where it is locally abundant. The early authors did not

279

GRAMINEAE

distinguish between this species and *P. incurva* (L.) C. E. Hubbard, but all specimens seen belong to *P. strigosa* and there is no evidence that *P. incurva* has ever occurred in Cambs. 41, [40]. **FA.** All round the coasts of the British Isles in suitable habitats, north to W. Lothian and Mull.

NARDUS L.

1. Nardus stricta L.   Mat-grass   Ray, 1660
Now known only at Gamlingay, but recorded by Relhan from Hildersham. 25, [54]. **G.** General on acid soils but rare in eastern England.

ECHINOCHLOA Beauv.

1. Echinochloa crus-galli (L.) Beauv.
Casual. 45, 55, *40*.

PANICUM L.

1. Panicum miliaceum L.   Millet
Casual. 45.

DIGITARIA P. C. Fabr.

1. Digitaria sanguinalis (L.) Scop.
Casual. *66*.

SETARIA Beauv.

1. Setaria viridis (L.) Beauv.   Relhan, 1786
An occasional casual of waste places and fields. 45, 40, [38, 46, 48, 67]. Uncommon in the British Isles; native of warm-temperate Europe and Asia.

(**S. lutescens** (Weigel) C. E. Hubbard (*S. glauca* sensu Evans). According to Evans (1939) this species was recorded from Sutton and Welches Dam by A. Fryer. In Fryer's notes Sutton Gault, 1880, is given, but the only record from Welches Dam refers to *S. viridis*.)

2. Setaria italica (L.) Beauv.
Rare casual; a constituent of cage-bird seed. 45, 58, 41, 50.

ZEA L.

1. Zea mais L.   Maize or Indian Corn
Casual. Occasionally grown as a crop in Cambs, the green corn being used for cattle fodder. [45.]

BRYOPHYTA

By H. L. K. Whitehouse

Introduction

In the eight years since M. C. F. Proctor's paper, 'A Bryophyte Flora of Cambridgeshire' (Proctor, 1956) went to press, knowledge of the bryophytes of the county has been considerably extended. About 30 additions have been made to the county list, and the frequency and distribution within the county of many species is now better known. For several species, the first record for the county has been put back over 50 years, by the finding of Cambridgeshire specimens in F. Y. Brocas's collection at Saffron Walden Museum, and through the correction of faulty identifications of specimens in the Cambridge University Herbarium. A few published records have been omitted when there is doubt about the identity of the plants. A start has been made with the recording of bryophytes within 10 km. squares of the National Grid, and this has led to collecting in new areas, with some unexpected finds. Details of the habitat, locality, collector and date for these records have been deposited in the library of the Botany School. I am particularly indebted to Mr P. J. Bourne, who has made numerous excursions to all parts of the county in search of bryophytes. His collecting has led to over 600 additions to the 10 km. square lists. Others who have helped are acknowledged on p. xiii.

The identifications of all additions to the county list, since the publication of Proctor's paper, have been confirmed by the referees of the British Bryological Society, and a majority of the new records have already been published in its *Transactions*. I should like to thank Mrs G. Spencer, Curator of the Saffron Walden Museum, for the loan of F. Y. Brocas's collection of bryophytes, and Miss J. M. Morris, Curator of the County Museum, Warwick, for the loan of H. E. Lowe's Cambridgeshire specimens.

Reproductive organs such as capsules and gemmae are referred to only when they have been observed on Cambridgeshire material of the plant. Hence, if there is no reference to them, it may be assumed that they are not present, or at least have not been observed. The Cambridgeshire bryophyte flora includes four species in which capsules are unknown (*Barbula nicholsonii, Trichostomum sinuosum, Riccia fluitans, Riccia rhenana*), and a further ten species in which, though known

281

SPHAGNACEAE

elsewhere, they have never been observed in the British Isles (*Dicranum montanum, Tortula papillosa, Barbula acuta, Tortella inclinata, Thuidium abietinum, T. philibertii, Scorpiurium circinatum, Entodon concinnus, Rhytidium rugosum, Ricciocarpus natans*). In addition, there are many species in which capsules are of very rare occurrence in the British Isles, for example, *Ditrichum flexicaule, Encalypta streptocarpa, Tortula latifolia, Barbula trifaria, Rhodobryum roseum, Aulacomnium androgynum, Climacium dendroides, Leucodon sciuroides, Campylium chrysophyllum, C. elodes, Acrocladium giganteum, Brachythecium glareosum, Cirriphyllum piliferum, Pleurozium schreberi* and *Lunularia cruciata*. In Cambridgeshire and adjoining counties, there is evidence that a number of species fruit less often than they did a century or more ago. Moreover, several epiphytic species, such as *Orthotrichum lyellii, Ulota crispa, Cryphaea heteromalla* and *Leucodon sciuroides*, are apparently rarer than formerly. Increased atmospheric pollution may be responsible for these changes.

MUSCI (Mosses)

SPHAGNALES

SPHAGNACEAE

SPHAGNUM L.

1†. Sphagnum palustre L. T. Martyn, 1763

Formerly in acid bogs at Hinton and Teversham Moors and Gamlingay. Last recorded in the county from Gamlingay Heath Wood in 1930. *25*, [45, 55]. General.

2†. Sphagnum squarrosum Pers. ex Crome Relhan, *c.* 1796

Formerly at Gamlingay, but no records except Relhan's. [25.] General.

3†. Sphagnum tenellum Pers. Relhan, *c.* 1796

Formerly at Gamlingay, but no records since Relhan's. [25.] Widely distributed in the British Isles, but rare in the south-eastern half of England.

4†. Sphagnum cuspidatum Ehr. ex Hoffm. Relhan, 1820

Formerly at Gamlingay Bog and Heath Wood, but not seen since 1930. *25*. General.

5†. Sphagnum subsecundum Nees var. **inundatum** (Russ.) C. Jens.

C. E. Moss, 1910

Formerly at Gamlingay, but no record since 1910. [25.] Widely distributed in the British Isles.

†Var. **auriculatum** (Schp.) Lindb.   Henslow, 1857

Formerly at Gamlingay Heath, but not seen since 1930. *25.* Widely distributed in the British Isles.

6†. Sphagnum plumulosum Röll.   Shrubbs, 1898

Formerly at Chippenham Fen and Gamlingay, but no records since 1913. [25, 66.] General.

POLYTRICHALES

POLYTRICHACEAE

ATRICHUM Beauv.

1. Atrichum undulatum (Hedw.) Beauv.   Relhan, 1785

Prefers lime-free soils. Abundant on the ground in woods on the Lower Greensand and Breck fringe, and occurs in most of the boulder-clay woods. Recently found in two places in Wicken Fen. Capsules frequent. 25, 26, 35, 36, 43, 54, 55, 57, 64, 65, 76. General.

POLYTRICHUM Hedw.

1†. Polytrichum nanum Hedw.   Relhan, 1802

Formerly grew on sandy soil on the Gog Magog Hills and at Gamlingay. Last record, Gamlingay, 1829, with capsules. [25, 55.] Widely distributed in the British Isles, though there are few records from W. and C. Ireland.

2†. Polytrichum aloides Hedw.   Henslow, *c.* 1830

Recorded by Henslow from Gamlingay, with capsules. A plant of lime-free sandy banks. No recent records. [25.] General, but rare in E. England.

3. Polytrichum piliferum Hedw.   Relhan, 1820

First record since Henslow: H. Godwin, 1944. Formerly at Gamlingay Heath, but not recorded there since 1827. The only recent records

283

POLYTRICHACEAE

are from the site of a former raised bog at Sutton Meadlands, and from cinders on waste ground. 37, 45, *47*, [25]. General.

4. Polytrichum juniperinum Hedw.   Relhan, 1785

Occurs on lime-free sandy soils on the Breck fringe, on the Gamlingay Lower Greensand, and on glacial gravel at Hildersham Furze Hills. Also occurs on railway embankments, etc. Capsules infrequent. 25, 36, 37, 54, 64, 65, *66, 67,* 76, [56]. General.

5. Polytrichum aurantiacum Sw.   R. S. Adamson, 1913
P. gracile Sm.

First certain record: E. W. Jones and P. W. Richards, 1934.

Occasional in woods on the Lower Greensand at Gamlingay, and rare at the foot of Beeches and on rotten wood on the chalk. Recorded from one boulder-clay wood. 25, 45, 55. GG. General, but not recorded from a number of vice-counties, particularly in N. Scotland and W. Ireland.

6. Polytrichum formosum Hedw.   Mr Tozer in Relhan, 1820

First record since Henslow: P. W. Richards, 1938. Frequent in woods on the Lower Greensand at Gamlingay, rare in boulder-clay woods. On clinker of sewage works, Madingley Hall. Capsules rather infrequent. 25, 35, 36, 54, *65*. General.

7†. Polytrichum commune Hedw.   Ray, 1660

Formerly grew in acid bogs at Hinton Moor and Gamlingay, with capsules. Last record: Gamlingay, 1881. Recorded from Chesterton Ballast Pits, Cambridge, about 1935, but the habitat has now been destroyed. *46*, [25, 45]. General.

FISSIDENTALES

FISSIDENTACEAE

FISSIDENS Hedw.

1. Fissidens viridulus (Web. & Mohr.) Wahl.   Henslow, 1829

On ditch-sides and clay banks, usually in woods, frequent. Capsules abundant. 15, 24, 25, 29, 34, *35*, 36, 44–7, 49, 54–6, 75. CH. Frequent in England and Wales, rare in Scotland and Ireland.

FISSIDENTACEAE

2. Fissidens minutulus Sull. var. **tenuifolius** (Boulay) Norkett
F. minutulus Sull. sec. Braithw.
D. E. Coombe and M. C. F. Proctor, 1951
On calcareous stone- or brick-work in shady places, rare. Capsules frequent. 45, 47, 55. *F. minutulus* occurs throughout the British Isles, but there are few records from S.E. Ireland or N. Scotland. The separate distributions of the var. *minutulus* (*F. pusillus* Wils. ex Milde), which is a plant of wet non-calcareous rocks, and the var. *tenuifolius* of calcareous rocks are incompletely known, but var. *tenuifolius* has not yet been found in Scotland.

3. Fissidens bryoides Hedw.   ?T. Martyn, 1763 or Relhan, 1820
First certain record: P. C. Hodgson, 1931.
Prefers lime-free soils, and hence rare in Cambs. Frequent in Gamlingay Wood and recorded from three woods wholly on boulder clay, and from ditch-sides on the Jurassic clays. Capsules abundant. 25, 29, 35, 54. General.

4. Fissidens incurvus Starke ex Web. & Mohr
First certain record: G. D. Haviland and J. J. Lister, 1881
Abundant in boulder-clay woods, often growing mixed with *F. taxifolius*. Occasional in arable fields and ditch-sides. Capsules abundant. 15, 25, 34–8, 43, 45–8, 54–8, 64, 65, 76, 20. Frequent in England, rare in Scotland, Wales and Ireland. Not recorded from N.W. Scotland, and most of W. Ireland.

5. Fissidens crassipes Wils.   P. W. Richards, 1928
On wood or stonework submerged in the River Cam, and on stonework near the R. Ouse, apparently rare. Capsules recorded. 37, 45, 56. Frequent in England, rare in Scotland, Wales and Ireland, and not recorded from N. and W. Scotland.

6. Fissidens exilis Hedw.   T. G. Tutin and P. W. Richards, 1932
A minute species of lime-free soils. Frequent in the non-calcareous part of Gamlingay Wood. Recorded from five woods wholly on boulder clay and from a temporary pasture. Capsules abundant. 25, 35, *36*, 43, *64*, 65. Throughout England, but apparently absent from large areas of Scotland, Ireland and Wales.

FISSIDENTACEAE

7. Fissidens taxifolius Hedw.   T. Martyn, 1763

On the ground in woods, ditches, hedgebanks, etc. One of the most abundant mosses in the county. Capsules frequent. All squares except 39, 67, 30. General.

8. Fissidens cristatus Wils.   F. Y. Brocas, 1874

In undisturbed chalk grassland, frequent. Capsules rare. 24, 33, 35, 45, 54–6, 65, 66. General.

9. Fissidens adianthoides Hedw.   T. Martyn, 1763

In grassland, frequent in fens, and occasional on the chalk. Capsules occasional. 35, 44, 55–7, 66, [45, 40]. W. General.

DICRANALES

ARCHIDIACEAE

ARCHIDIUM Brid.

1†. Archidium alternifolium (Hedw.) Schp.   Relhan, 1802

On damp lime-free soils. Formerly at Gamlingay. No recent record. [25.] Widely distributed in the British Isles, but apparently absent from most of eastern England between the Humber and the Thames.

DICRANACEAE

PLEURIDIUM Brid.

1. Pleuridium acuminatum Lindb.   ?Relhan, 1802

First certain record: E. W. Jones, 1934

P. subulatum sensu Dix.

A plant of lime-free soils, recorded from a path on the Breck fringe and from a stubble field near Gamlingay. Capsules abundant. 25, *66*. Widespread in the British Isles, but rare in Ireland and N.W. Scotland.

2. Pleuridium subulatum (Hedw.) Lindb.   P. W. Richards, 1946

P. alternifolium (Dicks.) Brid.

Calcifuge. A rare plant in Cambridgeshire, recorded only from the lime-free parts of Gamlingay Wood and from Little Widgham Wood. Capsules abundant. 25, *65*. Of widespread occurrence in the British Isles, but rare or absent from much of Ireland and N. and W. Scotland.

DICRANACEAE

DITRICHUM Hampe

1. Ditrichum cylindricum (Hedw.) Grout
P. F. Lumley, 1960; P. J. Bourne, 1960
A plant of lime-free arable fields, recorded from Oxford Clay and fen peat. Gemmae (on rhizoids) frequent. 26, 49, 20. Widely distributed in the British Isles.

2. Ditrichum flexicaule (Schleich.) Hampe P. W. Richards, 1929
Locally abundant in long-established chalk grassland. *35, 55, 66.* F. General.

CERATODON Brid.

1. Ceratodon purpureus (Hedw.) Brid. Relhan, 1785
Abundant in all lime-free habitats, such as thatched or tiled roofs, walls, tarmac, rotten wood, and on the ground at the foot of beech trees on the chalk; particularly abundant on the ground on the Lower Greensand and on lime-free sandy soils fringing the Suffolk Breckland. Capsules frequent. All squares. General.

SELIGERIA B. & S.

1. Seligeria paucifolia (Dicks.) Carruth. E. W. Jones, 1932
This minute species is of frequent occurrence in woods and grassland on the chalk, but appears to be strictly confined to detached lumps of chalk. Capsules abundant. 33, 35, 45, 55, 56, 65, 66. **CH.** Distribution in the British Isles restricted to S.E. England, east of a line from Portland Bill to Flamborough Head, and on the Continent known only from France and Italy.

2. Seligeria calcarea (Hedw.) B. & S. Mr Dickson in Relhan, 1802
This minute species is frequent on the vertical faces of abandoned chalk-pits, but appears not to colonize the chalk for many years after it is first exposed. It also occurs on detached lumps of chalk in shade, but is less common in this habitat than *S. paucifolia*. Capsules abundant. 35, 45, 55, 56, 65, [66]. **CH.** Of restricted distribution in the British Isles, having been recorded only from England south and east of a line from the Severn estuary to Flamborough Head, and from a limited area in N.E. Ireland.

DICRANELLA Schp.

1. Dicranella schreberana (Hedw.) Dix.
B. Reeve, 1959; H. L. K. Whitehouse, 1959
Sides of ditches, recently disturbed soil in woods, and arable fields,

DICRANACEAE

rare. Gemmae (on rhizoids) frequent. 25, 26, 36, 39, 49, 55, 64. Widespread in the British Isles, but few records from E. England, Wales, N. Scotland and much of Ireland.

2. Dicranella varia (Hedw.) Schp.   Relhan, 1785
var. **varia**
Abundant on the floor of chalk-pits, in arable fields and on ditch-sides. Gemmae (on rhizoids) and capsules frequent. All squares except 37, 67, 68, 30, 40, 41. General, but few records from N. and W. Scotland.

var. **callistoma** (Turn.) B. & S.   C. D. Pigott, 1951
Occasional, mixed with var. *varia*, in chalk-pits. Capsules abundant, but without viable spores. 35, 45, 54, 64. Recorded from scattered localities in the British Isles.

3. Dicranella rufescens (Sm.) Schp.
                                      P. J. Bourne and H. L. K. Whitehouse, 1960
Recorded from non-calcareous arable fields (on Oxford Clay and fen peat). Gemmae (on rhizoids) abundant. 26, 49, 20. Widely distributed in the British Isles, but few records from Ireland and E. England.

4. Dicranella heteromalla (Hedw.) Schp.   Relhan, 1785
Calcifuge. Abundant on the ground in woods on Lower Greensand at Gamlingay, occasional in woods elsewhere, chiefly at the foot of trees, or on much-decayed wood. Capsules uncommon except on the Lower Greensand. 25, 35, 43–5, 49, 54–7, 59, 64, 65, 67, 68, 76. General.

DICRANOWEISIA Lindb.

1. Dicranoweisia cirrata (Hedw.) Lindb.   Henslow, *c.* 1830

First record after Henslow's: P. W. Richards, 1929. Abundant on tree-trunks and wooden palings. On sandstone, Gamlingay church. Capsules and gemmae frequent. 15, 23, 25, 26, 34–6, 39, 43–6, *47*, 48, *49*, 54–9, 64–6. Widespread in the British Isles, but not recorded from much of W. Scotland and W. and C. Ireland. In the west it occurs more often on rocks than on trees.

DICRANUM Hedw.

1. Dicranum montanum Hedw.   C. C. Townsend, 1955
Locally abundant on tree-trunks in White Wood, Gamlingay. 25. Occurs in parts of S.E. England, the Midlands, N. England and a few

localities in Scotland. Not recorded from Ireland, Wales or S.W. England.

2†. Dicranum majus Turn. Relhan, 1820

Formerly in Eversden Wood, but no recent records. [35.] General, except that it becomes progressively rarer eastward in England and is largely absent from East Anglia.

3. Dicranum scoparium Hedw. Relhan, 1785

Occasional on the bark of trees in woods, and on the ground at the base of trees. Frequent in grass-heaths on the Lower Greensand at Gamlingay, on glacial gravel at Hildersham Furze Hills and on Breck sands near Kennett. Rare on the ground elsewhere. 25, 35, 36, 43–6, 54, 55, 64, *65*, 66, 67, 76, 31. General.

CAMPYLOPUS Brid.

1. Campylopus pyriformis (Schultz) Brid. P. W. Richards, 1930

Frequent on lime-free sandy soil and rotten wood in woods on the Lower Greensand at Gamlingay; occasional on rotten wood elsewhere. Capsules rare. 25, 46, 54, 66. General.

2. Campylopus flexuosus (Hedw.) Brid. P. W. Richards, 1930

On rotten stumps and tree-bases, rare. *25*, 45, 66. General.

LEUCOBRYUM Hampe

1. Leucobryum glaucum (Hedw.) Schp. H. L. K. Whitehouse, 1941

In old coppiced stump, Little Widgham Wood. *65*. General in woods and on heaths where the soil is free of lime.

POTTIALES

ENCALYPTACEAE

ENCALYPTA Hedw.

1. Encalypta vulgaris Hedw. Relhan, 1785

In beech woods and grassland on the chalk, rather rare. Capsules abundant. 45, 55, 66. **GG.** Widespread in the British Isles, but few records from W. and N. Scotland.

ENCALYPTACEAE

2. Encalypta streptocarpa Hedw. M. C. F. Proctor, 1951
Locally abundant in a few beech woods and one piece of grassland on the chalk. Gemmae abundant. 45, 55, 66. GG. General, except East Anglia.

POTTIACEAE

TORTULA Hedw.

1. Tortula ruralis (Hedw.) Crome T, Martyn, 1763
Frequent in a wide range of habitats but usually where well-drained and not heavily shaded. Most characteristic of thatched and tiled roofs, where it may form huge cushions. Occurring also at the base of trees, on chalk banks, and on tarmac. Capsules rather uncommon. 23, 24, *25*, 33–7, 39, 44–7, 49, 54–7, 59, 64–8, 76. General.

2. Tortula ruraliformis (Besch.) Dix. G. Halliday, 1955
A plant of sandy soils, abundant on the Suffolk Breckland, and recorded in Cambridgeshire from the Breck fringe and from a bunker on the Gog Magog golf-course. 45, 76. All round the British coast and in a few inland localities.

3. Tortula intermedia (Brid.) Berk. L. J. Sedgwick, 1905
Frequent on brick and limestone walls. Capsules rare. 24, 26, 29, 33–9, 43–8, 54–7, 66, 67, 76, 20, 41. Throughout England, Wales and Ireland, but rare in N. Scotland.

4. Tortula laevipila (Brid.) Schwaegr. Skrimshire, 1795
Frequent on the bases of trees and the bark of Elders where not heavily shaded. Thus it is frequent on trees in hedges but rare in woods. Gemmae and capsules frequent. 15, 23, 25, 34–7, 43–6, 48, 54–7, 65, 66, 69, 75, 76, [40]. Throughout England, Wales, Ireland and S. Scotland. Rare in N. Scotland.

5. Tortula papillosa Wils. ex Spruce R. E. Parker, 1952
A rare plant occurring on the bark of trees (including Elders) and on walls. Gemmae abundant. 54, 64–6. Widely distributed in the British Isles, but not recorded from S. Wales, nor from much of N. Scotland and W. and C. Ireland.

6. Tortula latifolia (Brich.) Hartm. Henslow, 1829
Frequent on river-banks where liable to be submerged in times of flood, usually growing in the silt on the roots of river-side trees. Gemmae abundant. 34–7, 44, 45, *46*, 48, 54, 55, 57, *66*, 69, 75. Frequent in

POTTIACEAE

England, rare in Wales, Scotland and Ireland and absent from N.W. Scotland and W. Ireland.

7. Tortula subulata Hedw.   T. Martyn, 1763

Frequent on the ground in beech woods on the chalk. Occasional in woods elsewhere. Capsules abundant. *25, 34, 35, 45,* 54–6, 65, 66, 76, 40. GG. General.

8. Tortula muralis Hedw.   T. Martyn, 1763

var. **muralis**

Abundant on brick or stonework, especially on wall-tops; frequent on bare chalk in chalk-pits; rare on wooden palings and trees. Calcicole and hence absent from tarmac. Capsules abundant. All squares. General.

var. **aestiva** (Beauv.) Brid.   R. S. Adamson, 1913

Frequent in deep shade on walls and on lumps of chalk, often mixed with var. *muralis.* Capsules abundant. 25, 36, 44, 45, 47, 54, 55, 65. Widely distributed in England, and a few records from N. Wales and S. Scotland.

9. Tortula marginata (Bry. Eur.) Spruce   P. W. Richards, 1949

Recorded from brickwork, stones and lumps of chalk in deep shade, usually by water; probably frequent but much overlooked. Capsules abundant. 36, *45,* 55, 64, *65,* 75. A Mediterranean species which is widespread in England, except the south-west and extreme north, but otherwise known only from scattered localities in N. Wales, the Scottish lowlands, and N.E. and W. Ireland.

10. Tortula vahliana (Schultz) Wils.   H. N. Dixon, 1882

On chalk-banks in shade, e.g. under bushes, in sheltered places; rare. Capsules frequent. 44, 45, 56, [66]. **CH.** A Mediterranean species which is recorded from scattered localities in England and Wales south-east of a line from Glamorgan to the Humber, and from N. and E. Ireland.

Aloina Kindb.

1. Aloina rigida (Hedw.) Kindb.   H. L. K. Whitehouse,1950

Undisturbed chalky soil, rare. Capsules abundant. 45, 55, 64. Widespread in England, rare in Wales, Scotland and Ireland. Not recorded from N.W. Scotland and N.W. Ireland.

2. Aloina ambigua (B. & S.) Limpr.   Skrimshire, 1796

Undisturbed calcareous soil, such as on the floor of chalk or gravel-pits, frequent. Capsules abundant. 34, 35, 44, 45, 54, 55, *56,* 57, 65–7,

POTTIACEAE

[40]. CH. Throughout England, Wales and Ireland, but rare in Scotland where it occurs chiefly on the east side.

3. **Aloina aloides** (Schultz) Kindb.   ?Rev. Hemsted in Relhan, 1802
First certain record: H. L. K. Whitehouse, 1950.
Undisturbed chalky soil, rare. Capsules abundant. 34, 45, 54. Throughout England, Wales, Ireland and S. Scotland, but not recorded from N.W. Scotland.

PTERYGONEURUM Jur.

1. **Pterygoneurum ovatum** (Hedw.) Dix.   Relhan, 1802
Formerly on mud walls, e.g. in Parker's-piece Lane (H. E. Lowe, 1837). Occasional on chalk banks, such as occur in chalk pits and field margins. Capsules abundant. 34, 35, 45, 54–6, 66. Of widespread occurrence in England and E. Scotland, but rare in Wales and not recorded from W: Scotland and most of Ireland.

2. **Pterygoneurum lamellatum** (Lindb.) Jur.   M. C. F. Proctor, 1951
A rare species of well-drained undisturbed soil on the chalk, such as occurs in chalk-pits or at the margins of arable fields. Recorded recently from three localities on the Gog Magogs. Capsules abundant. 45. Recorded from scattered localities in England (as far north as Yorkshire), from the Scottish lowlands and from E. Ireland.

POTTIA Fürnr.

1. **Pottia lanceolata** (Hedw.) C.M.   Skrimshire, c. 1796
Frequent in chalk-pits and occasional in grassland and arable fields on the chalk. Rare on banks on other soils. Capsules abundant. 34, 35, 44, 45, 54, 55, 66, 76, 31, [40]. Widespread in England and E. Scotland, but apparently absent from most of Wales, Ireland and W. Scotland.

2. **Pottia intermedia** (Turn.) Fürnr.   R. S. Adamson, 1913
First localized record: M. C. F. Proctor, 1951. Apparently rare in Cambs, and recorded chiefly from sandy arable fields. Capsules abundant. 44–6, 54. Widespread in England, but less frequent in Wales, Scotland and Ireland.

3. Pottia truncata (Hedw.) Fürnr. ?T. Martyn, 1763
First certain record: E. W. Jones, 1934
Rare in Cambs, owing to the scarcity of lime-free soils. Recorded chiefly from arable fields on the Jurassic clays, Lower Greensand and Breck fringe. Capsules abundant. 25, 26, 35–7, 45, 65, 66, 20, 40. General.

4. Pottia davalliana (Sm.) C. Jens. Henslow, 1828
A small annual species which is abundant in arable fields on the chalk and boulder clay. Capsules abundant. 23, 24, *25*, 34–6, 38, 44–6, 48, 54–6, 59, 65, 66, 41. Throughout England, but rare in Wales, Scotland and Ireland. Not recorded from N. Scotland and S.E. Ireland.

5. Pottia bryoides (Dicks.) Mitt. Relhan, 1820
First record since Henslow: D. G. Catcheside, 1942. Occurs on roadsides, edges of paths and in grassland, particularly on the chalk, apparently uncommon. Capsules abundant. 36, 45, *54*, *55*, 56, 65, 66, 76, 31. Widespread in England, but few records from Wales, Scotland and Ireland. Not recorded from W. Scotland and W. Ireland.

6. Pottia recta (Sm.) Mitt. Skrimshire, 1796
A minute winter-fruiting annual of frequent occurrence in chalk grassland and in arable fields on the chalk. Capsules abundant. 23, 44, 45, 54–6, 66. A Mediterranean species of widespread occurrence in England, but known from only a few localities in Wales, Scotland and Ireland

PHASCUM Hedw.

1. Phascum curvicollum Hedw. Relhan, 1802
A minute winter-fruiting annual of frequent occurrence on undisturbed chalky soils. Capsules abundant. 35, 45, 54–6, 66, 41. Widespread in England, but absent from most of Wales, Scotland and Ireland.

2. Phascum cuspidatum Hedw. T. Martyn, 1763
var. **cuspidatum**
An abundant annual moss of arable fields and other disturbed ground, fruiting chiefly in winter. Capsules abundant. All squares except 29, 68. Throughout England, Wales and S.E. Scotland. Rare in Ireland and N.W. Scotland.

POTTIACEAE

†var. **maximum** Web. & Mohr   Relhan, 1820
var. *schreberianum* (Dicks.) Brid.
Formerly on Gamlingay Heath. No recent record. [25.] Recorded from
scattered localities in England.

var. **bivalens** Springer   J. G. Hughes, 1954
In arable fields with var. *cuspidatum*, rare. Presumably arises by
regeneration of gametophytes from immature sporophytes. Apogamous
capsules develop at the tips of the leaf-nerves. 26, 45. Not recorded
elsewhere in the British Isles.

3. Phascum floerkeanum Web. & Mohr
                    H. N. Dixon in P. G. M. Rhodes, 1911
A minute annual of arable fields on the chalk, fruiting in autumn.
Abundant after wet summers. Capsules abundant. 34, 35, 44–6, 54–6,
66. Known from S.E. England (S. and E. of a line from the Severn to
the Wash), N.E. England (Yorkshire to Northumberland) and N. Wales.

ACAULON C. M.

1. Acaulon muticum (Brid.) C. M.   Skrimshire, 1796
First record since Skrimshire: P. J. Bourne and H. L. K. Whitehouse,
1960. A minute annual of lime-free arable fields and banks, fruiting in
winter. Rare. Capsules recorded. 26, 49, 54, [40]. Widespread in
England, rare in Scotland, Wales and Ireland.

CINCLIDOTUS Beauv.

1. Cinclidotus fontinaloides (Hedw.) Beauv.
                    H. L. K. Whitehouse, 1957; J. W. Cowan, 1957
In Cambs this species appears to be confined to the River Ouse and the
Old Bedford River. All the records refer to stonework where liable to
flooding, but it may be expected to occur on woodwork in similar
situations. It has been known from the Ouse in Bedfordshire for over
160 years, and its apparent absence from the River Cam is remarkable.
It perhaps requires to be submerged in winter but not in summer, and
the River Cam may have insufficient seasonal variation in water level.
37, 48, 69. General, except East Anglia.

2. Cinclidotus mucronatus (Brid.) Moenk. & Loeske

P. W. Richards, 1941

On alluvium on tree-roots by the River Granta (or Bourne) near Hildersham, and in similar situations by the River Kennett. 54, *66*. A Mediterranean species which is rare in Britain. It occurs over most of southern England and S. Wales, and at scattered localities in N. Wales, N. England and S. E. Scotland.

BARBULA Hedw.

1. Barbula convoluta Hedw.   Rev. Hemsted in Skrimshire, 1796

var. **convoluta**

Abundant in a wide range of habitats where the substratum is well drained, e.g. chalky soil in beech woods and on stones, banks and walls. Gemmae (on rhizoids) frequent. Capsules uncommon. All squares except 29, 39, 58, 68. General, but rare in N.W. Scotland.

var. **commutata** (Jur.) Husnot   E. W. Jones, 1935

Recorded from roadsides in several places, with capsules. 36, *56*, 65. Most frequent in S. England, becoming rarer northwards and not recorded from much of N. Scotland.

2. Barbula unguiculata Hedw.   Skrimshire, 1797

Abundant on the ground in many well-drained habitats, e.g. chalk- and gravel-pits, chalky banks, beech woods on the chalk, arable fields, edges of paths and roadsides. Capsules frequent. All squares. General.

3. Barbula revoluta Brid.   E. W. Jones, 1932

Frequent on mortar or brick near the base of walls, where shaded. Occasional on shaded rockery stones. Gemmae frequent. 25, *34*, 36, 45, 54, 55, *56*, 66. General, but rare in N. Scotland.

4. Barbula hornschuchiana Schultz   E. W. Jones, 1934

Abundant on paths, floors of chalk-pits and other relatively undisturbed, but not shaded, habitats. Capsules occasional. 23–5, 33–6, 44–7, 54–8, 64–7, 75, 76, 30, 41. Probably general, but few records from Ireland.

5. Barbula acuta (Brid.) Brid.   H. L. K. Whitehouse, 1956

Occasional on calcareous soil, such as in chalk- and gravel-pits and arable fields. It usually occurs in small quantity mixed with other

295

POTTIACEAE

mosses, such as *Barbula fallax*. Gemmae frequent. 34, 44, 45, 55, 57, 66. Recorded from most of the western English counties, but very few records from the rest of the British Isles.

6. Barbula fallax Hedw.   H. E. Lowe, 1827

Abundant in chalk grassland, chalk-pits and the edges of paths on the chalk. Occasional on paths and woodland rides elsewhere. Capsules occasional. 15, 23, 24, *25*, 29, 33–6, 38, 39, 43–9, 54–7, 64–7, 75, 76, 30. General, but rare in N.W. Scotland.

7. Barbula rigidula (Hedw.) Mitt.   L. J. Sedgwick 1911

Occasional on walls. Gemmae abundant. *25*, 36, 45, 54, 65, 75, [46]. General.

8. Barbula nicholsonii Culmann
                              E. F. Warburg and P. W. Richards, 1945

Normally a plant of stone- or brick-work by water, but in Cambs known only from the top of a brick wall at Gamlingay. 25. Recorded from a few scattered localities, particularly in S. England, but also S. Wales, S. Ireland, N.E. England and S.E. Scotland. Not known outside the British Isles.

9. Barbula trifaria (Hedw.) Mitt.   E. F. Warburg, 1941

Frequent in a wide range of well-drained habitats, such as on the ground in beech woods on the chalk; on banks or at the base of trees in chalk-pits; on walls or rockery stones; and on brick-, stone- or wood-work by water. 24, 25, 34–6, 44–7, 54, 55, 57, 64, 66. Throughout England and Wales, but rare in Ireland and absent from most of Scotland.

10. Barbula tophacea (Brid.) Mitt.   P. W. Richards, *c.* 1930

Abundant on earth, wood or stonework by water. Capsules frequent. 29, 34–6, 38, 39, 44–6, 49, 54–6, *57*, 59, 64, 40. General.

11. Barbula cylindrica (Tayl.) Schp.   R. S. Adamson, 1913

Occasional in sheltered places on earth, wood or stone, usually by water; also occurs about the roots of beeches on the chalk. 35, 36, 44, 45, 54, 55. General.

12. Barbula vinealis Brid.   Skrimshire, 1797

First record since Skrimshire: E. F. Warburg, 1941. Abundant on walls. Occasional on well-drained earth in chalk-pits and on silt-covered tree-roots by water. 23–6, 29, 34–6, 38, 43–7, 49, 54–7, 64–7, 75, 20, 30, 41,

296

[40]. Throughout England, but apparently less frequent in Wales, Scotland and Ireland.

13. Barbula recurvirostra (Hedw.) Dix.

G. D. Haviland and J. J. Lister, 1880

Abundant in shady places on the chalk, especially on the ground or on tree-roots in beech woods. Occasional elsewhere. Capsules abundant. 23, 25, 34, 35, 43–5, 49, 54–6, 65, 66. General.

GYROWEISIA Schp.

1. Gyroweisia tenuis (Hedw.) Schp. M. C. F. Proctor, 1951

A minute species of wet calcareous brick or stonework in deep shade, probably frequent, but much overlooked. Gemmae (on protonema) frequent. 36, 45, 65, 75. Widespread in the British Isles, but rare in Ireland, N. Scotland, Wales, S.W. and E. England.

EUCLADIUM B. & S.

1. Eucladium verticillatum (With.) B. & S. E. W. Jones, 1932

On brick or stone, where permanently wet with calcareous water, rare. 45, 66. General, but rare in East Anglia and the East Midlands.

TORTELLA (C.M.) Limpr.

1. Tortella tortuosa (Hedw.) Limpr.

Rev. Hemsted in Skrimshire, 1796

First record since Henslow: E. W. Jones, 1934. A plant of calcareous substrata, known in Cambs only from a limited stretch of chalk grassland on the Devil's Dyke. 66. General, except that it is rare in the eastern half of England and not recorded from Norfolk, Suffolk or Essex.

2. Tortella inclinata (Hedw. f.) Limpr. P. G. M. Rhodes, 1911

Recorded from chalk grassland over a limited stretch of the Devil's Dyke. 66. A rare plant of calcareous grassland and rocks, known only from a few scattered localities in the British Isles.

3. Tortella inflexa (Bruch) Broth. E. F. Warburg, 1961

On lumps of chalk on the Devil's Dyke. 66. A Mediterranean species first recorded in the British Isles on the South Downs in 1951, but now widespread in S.E. England.

POTTIACEAE

TRICHOSTOMUM Bruch

1. Trichostomum sinuosum (Wils.) Lindb.   P. W. Richards, 1929
Frequent on tree-roots by streams where liable to flooding. Occasional on the roots of beeches on the chalk. Rare in other calcareous habitats, e.g. limestone wall-tops. 34–6, *37*, 44, 45, 54, 55, 64–6. Widely distributed in S. England, but becoming rarer northward and absent from most of Scotland and N. Ireland.

2. Trichostomum crispulum Bruch   M. C. F. Proctor, 1950
Known in Cambs only from chalk grassland over a limited area of the Devil's Dyke. 66. Widespread in the north and west of the British Isles, but becoming rarer eastward and apparently absent from Norfolk Suffolk and Essex.

3. Trichostomum brachydontium Bruch   E. F. Warburg, 1944
Occurs in chalk grassland over a limited area of the Devil's Dyke. 66. Throughout the north and west of the British Isles, but only in scattered localities in E. England.

WEISSIA Hedw.

1. Weissia controversa Hedw.   Relhan, 1785
First record since Henslow: D. E. Coombe, 1957. Formerly at Gamlingay, but the only recent record is from chalk grassland at Quy Fen. Capsules abundant. 56, [25]. General.

2. Weissia microstoma (Hedw.) C. M.   H. E. Lowe, 1834
First record since Henslow: E. W. Jones, 1932. Abundant in chalk grassland. Capsules abundant. 35, 44, 45, 54–6, 64, 66. General in Britain but only a few records from Ireland.

3. Weissia rostellata (Brid.) Lindb.   H. L. K. Whitehouse, 1957
Sides of ditches and ruts in Gamlingay Wood, and in an arable field at Hatley St George. Capsules present. 25. Known in the British Isles only from scattered localities in S.E. England, the Midlands, N. England and N.E. Ireland.

4. Weissia crispa (Hedw.) Mitt.   Henslow, 1827
var. **crispa**
Frequent in chalk grassland. Capsules abundant. 45, 55, 56, 66. Widely distributed in England, but of very local occurrence in Wales and Ireland, and not recorded from Scotland.

var. **aciculata** (Mitt.) Braithw.   H. L. K. Whitehouse, 1952
Frequent in arable fields on the non-calcareous Jurassic clays, and
occasional in arable fields and pastures on boulder clay and non-
calcareous gravel. Capsules frequent. 25, 26, 35, 36, 45–7. Recorded
from scattered localities in England.

5. Weissia sterilis Nicholson   H. L. K. Whitehouse, 1940
In undisturbed chalk grassland, rare. Capsules recorded. 55, 56, 66.
Recorded from chalk and limestone in southern England. Not known
outside the British Isles.

GRIMMIALES

GRIMMIACEAE

GRIMMIA Hedw.

1. Grimmia apocarpa Hedw.   ?Relhan, 1785
First certain record: Henslow, 1827.
On wall-tops, apparently rather uncommon. Capsules abundant. 33,
34, 39, 43, 45, 47, 54, 55, 65, 66, [25]. General.

2. Grimmia pulvinata (Hedw.) Sm.   T. Martyn, 1763
Abundant on brick and limestone wall-tops and on tiled roofs, and three
records from wood. Capsules abundant. All squares. General.

3. Grimmia trichophylla Grev.   R. E. Parker, 1953
On shady brick and sandstone walls, rare. 54. Widespread in Ireland,
Scotland, Wales and N. and W. England, but rare in S. and E. England,
where it appears to be absent from large areas.

RHACOMITRIUM Brid.

1. Rhacomitrium heterostichum (Hedw.) Brid.   S. J. P. Waters, 1961
On clinker of filter bed at sewage works, Madingley Hall. 36. General
in north and west of British Isles, but absent from most of E. England.

2. Rhacomitrium canescens (Hedw.) Brid.   Relhan, 1820
First record since Henslow: S. J. P. Waters, 1961. Formerly at
Gamlingay Heath. On clinker of filter bed at sewage works, Madingley
Hall. 36, [25]. General, but rare or absent from much of E. England.

GRIMMIACEAE

3. Rhacomitrium lanuginosum (Hedw.) Brid.   S. J. P. Waters, 1961
On clinker of filter bed at sewage works, Madingley Hall. 36. General in north and west of British Isles, but absent from most of E. England.

FUNARIALES

FUNARIACEAE

FUNARIA Hedw.

1. Funaria hygrometrica Hedw.   Ray, 1660
On sites of fires, paths, in flower-pots, etc., abundant but seems always to be associated with human activity. All squares except 39, 41. General.

PHYSCOMITRIUM Brid.

1. Physcomitrium pyriforme (Hedw.) Brid.   Skrimshire, *c.* 1795
A summer-fruiting annual of mud thrown out of ditches, and of earth by streams where liable to flooding. Sporadic in appearance, but sometimes locally abundant. Capsules abundant. 44–6, 49, 56, 58, 59, 65, 67, 40, [55]. General, but few records from N. Scotland.

PHYSCOMITRELLA B. & S.

1. Physcomitrella patens (Hedw.) B. & S.   J. Carpenter, 1941
A summer-fruiting annual of stream, ditch and pond margins, and cart-ruts, where liable to flooding, apparently rare. Capsules abundant. 24, *25*, 37, 45, 48, 54, 55, 20. Widely distributed in England, but very few records from Wales, Scotland and Ireland. Not recorded north of Edinburgh.

EPHEMERACEAE

EPHEMERUM Hampe

1. Ephemerum recurvifolium (Dicks.) Boul.   E. W. Jones, 1934
A minute winter annual of arable fields on calcareous clay soils. Apparently rare, but probably much overlooked. Capsules recorded. *34*, *35*, *36*, 45, 46. Confined to S.E. England, south and east of a line from Somerset to Lincolnshire, apart from isolated records from Cheshire and Durham.

EPHEMERACEAE

2. Ephemerum serratum (Hedw.) Hampe var. **minutissimum** (Lindb.)
Grout   ?Rev. Hemsted in Relhan, 1802
var. *angustifolium* B. & S.   First certain record: G. Halliday, 1957
A minute autumn-fruiting annual of lime-free soil. Recorded recently
from Gamlingay Wood and Ditton Park Wood, and from arable fields
on the Jurassic clays. Capsules abundant. 25, 26, 36, 54, 65, [?66].
Var. *minutissimum* is a plant of S. and W. Europe which is widespread
in S. and E. England, becoming rarer to the north and west, with only
a few records from Scotland and Ireland. (Var. *serratum* is known from
scattered localities throughout the British Isles, but appears to be less
frequent than var. *minutissimum* in southern England.)

SPLACHNACEAE

SPLACHNUM Hedw.

1†. Splachnum ampullaceum Hedw.   T. Martyn, 1763
This moss of dung on acid heaths, formerly grew at Hinton, Teversham
and Sawston Moors and Gamlingay Bog, but has not been recorded
since about 1820. [25, 44, 45, 55.] Widely distributed in the British
Isles, but rare in lowland parts of England.

TETRAPHIDALES

TETRAPHIDACEAE

TETRAPHIS Hedw.

1. Tetraphis pellucida Hedw.   T. Martyn, 1763
First certain localized record: M. C. F. Proctor, 1951. Only known
from stumps in Gamlingay Heath Wood. Gemmae present. This species
favours markedly acid conditions, and therefore, in the absence of a
specimen, Relhan's record from the bank of the ditch between Coe Fen
and Peterhouse Garden is in doubt. 25, [?45]. General, but rare in
eastern England.

BRYACEAE

EUBRYALES

BRYACEAE

ORTHODONTIUM Schwaegr.

1. Orthodontium lineare Schwaegr. P. W. Richards, 1947

On rotten wood; abundant in woods on the Lower Greensand, occasional in woods on the boulder clay and chalk. Gemmae (on protonema) and capsules abundant. This species is a recent introduction and is probably still spreading in the county. 25, 26, 35, 36, 45, 54, 55, 66. G. A plant native to the temperate regions of the southern hemisphere, first recorded in the British Isles near Manchester in 1920. It has been spreading steadily since then. By 1939 it was widespread in N. England. During the 1940's it reached E. Ireland, the English Midlands, East Anglia, Belgium, Holland and N. Germany, and it has now spread to S. and E. Scotland, N.E. Wales, southern England and Denmark.

LEPTOBRYUM Wils.

1. Leptobryum pyriforme (Hedw.) Wils. Relhan, 1802

Relhan's record was from the Botanic Garden hot-house and the first certain record in a natural habitat was by P. W. Richards, 1930. Frequent in flower-pots. Occasional, or sometimes abundant, on the sites of fires. Occasional on ditch-sides and pond-margins. Gemmae (on rhizoids) and capsules frequent. 25, 29, 34, 45, 57, 20, 40, 41, 50. General.

POHLIA Hedw.

1. Pohlia nutans (Hedw.) Lindb. P. G. M. Rhodes, 1911

Of frequent occurrence in lime-free habitats, such as on rotten stumps and at the base of trees, particularly beeches, in woods and plantations. Capsules frequent. 23, 25, 26, 34, 36, 44, 45, *46*, 49, 54–6, *58*, *64*, *65*, 66, 68, 75, 31. General.

2. Pohlia wahlenbergii (Web. & Mohr) Andr. P. W. Richards, 1938
P. albicans (Wahl.) Lindb.

On damp clay soil in woodland rides, rare. *25, 35*. Throughout the British Isles except East Anglia and the E. Midlands, where there are very few records.

3. Pohlia delicatula (Hedw.) Grout T. Martyn, 1763

An abundant species of river-banks, ditch-sides, and margins of ponds, throughout the county. Capsules frequent. 15, 24–6, 29, 34–6, 38, 43–9,

54–7, 59, 64–7, 75, 76, 50. General, but not recorded from a number of Irish vice-counties.

BRYUM Hedw.

1. Bryum inclinatum (Brid.) Bland. R. S. Adamson, 1913
The only record is by Adamson without precise locality. A plant of dry heaths and walls. General, but few records from E. England.

2. Bryum pallens (Brid.) Rohl M. C. F. Proctor, 1953
By water, apparently rare. 45. General, but few records from E. England.

3. Bryum pseudotriquetrum (Hedw.) Schwaegr.

†var. **pseudotriquetrum** Relhan, 1785
Recorded from Hinton, Teversham and Shelford Moors, but no recosds since Relhan's. [45, 55.] General.

var. **bimum** (Brid.) Richards & Wallace Relhan, 1820
First record since Relhan's: P. W. Richards, 1929. Abundant in marshy places in the Fens, as at Wicken. Occasional in damp places elsewhere. - Capsules frequent. 44, 54, 56, 57, 66, [45, 55]. General.

4. Bryum creberrimum Tayl. B. Reeve, 1959
B. affine Lindb. & Arnell.
On damp sandy soil, Gray's Moor Pits, Chainbridge, near March, with capsules. 40. Scattered records, chiefly from C. and N. England.

5. Bryum intermedium (Ludw.) Brid. R. S. Adamson, 1913
First localized record: B. Reeve, 1959. On marshy ground in gravel-pits, rare. Capsules present. 49. Widely distributed in the British Isles, but few records from W. Scotland, W. and C. Ireland and E. England.

6. Bryum caespiticium Hedw. Relhan, 1785
Abundant on walls. Frequent on banks in gravel- and chalk-pits and on paths and railway ballast. Occasional on stumps. Capsules frequent. 23–5, 29, 33–9, 45, 46, 48, 49, 54, 57, 59, 65, 67, 68, 76, 20, 30, 31, 40, 41, 50. General.

7. Bryum argenteum Hedw. T. Martyn, 1763
var. **argenteum**
Abundant on paths and roadsides (including tarmac) and on roofs and walls. Occasional in a wide range of other habitats, but usually where

BRYACEAE

there has been recent disturbance by man. Capsules frequent in some years, rare in others. All squares. General.

Var. **lanatum** (Beauv.) B. & S.   P. J. Bourne, 1960

On paths and in gravel-pits, rare. Capsules recorded. 33, 34, 45, 66. Widely distributed in the British Isles.

8. Bryum bicolor Dicks.   G. D. Haviland and J. J. Lister, 1878

Abundant in chalk- and gravel-pits and on paths. Occasional in arable fields and on earth on walls. Gemmae abundant, especially in damp habitats. Capsules frequent. All squares. General, but few records from N. Scotland and C. Ireland.

9. Bryum radiculosum Brid.   R. E. Parker and M. C. F. Proctor, 1952

B. murale Wils. ex Hunt

Recorded only from chalk grassland on the Devil's Dyke. Gemmae (on rhizoids) present. 56. Widespread in England, Wales and Ireland, but rare in Scotland, where it is recorded only from scattered localities on the west side.

The following five species are closely allied and in the past have all been referred to *B. erythrocarpon* Schwaegr. nom. illegit. Species 12 and 13 have been recently described in *Bot. Notiser*, **116**, 94–5 (1963).

10. Bryum rubens Mitt.   Henslow, 1834

One of the most abundant mosses in the county, occurring in arable fields and other disturbed ground on all soils. Gemmae (on stems and rhizoids) abundant. 23, 25, 35, 36, 39, 44–7, 49, 54–7, 59, 64, 65, 30. Widely distributed in the British Isles.

11. Bryum microerythrocarpum C. Müll & Kindb.

D. F. Chamberlain, 1960

Recorded from undisturbed soil on the Jurassic clay and Breck sands on the extreme west and east borders of the county, respectively. Gemmae (on rhizoids) abundant. Capsules recorded. 15, 66. Widely distributed in the British Isles on non-calcareous soils.

12. Bryum ruderale Crundw. & Nyh.   H. L. K. Whitehouse, 1957

Abundant on roadsides and paths. Gemmae (on rhizoids) abundant. 34, 44, 45, 47, 54, 66, 30. Widely distributed in the British Isles.

13. Bryum violaceum Crundw. & Nyh.

P. J. Bourne and H. L. K. Whitehouse, 1960

Frequent in arable fields. Gemmae (on rhizoids) abundant. 23, 44–6, 54, 20. Widely distributed in the British Isles.

14. Bryum klinggraeffii Schp.   H. L. K. Whitehouse, 1956
Frequent in arable fields on all soils. Gemmae (on rhizoids) abundant.
23, 26, 34–6, 44–7, 49, 54–7, 59, 65, 67, 20. Widely distributed in the
British Isles.

15. Bryum capillare Hedw.   Relhan, 1785
Abundant on stumps and tree-bases in woods and hedges, the floor of
beech woods, walls, roofs, and banks in chalk- and gravel-pits. Gemmae
(on rhizoids) and capsules frequent. All squares. General.

RHODOBRYUM (Schp.) Limpr.

1. Rhodobryum roseum (Hedw.) Limpr.   Relhan, 1820
A plant of grassland on sandy soils, known from a field at Gamlingay
and from Hildersham Furze Hills. 54, [25]. Widespread in the British
Isles, but rather few records from Ireland and N. Scotland.

MNIACEAE
MNIUM Hedw.
1. Mnium hornum Hedw.   Relhan, 1802
A plant of lime-free shady soils. Abundant on the ground in woods on
the Lower Greensand at Gamlingay, and occasional in boulder-clay
woods. In beech woods on the chalk, it occurs chiefly at the foot of
trees. Capsules frequent. 25, 26, 35, 36, 43, 45, 46, 54, 55, 57, 64–7, 76.
General.

2. Mnium cuspidatum Hedw.   Relhan, 1785
On the ground in woods and in lawns, rare. 45, 54. **GG.** General.

3. Mnium longirostrum Brid.   Relhan, 1820
On the ground in woods and in lawns, rare. Capsules occasional. 25, 36,
45, 54, 66. General.

4. Mnium affine Bland.   R. S. Adamson, 1913
Frequent on the ground in grassland and in woods, particularly where
the soil is sandy or liable to flooding by streams. 15, 25, 36, 44–7, 54,
56, 57, 64–7, 76, 31. General, but rather few records from Ireland,
W. Scotland and W. Wales.

5. Mnium seligeri (Jur. ex Lindb.) Limpr.   P. G. M. Rhodes, 1911
On the ground in marshes and fens, locally abundant. One of the few
bryophytes characteristic of fen communities. 34, 57, 66. **W.** Recorded

MNIACEAE

from scattered localities throughout the British Isles, but few records from Ireland.

6. Mnium undulatum Hedw.   T. Martyn, 1763

On the ground in woods, abundant. Occasional in damp grassland, including lawns. Capsules very rare. 15, 25, 26, 34–6, 39, 43–7, 49, 54, 55, 57, 64–7, 75, 76, 41. General.

7. Mnium punctatum Hedw.   Relhan, 1802

First record since Henslow: R. A. Lewin, 1941. On rotten wood or on the ground, in fen-woods or other damp woodland, occasional and usually stunted and only in small quantity. 15, 25, 44, 45, 54, 55, 66. General.

AULACOMNIACEAE

AULACOMNIUM Schwaegr.

1†. Aulacomnium palustre (Hedw.) Schwaegr.   T. Martyn, 1763

Formerly on Hinton and Shelford Moors and Gamlingay Bogs. Last record: Gamlingay, 1881. [25, 45.] General.

2. Aulacomnium androgynum (Hedw.) Schwaegr   ?Henslow, 1829
                                          First certain record: R. S. Adamson, 1913.

Frequent on stumps and rotten wood, occasional on peaty soil in damp woodland. Gemmae abundant. This species appears to have increased in frequency in recent years, as, until 1940, it was known only from the neighbourhood of Gamlingay and from the Breck fringe. 15, 25, 26, 29, 34–6, 44–6, 49, 54–6, 58, 59, 65–8, 30. Widespread in E. England and E. Scotland, becoming rarer westward. Absent from most of Ireland except the north-east, and from much of S.W. England, W. Wales and W. Scotland.

BARTRAMIACEAE

BARTRAMIA Hedw.

1†. Bartramia pomiformis Hedw.   Relhan, 1802

On shady lime-free banks. Formerly at Gamlingay, but not recorded since 1830. [25.] General, but rare in E. England.

BARTRAMIACEAE

PHILONOTIS Brid.

1†. Philonotis fontana (Hedw.) Brid.   T. Martyn, 1763
Formerly grew on Hinton and Sawston Moors and in Gamlingay Bog.
Last record, Gamlingay, 1827. [25, 44, 45.] General.

2†. Philonotis calcarea (B. & S.) Schp. W. Watson between 1907 and 1926
Watson's record is without locality. No other record. A plant of
marshes and springs. Widely distributed in the British Isles, but absent
from much of the eastern half of England and few records from
W. Scotland and W. Ireland.

ISOBRYALES

ORTHOTRICHACEAE

ZYGODON Hook. & Tayl.

1. Zygodon viridissimus (Dicks.) R.Br.
British Bryological Society Census Catalogue, 1907
First localized record: P. W. Richards, 1928. A frequent epiphyte,
occurring on tree-trunks and on the bark of Elders, especially where
sheltered and not too heavily shaded. All the Cambs specimens
examined have belonged to var. **viridissimus.** Gemmae abundant. 15,
26, 34–6, 43–5, *46*, 54, 55, 65–7, 76. **CH.** General.

ORTHOTRICHUM Hedw.

1. Orthotrichum anomalum Hedw.   Relhan, 1820
On limestone wall-tops, frequent. Capsules abundant. 24, *35*, 36, 44–6,
48, 54, 65. General.

2. Orthotrichum cupulatum Brid.   Henslow, 1829
First localized record: A. O. Chater, 1955. On stonework of bridge at
Murrow, with capsules. 30. Widely distributed in the British Isles, but
rare in E. England.

3. Orthotrichum affine Brid.   ?T. Martyn, 1763
First certain record: G. D. Haviland, 1881.
Frequent on bark, particularly of Elders, where sheltered and not too
heavily shaded, e.g. in abandoned chalk-pits. Capsules abundant. *34*,
35, 44, 45, *46*, 48, 54, 55, 57, 58, 65–7, 31. General.

ORTHOTRICHACEAE

4†. Orthotrichum lyellii Hook. & Tayl.   P. G. M. Rhodes, 1911
An epiphytic species, only once recorded from the county. [45.]
Widespread in the British Isles, but rare in E. England, where it appears
to have decreased in frequency during the last hundred years, perhaps
through increased atmospheric pollution. Not recorded from much of
W. Ireland and W. Scotland.

5. Orthotrichum diaphanum Brid.   Henslow, 1829
Abundant on the bark of Elders and other trees where not too heavily
shaded, and on walls. Capsules and gemmae abundant. 15, 23–6, 33–8,
44–8, 54–9, 64–8, 75, 76, 20, 30. General.

ULOTA Brid.

1†. Ulota crispa (Hedw.) Brid.   Relhan, 1802
Formerly recorded, with capsules, from trees in a number of localities,
but no record since 1881. The reduction in frequency of this and a
number of other epiphytic species during the last century may be due
to increased atmospheric pollution. [35, 36, 65.] General, but rare in
E. England.

FONTINALACEAE

FONTINALIS Hedw.

1. Fontinalis antipyretica Hedw.   Ray, 1663
Attached to wood or stone in rivers, streams, ditches and ponds,
frequent. Capsules rare. 25, 34, 36, 37, 44, 45, 48, 54, 56, 57, 65, 68,
69, 75, 20. General.

CLIMACIACEAE

CLIMACIUM Web. & Mohr

1. Climacium dendroides (Hedw.) Web. & Mohr
D. Vernon in Ray, 1696
First record since Henslow: P. W. Richards, 1927. Known from the
Breck fringe (the plant is common in damp places in the Suffolk
Breckland), from marshy ground adjoining Wicken Fen, and from two
boulder-clay woods. *35, 45, 57, 67*, [25]. General.

308

CRYPHAEACEAE

CRYPHAEACEAE

CRYPHAEA Mohr

1. Cryphaea heteromalla (Hedw.) Mohr   T. Martyn, 1763

On the bark of trees, usually Elders, in sheltered places, rare. Has apparently decreased in abundance since last century. Capsules uncommon. 54, 55, 65, 66, [25, 35, 45, 46]. A species of S. and W. Europe, which occurs throughout the British Isles except the northern half of Scotland, but rare in E. England, and no records from E. Wales.

LEUCODONTACEAE

LEUCODON Schwaegr.

1. Leucodon sciuroides (Hedw.) Schwaegr.   T. Martyn, 1763

On bark of isolated trees and on walls, rare and apparently decreasing. *25, 34, 35*, 55, *65*, [45]. Throughout England and Wales, but largely ·confined to the south and east of Scotland and Ireland.

NECKERACEAE

NECKERA Hedw.

1. Neckera complanata (Hedw.) Hüben.   T. Martyn, 1763

On the bark of trees and on stumps, both in woods and hedges, frequent. On the ground in chalk grassland, rare. Capsules rare. 23–5, *35*, 36, 43–6, 54, 55, 57, 64–6, 76. General.

OMALIA (Brid.) B. & S. (*Homalia*)

1. Omalia trichomanoides (Hedw.) B. & S.   Relhan, 1802

On the bark of trees and on stumps, chiefly in woods, frequent. Capsules frequent. 24, *25*, 35, 36, 43–6, 54, 64–6. General, but few records from N. Scotland.

THAMNIUM B. & S.

1. Thamnium alopecurum (Hedw.) B. & S.   T. Martyn, 1763

Abundant on the ground in woods, especially those on the boulder clay. Bearing nematode galls, Great Abington, February 1940. Infected with the fungus *Pachybasium tilletii* (Desm.) Oudem (det. N. F. Robertson) at Dry Drayton, November 1956. Capsules rare. 15, 24–6, 34–6, 43–7, 54–8, 64–6, 76. General.

HYPNOBRYALES

LESKEACEAE

LESKEA Hedw.

1. Leskea polycarpa Hedw.

P. G. M. Rhodes and L. J. Sedgwick, 1911

On tree-bases or stonework by streams and rivers, where liable to flooding, frequent. Capsules frequent. 34, *35*, *36*, 37, 45, 48, 54, 55, 57, 58, [46]. Apparently absent from the northern half of Scotland, otherwise general.

THUIDIACEAE

ANOMODON Hook. & Tayl.

1. Anomodon viticulosus (Hedw.) Hook. & Tayl.   T. Martyn, 1763

Frequent on tree-bases and on the ground in beech woods on the chalk and occasional in woods on the chalky boulder clay. Frequent on silt-covered tree-roots by streams. 25, *35*, 36, 45, 46, 54, 55, *64*, 65, 75, [66]. GG. General, but few records from N. Scotland.

THUIDIUM B. & S.

1. Thuidium abietinum (Brid.) B. & S.   Relhan, 1802

Locally frequent in undisturbed chalk grassland where the turf is short. 44, 45, 55, 56, 66, 67. GG. Erratically distributed in the British Isles, occurring in a number of widely scattered areas, and not recorded from S. Ireland and N.W. Scotland. This distribution pattern can probably be related to inefficient dispersal, as capsules have never been found in the British Isles.

2. Thuidium tamariscinum (Hedw.) B. & S.   Ray, 1663

Locally abundant on the ground in woods on the boulder clay. Capsules very rare. 25, 35, 36, 43, *44*, 45, 46, 54, 64–6. General.

THUIDIACEAE

3. Thuidium philibertii Limpr.

C. D. Pigott and M. C. F. Proctor, 1951

A local plant of calcareous grassland, known in Cambs only from the lawn at Hildersham Hall. 54. Widely distributed in the British Isles, but distribution rather erratic perhaps owing to the absence of capsules.

HYPNACEAE
CRATONEURON (Sull.) Roth.

1. Cratoneuron filicinum (Hedw.) Roth   T. Martyn, 1763
var. **filicinum**

Abundant in damp grassland, particularly by water. Forming large cushions in the splash zone at weirs in the Cam and at the margins of fen lodes. Capsules rare. 24, 25, 29, 34–7, 44–6, 48, 54–7, 64, 65, 66, 67, 75, 30. General.

var. **fallax** (Brid.) Moenk.   P. G. M. Rhodes, 1911
Submerged in pools, rare. 66, [55]. Widely distributed in the British Isles.

2. Cratoneuron commutatum (Hedw.) Roth   R. S. Adamson, 1913
First localized record: P. W. Richards, 1929. A plant of permanently wet calcareous substrata. Rare in Cambs, and known only from relics of undrained fens. 66, [44, 57]. C. General, except S.E. England, where it is rare and apparently absent from a number of vice-counties.

CAMPYLIUM (Sull.) Mitt.

1. Campylium stellatum (Hedw.) J. Lange & C. Jens.   Relhan, 1802
Frequent on the ground in Wicken and other fens, occasional in marshes elsewhere, e.g. in gravel-pits. Capsules rare. 29, 44, 54, 56, 57, 66, [25, 45, 55]. General.

2. Campylium protensum (Brid.) Kindb.
G. D. Haviland and J. J. Lister, 1881
Frequent on the ground in the relics of undrained fen; rare in chalk grassland and in woodland rides on boulder clay. 25, 29, 44, 45, 56, 57, 66. GG. General, but few records from S. Ireland.

3. Campylium chrysophyllum (Brid.) Bryhn.   P. G. M. Rhodes, 1908
Abundant in chalk grassland. A single record from a clover field on boulder clay. 24, 34, 35, 44, 45, 55, 56, 66. General, but few records from W. Scotland.

311

HYPNACEAE

4. Campylium polygamum (B. & S.) J. Lange & C. Jens.

R. S. Adamson, 1913

First localized record: P. J. Chamberlain, 1949. Frequent on the ground in undrained fens; occasional in gravel-pits and pastures. Capsules frequent. 45, 56, 57, 40. Widely distributed in the British Isles.

5. Campylium elodes (Spruce) Broth. L. J. Sedgwick in Rhodes, 1911

On the ground in undrained fens, rare. 56, 57, 66, [55]. **W.** Widely distributed but rare in the British Isles.

6. Campylium calcareum Crundw. & Nyh.   P. W. Richards, 1940

C. sommerfeltii auct. eur.

Frequent on the bases of trees or on the ground in woods on the chalk. Capsules abundant. 45, 55. **GG.** Widely distributed in the British Isles, but not recorded from W. Scotland or S. Ireland.

LEPTODICTYUM (Schp.) Warnst.

1. Leptodictyum riparium (Hedw.) Warnst.   T. Martyn, 1763

On earth, wood or stone by ponds and streams, abundant. Capsules abundant. 15, *24*, 25, 26, 34–7, 43–6, 48, 49, 54–7, 64–7, 75, 40. General, but rather few records from N. Scotland.

AMBLYSTEGIUM B. & S.

1. Amblystegium serpens (Hedw.) B. & S.   T. Martyn, 1763

Abundant on tree-stumps and rotten wood in woods and hedges, less frequently on the ground. Frequent on damp shady brick or stone-work. Capsules abundant. All squares. General.

2. Amblystegium juratzkanum Schp.   M. C. F. Proctor, 1950

Occasional by water, occurring either on tree-bases, rotten stumps and woodwork, or less commonly on stones or on the ground. Capsules frequent. 35, 44, 45, 54, 57. Widespread in England and E. Scotland, becoming rarer westward. Very few records from W. Scotland, W. Wales and Ireland.

3. Amblystegium varium (Hedw.) Lindb.   P. W. Richards, 1929

At the base of trees or on decaying wood, by water, apparently rare, but perhaps overlooked. Capsules recorded. 25, 43, 45, [46, 57]. Widespread in England and N. Ireland, but few records from Scotland, Wales and S. Ireland.

DREPANOCLADUS (C. M.) Roth

1. Drepanocladus aduncus (Hedw.) Warnst.   Relhan, 1785
Abundant in or near ponds and ditches. Capsules rare. 25, 29, 34, 36, 38, 39, 44–7, 49, 54–7, 59, 66, 67, 69, 20, 30, 31, 40, 50. General.

2†. Drepanocladus fluitans (Hedw.) Warnst.   Relhan, 1802
A plant of relatively acid marshes and pools. A number of records, between 1802 and 1904, but all may refer to *D. aduncus*. [25, 45.] General.

3†. Drepanocladus revolvens (Sm.) Warnst.   Henslow, 1827
Formerly at Gamlingay. No record since Henslow's. [25.] Absent from much of S.E. England, otherwise general.

SCORPIDIUM (Schp.) Limpr.

1. Scorpidium scorpioides (Hedw.) Limpr.   Relhan, 1785
First record since Henslow: H. L. K. Whitehouse, 1955. Formerly in bogs at Hinton, Shelford, Sawston and Fowlmere and probably abundant in the Fens before they were drained. The only recent record is from the coprolite pit at Quy Fen. 56, [44, 45]. General, except in the English Midlands and parts of S.E. England.

ACROCLADIUM Mitt.

1. Acrocladium giganteum (Schp.) Richards & Wallace
                                                             T. G. Tutin, 1942
In wet hollow, Dernford Fen. *45*. Widely distributed in the British Isles, but few records from S. Ireland and S.W. England.

2. Acrocladium cuspidatum (Hedw.) Lindb.   Ray, 1696
Abundant amongst grass, particularly in marshes, at pond margins, in woodland rides and in chalk grassland. Capsules rare. 15, 25, 26, 29, 34–9, 43–8, 54–7, 64–6, 75, 30, 40, 41. General.

ISOTHECIUM Brid.

1. Isothecium myurum (Brid.) Brid.   P. W. Richards, 1932
On stumps and tree-trunks in woods, frequent. Capsules rare. 24, 25, 29, 35, 39, 43, 45, 46, *49*, 54, 57, 64–6. General.

HYPNACEAE

2. Isothecium myosuroides Brid.   T. Martyn, 1763
On stumps or tree-trunks in woods, occasional. 25, 26, 35, *36*, 45,
54, 55, 64, [66]. Throughout the British Isles, very abundant in the
west but becoming rarer eastward.

SCORPIURIUM Schp.

1. Scorpiurium circinatum (Brid.) Fleisch. & Loeske
                                         J. M. Lock, 1961
On limestone in rockery, Hildersham Hall, where it was presumably
introduced with the stones, which were brought from Killarney about
1936. 54. A Mediterranean species which in the British Isles is always
sterile and confined to calcareous rock or earth in southern and south-
western districts. Its natural stations nearest to Cambs are in Kent and
Surrey.

CAMPTOTHECIUM B. & S.

1. Camptothecium sericeum (Hedw.) Kindb.   T. Martyn, 1763
Abundant on walls and roofs, and the bases of trees, extending on to
humus in beech woods on the chalk. Capsules uncommon. All squares
except 59, 68. General.

2. Camptothecium lutescens (Hedw.) Brid.   Relhan, 1802
Abundant on the chalk, particularly in grassland. Rare in hedgebanks
and the edges of woods on the boulder clay. Capsules rare. 23, 24, *25*,
33–6, 43–5, 54–7, 64–7, 76. General.

BRACHYTHECIUM B. & S.

1. Brachythecium albicans (Hedw.) B. & S.   Skrimshire, 1797
Abundant on lime-free sandy soils, such as on the Lower Greensand
near Gamlingay and on the Breck fringe near Kennett. Frequent on
roadsides and on thatch throughout the county. 23, 25, 33, 34, 38, 39,
44–7, 54–6, 58, 59, 64–7, 76, 31, 40. General, but not recorded from
much of C. Ireland.

2. Brachythecium glareosum (Bruch) B. &. S.   F. Y. Brocas, 1874
In chalk-pits and on chalk banks, rare. 55, *56*, *64*, *65*, [44]. Throughout
England, but only scattered records from Wales, Scotland and Ireland.

3. **Brachythecium salebrosum** (Web. & Mohr) B. & S.
G. Halliday, 1958
On stumps and rotten logs, apparently rare. Capsules recorded. 34, 54.
Widely distributed in the British Isles.

4. **Brachythecium mildeanum** (Schp.) Milde   P. W. Richards, 1929
On the ground amongst grass near water, apparently rare. Capsules
recorded recently. 57, [45]. Widely distributed in the British Isles.

5. **Brachythecium rutabulum** (Hedw.) B. & S.   J. Ray, 1660
One of the most abundant mosses in the county. Particularly abundant
on rotten wood, but frequent also on the ground in woods and fields,
and on shady walls. Capsules abundant. All squares. General.

6. **Brachythecium rivulare** (Bruch) B. & S.   R. S. Adamson, 1913
First localized record, R. E. Parker, 1953. On earth by streams or
springs, occasional. 24, 45, 54, 57. General, but rather rare in E.
England.

7. **Brachythecium velutinum** (Hedw.) B. &. S.   T. Martyn, 1763
Abundant on tree-bases and stumps. Occasional on humus or sandy soil
in beech plantations. Capsules abundant. 23, 24, *25*, 33–7, 43–7, 49,
54–7, 59, 64–8, 76, 20, 40. Throughout England, Wales and E. Scotland,
but few records from W. Scotland and C. Ireland.

8. **Brachythecium populeum** (Hedw.) B. & S.
J. Harding and P. D. Brown, 1960
In chalk grassland on the Devil's Dyke. 66. General, but uncommon in
E. England.

SCLEROPODIUM B. & S.

1. **Scleropodium caespitosum** (Wils.) B. & S.   R. E. Parker, 1953
At the base of trees by water, apparently rare but perhaps overlooked.
44–6, 54. A species of W. Europe from Holland to the Pyrenees,
known in the British Isles chiefly from S. England and the Midlands,
but with isolated records from Wales, Ireland and S. Scotland.

2. **Scleropodium tourretii** (Brid.) L. F. Koch   M. C. F. Proctor, 1953
S. illecebrum auct.
Under Beeches, in grounds of Hildersham Hall. 54. A species of S. and
W. Europe, most frequent in the British Isles near the south coast of

HYPNACEAE

England and Ireland, and, although known as far north as Perthshire, there are very few records north of a line from the mouth of the Shannon to Anglesey and the Wash.

CIRRIPHYLLUM Grout

1. Cirriphyllum piliferum (Hedw.) Grout   P. W. Richards, 1931
On the ground in woods, abundant. Occasional among grass on ditch banks and in lawns. 25, 35, 36, 43–6, *49*, 54, 57, 64–7, 76. General.

2. Cirriphyllum crassinervium (Tayl.) Loeske & Fleisch.
H. L. K. Whitehouse, 1951
Frequent on the ground in beech plantations on the chalk. Occasional on tree-bases, particularly when by water where liable to flooding. 25, 35, 36, 45, 46, 54, 65. GG. General, but few records from N. Scotland and East Anglia.

EURHYNCHIUM B. & S.

1. Eurhynchium striatum (Hedw.) Schp.   Relhan, 1820
First localized record: P. W. Richards, 1929. Abundant on the ground in well-established woodland, but usually absent from plantations or other woodland of recent origin. Rare in chalk grassland. Capsules rather uncommon. 23, 25, 26, 33, 35, 36, 43–6, 54, 55, 64–6. General.

2. Eurhynchium praelongum (Hedw.) Hobk.   T. Martyn, 1763
Abundant on tree-bases and rotten wood and on the ground in woods and damp grassland, particularly where the soil is free of lime. Capsules occasional. All squares. General.

3. Eurhynchium swartzii (Turn.) Curn.   Relhan, 1820
var. **swartzii**
One of the most abundant mosses in the county, occurring in woods and grassland on calcareous soils. Fruiting occasionally, particularly in damp habitats. At Paradise Island, Newnham, in February 1958, fruiting material was infected with nematode galls caused by *Tylenchus davainii* Bastian (det. D. L. Lee). All squares. General.

var. **rigidum** Boul.   P. W. Richards, 1946
In chalk grassland, apparently rare but perhaps overlooked. *55*, 66. Widely distributed in the British Isles.

316

4. Eurhynchium speciosum (Brid.) Milde

H. Godwin & P. W. Richards, 1929

On earth by rivers and in fens where liable to flooding, frequent. Capsules frequent. 25, 34, 37, 44, 45, 54, 56, 57, 66, 20. W. Widely distributed in the British Isles, but rare in Ireland and Scotland and apparently absent from N.W. Ireland and N.W. Scotland.

5. Eurhynchium riparioides (Hedw.) Jennings   Relhan, 1802

On wood or stone in fast-flowing water, such as on weirs and bridges in streams, frequent. Capsules frequent. 25, 34, *35*, 36, 45, 46, 54, *55*, 64, 65, *66*, 76. General.

6. Eurhynchium murale (Hedw.) Milde   R. S. Adamson, 1913

First localized record, E. W. Jones, 1932. Frequent on shaded wall-bases or stones, particularly by water. Occasional on tree-roots. Capsules frequent. 24, 36, 39, 43–5, 54, 57, 64–7, 75. Throughout England, Wales, S. Scotland and much of Ireland, but few records from N.W. Scotland and N.W. Ireland.

7. Eurhynchium confertum (Dicks.) Milde

G. D. Haviland and J. J. Lister, 1881

Abundant on tree-bases and stumps, bark of Elders, the bases of walls, and on humus in beech woods. Capsules abundant. All squares. General, but few records from N. Scotland.

8. Eurhynchium megapolitanum (Bland.) Milde

M. C. F. Proctor, 1951

On the ground, particularly on sandy soils, rather uncommon but perhaps overlooked. Capsules frequent. 23, 35, 45, 54, 55, 65, 66, 30, 40. A species of central and southern Europe, which is largely confined to England south-east of a line from the Severn estuary to the Humber, although there are isolated records from Scotland, Ireland and Wales.

RHYNCHOSTEGIELLA Limpr.

1. Rhynchostegiella pumila (Wils.) E. F. Warb.

T. G. Tutin *et al.*, 1943

R. pallidirostra (Brid.) Loeske

Of frequent occurrence on the ground in woods, often in deep shade. Capsules rare. 25, 35, 36, 44–6, 54, 55, 64, 65, 76. GG. General, but few records from N. Scotland and N.W. Ireland.

HYPNACEAE

2. Rhynchostegiella tenella (Dicks.) Limpr. Henslow, 1829
var. **tenella** D. E. Coombe and M. C. F. Proctor, 1951
On bare chalk and on mortar or brick near the base of walls, where
shaded, rare. Capsules frequent. 35, 45, 46, 55, 20. A species of
central and southern Europe, which occurs throughout the British
Isles, but is rare in N. Scotland and E. England.

var. **litorea** (De Not.) Richards & Wallace P. J. Bourne, 1960
On stumps and tree-bases, rare. Capsules present. 45, 64. A Mediterranean plant which is confined to S. England. The Cambs localities are
the most northerly known in the British Isles.

ENTODON C. M.

1. Entodon concinnus (De Not.) Paris F. Y. Brocas, 1874
E. orthocarpus (Brid.) Lindb.
In small quantity in long-established chalk grassland, rare. 45, 54, 55,
[44]. GG. Widely distributed in the British Isles, but absent from large
areas perhaps because of lack of spores or other means of dispersal.
Not recorded from S.W. England and very few records from Wales.

PSEUDOSCLEROPODIUM (Limpr.) Fleisch.

1. Pseudoscleropodium purum (Hedw.) Fleisch. T. Martyn, 1763
Abundant in chalk grassland and frequent amongst grass in woods
irrespective of soil. Capsules very rare. 15, 23–5, 33–6, 38, 43–7, 54–6,
64–7, 76, 41. General.

PLEUROZIUM Mitt.

1. Pleurozium schreberi (Brid.) Mitt.
G. D. Haviland and J. J. Lister, 1880
A rare plant in Cambs, as it is a strict calcifuge. Occurs in the remnants
of acid heathland on the Lower Greensand near Gamlingay, and it was
found in a boulder-clay wood (Hardwick Wood) 80 years ago. 25,
[35]. General.

ISOPTERYGIUM Mitt.

1. Isopterygium seligeri (Brid.) Dix. P. W. Richards, 1946
At the foot of a birch in a plantation near Madingley and on rotten
wood at Chippenham Fen. Capsules present. *46*, 66. A rare plant which

in the British Isles is confined to E. England, occurring particularly on introduced trees such as *Castanea sativa*, and hence possibly not native, although known in Britain for over a century.

PLAGIOTHECIUM B. & S.

1. Plagiothecium denticulatum (Hedw.) B. & S.
E. W. Jones and P. W. Richards, 1934
On stumps or on the ground in woods, frequent on the Lower Greensand at Gamlingay, occasional on other soils. Gemmae rare. Capsules frequent. 25, 35, 36, 44, 45, *54, 55,* 56, *64,* 65, 66. General.

2. Plagiothecium curvifolium Schliep. M. C. F. Proctor, 1953
P. denticulatum (Hedw.) B. & S. var. *aptychus* Spr.
On stumps or on the ground in woods, frequent on the Lower Greensand at Gamlingay, but apparently uncommon on other soils. Gemmae rare. Capsules frequent. 25, 35, 46, 54, 64, 65. Widespread, but chiefly in S.E. England and the Midlands.

3. Plagiothecium ruthei Limpr. C. C. Townsend, 1956
Chippenham Fen. Capsules present. 66. Recorded in the British Isles chiefly from southern England, in marshes and fens.

4. Plagiothecium sylvaticum (Turn.) B. & S. E. F. Warburg, 1941
On the ground and at the base of trees in woods, frequent. Gemmae frequent. Capsules occasional. 25, 26, 35, 36, 43, 45, 46, *49,* 54–6, 64–6. General.

HYPNUM Hedw.

1. Hypnum cupressiforme Hedw. T. Martyn, 1763
var. **cupressiforme**
Abundant in a great variety of habitats, but particularly so on the bark of trees and rotten wood and on walls and roofs. Capsules abundant. All squares except 29. General.

var. **resupinatum** (Wils.) Schp. British Bryological Society
Census Catalogue, 1907
First localized record: P. W. Richards, 1938. Frequent on the bark of trees and on fallen logs and stumps in woods. Capsules occasional. 25, 34–6, 45, 46, 54, 55, 57, 59, *64,* 65, 66. General.

HYPNACEAE

var. ericetorum B. & S.   H. L. K. Whitehouse, 1952
On lime-free soil. Recorded from White Wood, Gamlingay and
Hildersham Furze Hills. 25, 54. General.

var. tectorum Brid.   T. G. Tutin, 1932; P. W. Richards, 1932
On roofs and in lime-free grassland, apparently rare. Capsules
occasional. *35, 67*, 30. General.

var. lacunosum Brid.   P. G. M. Rhodes, 1911
Abundant in undisturbed chalk grassland. Capsules rare. 24, 45, 55,
56, 66, 67. Widely distributed in the British Isles.

CTENIDIUM (Schp.) Mitt.

1. Ctenidium molluscum (Hedw.) Mitt.   T. Martyn, 1763
Abundant on the chalk, particularly in undisturbed grassland, beech
woods and chalk-pits. Rare in fens and boulder-clay woods. Capsules
rare; no recent record. 24, 25, 34, 35, 44, 45, 55, 56, *57, 64,* 66. General.

RHYTIDIUM (Sull.) Kindb.

1. Rhytidium rugosum (Hedw.) Kindb.   P. W. Richards, 1930
Abundant on sandy calcareous soils in the Breckland, and in Cambs
known only from the Breck fringe. *67.* Widely distributed but local in
calcareous grassland in Scotland and N. England, otherwise known in
the British Isles only from the Breckland and from single localities in
N. Ireland, N. Wales and Herefordshire. Capsules unknown in the
British Isles.

RHYTIDIADELPHUS (Lindb.) Warnst.

1. Rhytidiadelphus triquetrus (Hedw.) Warnst.   T. Martyn, 1763
Locally abundant in most of the boulder-clay woods, and recorded
from two chalk-pits. Capsules very rare; no recent record. 25, 35, 36, 45,
46, 54, *64,* 65. General.

2. Rhytidiadelphus squarrosus (Hedw.) Warnst.   Relhan, 1785
In grassland, frequent on base-deficient sandy soils such as at Gamlingay and on the Breck fringe, and frequent also in lawns; occasional
in woodland rides. Recorded from one chalk-pit, but otherwise
absent from chalk grassland. Capsules rare. 25, 35, 36, 43, 45–7, 54,
64–6, *67.* General.

HYLOCOMIUM B. & S.

1. Hylocomium splendens (Hedw.) B. & S.   T. Martyn, 1763
On the ground in plantations, rare. Also known from one chalk-pit
in the county. 35, 36, 54, [45]. General.

HEPATICAE (Liverworts)

ANTHOCEROTALES

ANTHOCEROTACEAE

ANTHOCEROS L.

1. †Anthoceros punctatus L.   Henslow, 1835
A plant of arable fields on lime-free soils. Henslow's record, without
locality, is the only one for the county. Widely distributed in the
British Isles, but few records from N. Scotland and none from C. Ireland.

SPHAEROCARPALES

SPHAEROCARPACEAE

SPHAEROCARPOS Ludwig

1. †Sphaerocarpos michelii Bellardi   Relhan, 1802
Barnwell gravel-pit. No record since Relhan's. A rare plant, occurring
chiefly in lime-free arable fields. [45.] A plant of southern and western
Europe which in the British Isles is known only from the English
Midlands and East Anglia with isolated records from Yorkshire, S.W.
England and Kent.

MARCHANTIALES

MARCHANTIACEAE

CONOCEPHALUM Weber

1. Conocephalum conicum (L.) Dum.   Relhan, 1785
Locally abundant on earth, brick or stone at the sides of ditches and
streams. 34, 44, 45, 54, 55, 65–7. General.

MARCHANTIACEAE

LUNULARIA Adans.

1. Lunularia cruciata (L.) Dum. Skrimshire, 1795

Frequent on earth, paths and wall-bases in gardens, otherwise rare and usually by streams. Gemmae abundant. 25, *34*, *35*, 36, 44, 45, 49, 54, 56, 65, *66*, 67, [40]. General.

MARCHANTIA Raddi

1. Marchantia polymorpha L. Ray, 1660

Abundant at Wicken Fen after fires in 1930 and 1957; frequent in gardens; occasional on wall-bases by water, in gravel-pits and in woods after felling. Gemmae abundant. Antheridiophores and archegoniophores frequent. 25, 29, 45, 49, 57, 76, 20. General.

RICCIACEAE

RICCIA L.

1. Riccia glauca L. Relhan, 1802

First record since Henslow: E. F. Warburg, 1941. In arable fields, whether calcareous or not, occurring particularly in damp hollows, rare. Capsules abundant. 26, 44, 45, 66, [46]. Widely distributed in England, Wales, and S. Scotland, and a few records from Ireland.

2. Riccia sorocarpa Bisch. P. J. Bourne, 1960

A plant of lime-free arable fields, recorded with capsules from a stubble field on Oxford Clay. 26. Few records from Ireland and N. Scotland, otherwise general.

3. Riccia fluitans L. Henslow, 1835

Floating in ponds and ditches, or growing on earth at their margins, locally frequent. 44–6, 56, 57, 68, [66]. Widely distributed in England, and a few records from Wales and Ireland.

4. Riccia rhenana Lorbeer ex K. Müll.

E. A. George, 1959; B. Reeve, 1959

Recorded from Madingley brick-pits and from Gray's Moor gravel-pits at Chainbridge near March, floating and on earth at the margins of pools. These are the first records of this plant in semi-natural habi-

RICCIACEAE

tats in the British Isles. 46, 40. Widely distributed in continental Europe, but known previously in the British Isles only from aquaria and one garden pond.

5. Riccia crystallina L. Skrimshire, 1796

First record since Skrimshire's: R. E. Parker, 1952. On the bed of dried-up pools, rare. Capsules abundant. 44, 54, [40]. Rare but widely distributed in England and Wales. Very few records from Scotland and Ireland.

RICCIOCARPUS Corda

1. Ricciocarpus natans (L.) Corda ?Skrimshire, 1795

First certain record: Relhan, 1802.

Floating in ponds or ditches, or on earth at their margins, chiefly in the Fens, rare. 37, *45*, *46*, 48, 57, *58*, *66*, [44, 65, ?50]. Widely distributed in England and most frequent on the east side. A few records from N. Wales and Ireland.

METZGERIALES

RICCARDIACEAE

RICCARDIA S. F. Gray

1. Riccardia multifida (L.) S. F. Gray B. Reeve, 1959

Gray's Moor Pits, Chainbridge, near March. 40. General, but rare in the English Midlands.

2. Riccardia sinuata (Dicks.) Trev. E. W. Jones, 1934

Occasional in a variety of damp habitats, such as in marshes and springs, at ditch and pond margins, in hollows in arable fields, and in woodland rides. 25, *35*, *36*, 37, *45*, 48, 66. General.

3. Riccardia pinguis (L.) S. F. Gray Relhan, 1785

First record since Henslow: E. W. Jones, 1933. Frequent by pools in gravel-pits, occasional on ditch-sides and in damp hollows on clayey soils. Capsules occasional. 29, 34, 35, *36*, 37, 44, 45, 49, 54, 40, [25]. General.

PELLIACEAE

PELLIACEAE

PELLIA Raddi

1. Pellia epiphylla (L.) Corda   Relhan, 1802

First certain record since Henslow: P. D. Sell, 1957. Formerly in Gamlingay Bog. Calcifuge. Recorded recently from damp places on the Lower Greensand at Gamlingay Great Heath. Capsules frequent. 25. General.

2. Pellia fabbroniana Raddi   ?P. G. M. Rhodes, 1911

First certain record: P. W. Richards, 1929.

Abundant on damp calcareous soil, such as the sides of ditches, ponds and streams, and in woodland rides. Capsules occasional. 24, 25, 29, 34–6, 43–6, 48, 54–8, 64–6, 40. General.

METZGERIACEAE

METZGERIA Raddi

1. Metzgeria furcata (L.) Dum.   Relhan, 1802

A frequent epiphyte on the bark of Elders and other trees. It is tolerant of considerable shade and occurs in woods as well as on trees in hedges. Gemmae occasional. 15, 25, 34–7, 43–7, 54–6, 64–6, 76. General.

2. Metzgeria fruticulosa (Dicks.) Evans   R. E. Parker, 1953

Rotten tree stump, swampy wood, Hildersham. 54. Widely distributed in the British Isles.

FOSSOMBRONIACEAE

FOSSOMBRONIA Raddi

1. Fossombronia pusilla (L.) Dum.   Relhan, 1802

First certain record since Henslow: H. L. K. Whitehouse, 1957. Formerly on Gamlingay Heath. Calcifuge. Recorded recently from ruts in rides in Gamlingay Wood and Ditton Park Wood. Capsules frequent. 25, 65. Throughout England, Wales and S. Scotland, but rather few records from Ireland and N. Scotland.

JUNGERMANNIALES

PTILIDIACEAE

PTILIDIUM Nees

1. Ptilidium pulcherrimum (Weber) Hampe   P. W. Richards, 1946
On a birch in a plantation near Madingley, and on an oak in Hardwick
Wood. 35, 46. Recorded from scattered localities on the east side of
Britain, becoming rarer westward and apparently absent from most of
Ireland, S.W. England, Wales and W. Scotland.

CALYPOGEIACEAE

CALYPOGEIA Raddi

1. Calypogeia fissa (L.) Raddi   M. C. F. Proctor, 1953
Calcifuge; recorded from woods on the Lower Greensand at Gamlingay, and boulder clay at Hildersham Wood. Gemmae abundant.
25, 54. General.

LOPHOZIACEAE

LOPHOZIA (Dum.) Dum.

1†. Lophozia ventricosa (Dicks.) Dum.   Relhan, 1820
Formerly in Gamlingay Bog, but no record since 1835. [25.] General,
but rare in the English Midlands.

LEIOCOLEA (K. Müller) Buch

1. Leiocolea turbinata (Raddi) Buch   P. G. M. Rhodes, 1908
Frequent on damp soil on the chalk, such as the floor of chalk-pits, the
sides of ditches, and in chalk grassland. 33–5, 44, 45, 55, 56, 65, 66.
CH. A plant of southern and western Europe which is widely dis-
tributed in the British Isles, but with few records from W. Scotland and
East Anglia.

PLAGIOCHILACEAE

PLAGIOCHILACEAE
PLAGIOCHILA (Dum.) Dum.

1. Plagiochila asplenioides (L.) Dum. var. major Nees   Relhan, 1802
First record since Henslow: P. C. Hodgson, 1931. On the ground in boulder-clay woods, rather rare but locally abundant. Also recorded from Wicken Fen. 25, 35, 43, *57*, 64, *65*. General.

HARPANTHACEAE
LOPHOCOLEA (Dum.) Dum.

1. Lophocolea bidentata (L.) Dum.   T. Martyn, 1763
Frequent on the ground in damp shady grassland, such as in woods and lawns, particularly if soil is not calcareous. 15, 25, 26, 29, *35*, 36, 43–7, 54, 55, 64–6, 75, 31. General.

2. Lophocolea cuspidata (Nees) Limpr.   E. F. Warburg, 1941
Abundant on tree bases and rotten wood. Capsules abundant. 15, 25, 26, 35, 36, 43–6, 54–6, 64–6, 41. General.

3. Lophocolea heterophylla (Schrad.) Dum.   P. G. M. Rhodes, 1905
Abundant on tree-bases and particularly on rotten wood. Gemmae frequent. Capsules abundant. 15, 23–6, 29, 33–6, 39, 43–6, *49*, 54–7, 64–8, 31. Throughout England, Wales and S. Scotland. Not recorded from much of N. and W. Scotland, and few records from central Ireland.

CHILOSCYPHUS Corda

1. Chiloscyphus polyanthos (L.) Corda
              British Bryological Society Census Catalogue, 1930
All the Cambs material, formerly called *C. polyanthos*, seems in fact to be *C. pallescens*. The Census Catalogue record may therefore be in error. *C. polyanthos* usually occurs in or near running water and there are no Cambs records of *Chiloscyphus* in this habitat. Throughout the British Isles, but apparently rather uncommon in east England.

2. Chiloscyphus pallescens (Ehr.) Dum.   P. W. Richards, 1939
On the ground or decaying wood, in marshes and woods, occasional. Capsules occasional. 25, *35*, 36, 44, 45, *65*, 66. Throughout Britain, although rather few records from the English Midlands. Apparently rare in Ireland, where there are only scattered records.

CEPHALOZIELLACEAE

CEPHALOZIELLA (Spr.) Schiffn.

1. Cephaloziella rubella (Nees) Warnst.
P. W. Richards and E. W. Jones, 1934
Gamlingay Heath Wood. Capsules recorded. *25*. Widely distributed in England and Scotland, but very few certain records from Wales and Ireland.

2. Cephaloziella hampeana (Nees) Schiffn.   M. C. F. Proctor, 1952
On sandy soil, Hildersham Furze Hills, with perianths. *54*. Widely distributed in the British Isles.

3. Cephaloziella starkei (Funck) Schiffn.   ?E. W. Jones, 1934
First certain record: G. Halliday and H. L. K. Whitehouse, 1957.
On heathland at Gamlingay Cinques. An unidentified *Cephaloziella*, possibly this species, has been found in Gamlingay Heath Wood and on humus in a beech wood on the Gog Magog Hills. Gemmae and capsules recorded. *25, ?45*. General, but there are few records from C. Ireland.

CEPHALOZIACEAE

CEPHALOZIA (Dum.) Dum.

1. Cephalozia bicuspidata (L.) Dum.   Relhan, 1802
First record since Henslow: P. J. Chamberlain, 1949. A plant of base deficient substrata, which is frequent on the ground in Gamlingay Heath Wood (Lower Greensand), but is rare in other districts where it is recorded only from rotten wood. Capsules recorded. *25, 35, 54, 66*. General.

NOWELLIA Mitt.

1. Nowellia curvifolia (Dicks.) Mitt.   M. H. Martin, 1962
On rotten wood, Hayley Wood. *25*. Scattered records in S. England, more frequent in the north and west of the British Isles.

CEPHALOZIACEAE

ODONTOSCHISMA (Dum.) Dum.

1†. Odontoschisma sphagni (Dicks.) Dum.   Relhan, 1820
Formerly in Gamlingay Bogs, but not recorded since 1835. [25.]
General, but rare in the English Midlands.

RADULACEAE

RADULA Dum.

1. Radula complanata (L.) Dum.   T. Martyn, 1763
First record since Henslow: P. W. Richards, 1931. Occasional on the
trunks of trees. Tolerant of considerable shade and recorded chiefly
from woods and copses. Capsules rare. *25*, 35, *36*, 44, 45, *54*, 55, *64*,
65, 66, [40]. General.

PORELLACEAE

PORELLA L.

1. Porella platyphylla (L.) Lindb.   Relhan, 1785
A calcicole species which is abundant about the exposed roots of
beeches on the chalk. Occasional on stumps in boulder-clay woods.
Rare on wall-bases. *25*, 35, 36, 44–6, 54, 55, 65. GG. Throughout
England and Wales, but not recorded from a number of Scottish and
Irish vice-counties.

LEJEUNEACEAE

LEJEUNEA Lib.

1. Lejeunea cavifolia (Ehr.) Lindb.   E. F. Warburg, 1941
On stumps in boulder-clay woods, rare. *25*, *54*, 65. General, but rare
in E. England.

FRULLANIACEAE

FRULLANIA Raddi

1. Frullania dilatata (L.) Dum.   T. Martyn, 1763
Frequent on tree-trunks, especially Elders, where sheltered and not too
heavily shaded. *25*, 34–6, 44–6, 54–7, 66, 41, [40]. General.

KEY TO MAP

Key to numbered localities on map abbreviated in text

When an Ordnance Survey 1 km. Grid Reference (6 figures) is given, the locality is wholly or largely within that square. In some cases a more precise 100 m. Grid Reference (8 figures) can be usefully given.

1. **FA.**   **Foul Anchor,** GR. 53/4617. Maritime vegetation on the banks of the tidal River Nene. See page 30.

2. **WW.**   **Welney Washes,** GR. 52/5191. Fen and marsh plants. See page 22.

3. **R.**   **Roswell Pits,** GR. 52/5580. Water plants.

4. **W.**   **Wicken Fen,** GR. 52/5569 and 5570. For full account see pages 17–21.

5. **C.**   **Chippenham Fen,** GR. 52/6569. For full account see pages 21–2.

6. **K.**   **Kennett Heath,** GR. 52/7068. One of the few areas where Breckland species are found in Cambs.

7. **D.**   **Devil's Dyke.** Old earthwork stretching from GR. 52/632603 to GR. 52/569659 with a characteristic chalk grassland flora.

8. **H.**   **Hardwick Wood,** GR. 52/3557. Characteristic woodland flora of the western boulder clay. See page 28.

9. **Cam.**   **Cambridge.** Mainly for species occurring on old walls.

10. **CH.**   **Cherry Hinton.** Chalk species occurring in Lime Kiln Close (The Spinney), GR. 52/485560, and around the chalk-pits, GR. 52/4855.

11. **DPW.**   **Ditton Park Wood,** GR. 52/6656 and 6657. Characteristic woodland flora of the eastern boulder clay. See page 28.

12. **F.**   **Fleam Dyke.** Old earthwork (like Devil's Dyke) stretching from GR. 52/537556 to GR. 52/570524.

13. **G.**   **Gamlingay.** This village has the only well-developed greensand flora in the county. The best undisturbed areas are: The Cinques (GR. 52/226529), White Wood (GR. 52/215520), Great Heath Wood (GR. 52/2251), Little Heath (GR. 52/231513), and Gamlingay Wood (GR. 52/2453), the last of which is partly on the boulder clay. See page 29.

KEY

14. **HA.** Hayley Wood, GR. 52/2952. As Hardwick Wood.

15. **GG.** Gog Magog Hills. Golf course (GR. 52/4954) containing a characteristic chalk grassland flora; also the planted grounds of Wandlebury (GR. 52/4953).

16. **B.** Buff Wood, East Hatley, GR. 52/2850. As Hardwick Wood.

17. **DF.** Dernford Fen, GR. 52/4750. Marsh and fen plants. (Compare paper by Bishop's Stortford Coll. Nat. Hist. Soc. (1958).)

18. **T.** Thriplow, GR. 52/4547. Compare paper by G. Crompton (1959). Formerly had an acid bog and heath flora, now extinct.

19. **FH.** Hildersham Furze Hills, GR. 52/553488. See page 29.

330

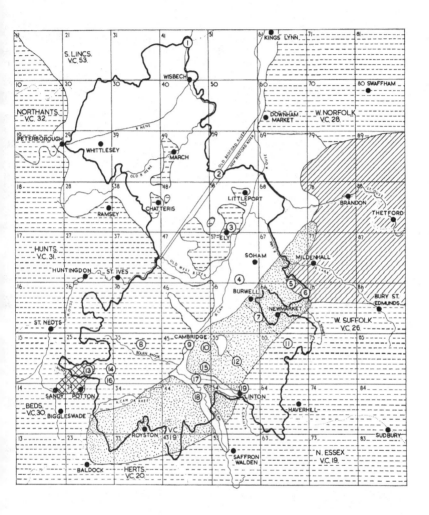

☐ Peat and silt   ▨ Breckland sands   ▦ Chalk

▦ Clays   ▩ Greensand

Key to map is on pp. 329–30.

BIBLIOGRAPHY

A number of additional less important papers on Cambridgeshire plants are referred to in Simpson (1960).

ABEYWICKRAMA, B. A. (1949). *A Study of the Variations in the Field Layer Vegetation of Two Cambridgeshire Woods.* (Unpublished thesis in University Library, Cambridge.)

ADAMSON, R. S. (1911). An ecological study of a Cambridgeshire woodland. *J. Linn. Soc., Lond.*, **40**, 339–387.

ARBER, A. (1920). *Water Plants.* Cambridge.

BABINGTON, C. C. (1848). *Orobanche picridis* F. W. Schultz. *Ann. Mag. Nat. Hist.* (Second Series), **2**, 149.

BABINGTON, C. C. (1849). *Poterium muricatum* Spach. *Bot. Gaz.*, **1**, 224.

BABINGTON, C. C. (1851). Note concerning *Anacharis alsinastrum*. *Bot. Gaz.*, **3**, 135.

BABINGTON, C. C. (1855). Letter in *Phytologist* (New Series), **1**, 190.

BABINGTON, C. C. (1857). *Senecio paludosus. Phytologist* (New Series), **2**, 303.

BABINGTON, C. C. (1860). *Flora of Cambridgeshire.* London. (Original manuscript and author's annotated copy in Cambridge University Botany School Library. Also copies annotated by A. Fryer, C. E. Moss, W. H. Mills, A. Shrubbs, and W. West, jun. A copy annotated by H. N. Dixon is in the library of Cambridge University Botanic Garden and two more in N. D. Simpson's library are annotated by J. S. L. Gilmour and A. H. Evans.)

BABINGTON, C. C. (1862). On the discovery of *Carex ericetorum* Poll. as a native of Britain. *J. Linn. Soc., Lond.*, **6**, 30–31.

BABINGTON, C. C. (1863). *Sturmia loeselii* [and *Senecio paludosus*]. *J. Bot., Lond.*, **1**, 57.

BABINGTON, C. C. (1863). In Smith, J. E., *Supplement to the English Botany*, **5**, t. 2971.

BABINGTON, C. C. (1867). On *Aster salignus* Willd. *J. Bot., Lond.*, **5**, 367–369.

BABINGTON, C. C. (1872). *Callitriche obtusangula* Legall. *J. Bot., Lond.*, **10**, 78.

BABINGTON, C. C. (1881). *Osmunda regalis* in Cambridgeshire. *J. Bot., Lond.*, **19**, 88.

BABINGTON, C. C. (1897). *Memorials Journal and Botanical Correspondence.* Cambridge.

BENNETT, A. (1880). *Potamogeton lanceolatus* in Cambridgeshire. *J. Bot., Lond.*, **18**, 276.

BENNETT, A. (1886). *Potamogeton coriaceus* Nolte. *J. Bot., Lond.*, **24**, 223.

BENNETT, A. (1899). *Selinum carvifolia* L. *J. Bot., Lond.*, **37**, 359.

BENNETT, A. (1899). Notes on Cambridgeshire plants. *J. Bot., Lond.*, **37**, 243–247.

BIBLIOGRAPHY

BENNETT, A. (1899). *Senecio paludosus* and *S. palustris* in East Anglia. *Trans. Norf. & Norw. Nat. Soc.*, **6**, 457–462.

BENNETT, A. (1905). Distribution of *Sonchus palustris* L. and *Atriplex pedunculata* L. in England. *Trans. Norf. & Norw. Nat. Soc.*, **8**, 35–43.

BENNETT, A. (1906). *Holosteum umbellatum* L., *Statice reticulata* L., and *Phleum boehmeri* Wibel. *Trans. Norf. & Norw. Nat. Soc.*, **8**, 231–238.

BENNETT, A. (1910). *Medicago sylvestris, M. falcata, Carex ericetorum* and *Psamma baltica* in England. *Trans. Norf. & Norw. Nat. Soc.*, **9**, 16–25.

BENTHAM, G. (1866). *Handbook of the British Flora*. London. (Copy annotated by Dr Venn.)

BISHOP'S STORTFORD COLLEGE NATURAL HISTORY SOCIETY (1958). *A Survey of Dernford Fen, Sawston, Cambridgeshire 1950–1958* (in place of the Society's Report, *Coturnix*, for 1958). Bishop's Stortford.

BLOMEFIELD, L. (1922). See JENYNS.

BREE, W. T. (1851). Recollections of a morning's ramble in the Whittlesea Fens. *Phytologist*, **4**, 98–105.

BRITTEN, J. (1881). *Sonchus palustris* in Cambridgeshire. *J. Bot., Lond.*, **19**, 152.

BRITTEN, J. (1899). *Gnaphalium luteo-album* in East Anglia. *J. Bot., Lond.*, **37**, 520.

BURKILL, I. H. (1893). Notes on the plants distributed by the Cambridge dust carts. *Proc. Camb. Phil. Soc.*, **8**, 91–95.

BURKILL, I. H. (1893). Cambridgeshire aliens. *J. Bot., Lond.*, **31**, 308–309.

BURKILL, I. H. (1894). *Erucastrum pollichii* in Cambridgeshire. *J. Bot., Lond.*, **32**, 21.

CAMBRIDGE NATURAL HISTORY SOCIETY'S *Card Index of Flowering Plants and Ferns*. (In the Herbarium, Botany School, Cambridge.)

CAMDEN, W. *Britannia* (*Cambridgeshire*). Trans. E. GIBSON. (1695), 416; (1722), 496; (1753), 496–500; (1772), 394–397; trans. R. GOUGH, ed. ii (1789), 144; (1806), 237. London.

CLAPHAM, A. R., TUTIN, T. G. & WARBURG, E. F. (1952). *Flora of the British Isles*. Cambridge.

CORBYN, S. (1656). *Catalogue of Cambridge Plants*. (See DRUCE, G. (1912). *J. Bot., Lond.*, **50**, 76–79.)

CROMPTON, G. (1959). The Peat Holes of Thriplow. *Nature in Cambridgeshire*, **2**, 25–34.

DANDY, J. E. (1958). *List of British Vascular Plants*. London.

DIXON, H. N. (1892). *Potentilla reptans* var. *microphylla* Trattinick. *J. Bot., Lond.*, **30**, 309.

DOBBS, F. D. (1934 for 1932, 1933). Wild flowers of Peterborough and District. *Peterb. Nat. Hist. Sc. Arch. Soc.*, 21.

DONY, J. G. (1953). *Flora of Bedfordshire*. Luton.

DRUCE, G. C. (1904). A hybrid Galeopsis. *J. Bot., Lond.*, **42**, 89.

D., J. (1831). Cornfield weeds. *Loudon's Mag. Nat. Hist.*, **4**, 442.

EVANS, A. H. (1911). The flora of the Fenland, as compared with that of the bogs, marshes, and mosses of Scotland. *Trans. Bot. Soc., Edinb.*, **74**, 164–170.

332

BIBLIOGRAPHY

EVANS, A. H. (1911). A Short Flora of Cambridgeshire, chiefly from an ecological standpoint, with a history of its chief botanists. *Proc. Camb. Phil. Soc.*, **16**, 197–284.

EVANS, A. H. (1913). Notes on additions to the Flora of Cambridgeshire. *Proc. Camb. Phil. Soc.*, **17**, 229–235.

EVANS, A. H. (1939). *A Flora of Cambridgeshire*. London.

EVANS, A. H. & MILLS, W. H. (1922). *Cirsium tuberosum* All. in Cambridgeshire. *J. Bot., Lond.*, **60**, 21.

FITCH, W. H. & SMITH, W. G. (1897). *Illustrations of the British Flora*. London. (Annotated copy in an unknown hand in the possession of F. A. Lees.)

FRYER, A. M. *Flora Hunts. and Cambs.* in Bot. Dept., Oxford, and annotated *London Catalogues of British Plants* in library of Botany School, Cambridge.

FRYER, A. (1883). *Liparis loeselii* in Cambridgeshire. *J. Bot., Lond.*, **21**, 316.

FRYER, A. (1883). *Potamogetons* new to Cambridge and Hunts. *J. Bot., Lond.*, **21**, 316.

FRYER, A. (1883). *Myosurus minimus*, native or colonist. *J. Bot., Lond.*, **21**, 280.

FRYER, A. (1883). *Ceratophyllum submersum* in Cambridgeshire and Hunts. *J. Bot., Lond.*, **21**, 375.

FRYER, A. (1883). *Senecio viscosus* in Cambridgeshire. *J. Bot., Lond.*, **21**, 346–347.

FRYER, A. (1883). *Limosella aquatica* in Cambridgeshire and Hunts. *J. Bot., Lond.*, **21**, 377.

FRYER, A. (1884). *Lepidium smithii* Hook. in Cambridgeshire. *J. Bot., Lond.*, **22**, 247.

FRYER, A. (1884). *Polygonum minus* Huds. in Cambridgeshire. *J. Bot., Lond.*, **22**, 28.

FRYER, A. (1884). *Agrostis nigra* in Cambridgeshire. *J. Bot., Lond.*, **22**, 125.

FRYER, A. (1884). *Bupleurum tenuissimum* Linn. inland in Cambridgeshire. *J. Bot., Lond.*, **22**, 28.

FRYER, A. (1884). *Juncus gerardii* Lois. in Cambridgeshire. *J. Bot., Lond.*, **22**, 151–152.

FRYER, A. (1884). Cambridgeshire Fumarias. *J. Bot., Lond.*, **22**, 279.

FRYER, A. (1885). *Carex paradoxa* Willd. in Cambridgeshire. *J. Bot., Lond.*, **23**, 221.

FRYER, A. (1886). *Epilobium angustifolium* L. in Cambridgeshire. *J. Bot., Lond.*, **24**, 345.

FRYER, A. (1886). *Potamogeton fluitans* in Cambridgeshire. *J. Bot., Lond.*, **24**, 306–7.

FRYER, A. (1889). *Polygala calcarea* in Cambridgeshire. *J. Bot., Lond.*, **27**, 119.

FRYER, A. (1889). *Gnaphalium uliginosum* L. var. *pilulare* Wahl. *J. Bot., Lond.*, **27**, 83–85.

BIBLIOGRAPHY

FRYER, A. (1890). On a new hybrid *Potamogeton* of the *Fluitans* group. *J. Bot., Lond.*, **28**, 321–326.

FRYER, A. (1892). *Potamogeton undulatus* Wolfgang, in Cambridgeshire. *J. Bot., Lond.*, **30**, 377.

FRYER, A. (1893). Notes on Pondweeds. *J. Bot., Lond.*, **31**, 353–355.

FRYER, A. (1894). *Potamogeton nitens* in Cambridgeshire. *J. Bot., Lond.*, **32**, 345.

FRYER, A. (1896). *Potamogeton nitens* Weber *f. involuta*. *J. Bot., Lond.*, **34**, 1–3.

FRYER, A. (1897). *Potamogeton trichoides* Cham. in Cambridgeshire. *J. Bot., Lond.*, **35**, 446–447.

GARDINER, J. S. & TANSLEY, A. G. (eds.) (1923–32). *The Natural History of Wicken Fen*, **1–6**. Cambridge.

GELDART, A. M. (1917). *Liparis loeselii*. *J. Bot., Lond.*, **55**, 292.

GIBSON, G. S. (1842, 1844). Flora of the neighbourhood of Saffron Walden. *Phytologist* (1842), **1**, 408–415; and (1844), **1**, 838–839 and 1123–1126.

GIBSON, G. S. (1848). Notice of the discovery of *Filago jussiaei* near Saffron Walden. *Phytologist*, **3**, 216.

GIBSON, G. S. (1848). Botanical notes for 1848. *Phytologist*, **3**, 308–310.

GIBSON, G. S. (1849). Botanical notes for 1849. *Phytologist*, **3**, 707–708.

GIBSON, G. S. (1862). *The Flora of Essex*. London.

GIBSON, G. S. (1868). On the discovery of *Potentilla norvegica* Linn. in England. *J. Bot., Lond.*, **6**, 302–303.

GILMOUR, J. S. L. & STEARN, W. T. (1932). Notes from the University Herbarium, Cambridge. *J. Bot., Lond.*, **70**, Suppl. 1–29.

GODWIN, H. (1931). Studies in the Ecology of Wicken Fen. I. The ground water level of the Fen. *J. Ecol.*, **19**, 449–473.

GODWIN, H. (1936). Studies in the Ecology of Wicken Fen. III. The establishment and development of Fen Scrub (Carr). *J. Ecol.*, **24**, 82–116.

GODWIN, H. (1938). Botany of Cambridgeshire. In *Victoria County History of Cambridgeshire and the Isle of Ely*, **1**, 35–76.

GODWIN, H. (1938). The Botany of Cambridgeshire. In *A Scientific Survey of the Cambridge District*. Ed. H. C. Darby for the British Association for the Advancement of Science. London.

GODWIN, H. (1941). Studies in the Ecology of Wicken Fen. IV. Crop-taking Experiments. *J. Ecol.*, **29**, 83–106.

GODWIN, H. & BHARUCHA, F. R. (1932). Studies in the Ecology of Wicken Fen. II. The Fen Water Table and its Control of Plant Communities. *J. Ecol.*, **20**, 157–191.

GOODE, G. (1898). Cambridgeshire Plants. *J. Bot., Lond.*, **36**, 400–401.

GOODE, G. (1908). *Prunella laciniata* in Cambridgeshire. *J. Bot., Lond.*, **46**, 266.

GOODE, G. (1914). The adventitious flora of a library court. *J. Bot., Lond.*, **52**, 46.

GORHAM, G. C. (1830). *Memoirs of John Martyn, F.R.S., and of Thomas Martyn, B.D., F.R.S., F.L.S.* (pp. 105–109). Cambridge.

BIBLIOGRAPHY

GRAHAM, R. A. (1954). Mint notes. V. *Mentha aquatica*, and the British Water Mints. *Watsonia*, 3, 109–121.

GREGORY, E. S. (1904). *Viola calcarea* as a species. *J. Bot., Lond.*, 42, 67–68 and 186.

HENSLOW, J. S. (1829). *A Catalogue of British Plants.* Cambridge. (Plants not found in Cambs are italicized.) Ed. ii (1835). (Cambs species marked with a small c.)

HENSLOW, J. S. (1829). On the leaves of *Malaxis paludosa. Loudon's Mag. Nat. Hist.*, 1, 441.

HENSLOW, J. S. (1832). On the varieties of *Paris quadrifolia*, considered with respect to the ordinary characteristics of monocotyledonous plants. *Loudon's Mag. Nat. Hist.*, 5, 429–433.

HIERN, P. (1867). On the occurrence of *Aster salignus* Willd., in Wicken Fen, Cambridgeshire. *J. Bot., Lond.*, 5, 306–307.

HOSKING, A. (1903). Notes on Cambridgeshire plants. *J. Bot., Lond.*, 41, 157–159.

HUDSON, G. (1762). *Flora Anglica.* London. (Interleaved copy annotated by I. Lyons and M. Tyson in library of Linnean Society.)

JACKSON, A. B. (1898). *Lathyrus aphaca* in Cambridgeshire. *J. Bot., Lond.*, 36, 353, 400.

JENYNS, L. (1846). *Observations in Natural History.* London.

JENYNS (BLOMEFIELD), L. (1922). *A Naturalist's Calendar*, ed. ii. Cambridge. (Ed. F. Darwin.)

JOHNSON, T. (1641). *Mercurii Botanici pars Altera*, pp. 15–36. London.

JONES, E. W. (1958). An annotated list of British Hepatics. *Trans. Brit. Bryol. Soc.*, 3, 353–374.

KASSAS, M. (1951). Studies in the Ecology of Chippenham Fen. *J. Ecol.*, 39: I. The Fen water-table, 1–18; II. Recent history of the Fen, from evidence of historical records, vegetational analysis and tree-ring analysis, 19–32.

KENT, D. H. (1955). Abstract of 'A taxonomic spectrum of the section eu-Callitriche in the Netherlands'. *Proc. Bot. Soc. Brit. Isles*, 1, 340–341.

KENT, D. H. (1956, 1960). *Senecio squalidus* L. in the British Isles. (1956): 1. Early records (to 1877). *Proc. Bot. Soc. Brit. Isles*, 2, 115–118. (1960): 2. The spread from Oxford (1879–1939). *loc. cit.*, 3, 375–379.

LAWSON, T. (before 1677). Manuscript notebook now in possession of the Linnean Society. (*See* RAVEN, C. E. (1948). Thomas Lawson's Notebook. *Proc. Linn. Soc., Lond.*, session 160, 3–12.)

LEES, F. A. (1899). The Cambridge and Lincoln *Selinum. J. Bot., Lond.*, 37, 326–327.

LEES, F. A. & MARSHALL, W. (1882). *Selinum carvifolia* in Cambridgeshire. *J. Bot., Lond.*, 20, 284.

LEWIN, R. A. (1948). *Sonchus* L. for Biological Flora of the British Isles. *J. Ecol.*, 36, 203–223.

LOUDON, J. C. (1835). *The Magazine of Natural History*, 8, 338. (Note on *Geranium pyrenaicum* in Cambs.)

335

BIBLIOGRAPHY

LYONS, I. (1763). *Fasciculus Plantarum circa Cantabrigiam nascentium.* London.

MARSHALL, E. S. (1911). *Erophila virescens* in Cambs. *J. Bot., Lond.,* **49,** 198.

MARSHALL, W. (1852). Excessive and noxious increase of *Udora canadensis* (*Anacharis alsinastrum*). *Phytologist,* **4,** 705–715.

MARSHALL, W. (1857). The American water-weed *Anacharis alsinastrum. Phytologist* (New Series), **2,** 194–197.

MARTYN, J. (1727). *Methodus Plantarum circa Cantabrigiam nascentium.* London. (24 printed pages of a second edition are to be found in the Cambridge University Botany School library, but they were never published. Also there is a copy of the main work annotated by T. Martyn.)

MARTYN, T. (1763). *Plantae Cantabrigienses.* London. (This work contains not only a list of Cambs plants but also a list of plants to be found in each of thirteen localities in the vicinity of Cambridge.)

MAYNARD, G. N. & MAYNARD, N. (19th century). Manuscript notes and specimens on the Thriplow area in the care of the County Archivist's Department, The Shire Hall, Cambridge.

MELVILL, J. C. (1871). *Siler trilobum* Scop. in England. *J. Bot., Lond.,* **9,** 211.

MILLER, S. H. (1889). *The Hand-book to the Fenland* [Botany by C. C. Babington]. Ed. ii (1890). London, Wisbech.

MILLER, S. H. & SKERTCHLY, S. B. J. (1878). *The Fenland, Past and Present.* Wisbech.

MOSS, C. E. (1914–20). *The Cambridge British Flora,* **2** and **3.** Cambridge.

NELMES, E. (1947 for 1945). Two critical groups of British sedges. *Rep. Bot. Soc. & Exch. Cl. Brit. Isles,* **8,** 95–105.

PALEY, F. A. (1860). *A List of Four Hundred Wild Flowering Plants being a contribution to the Flora of Peterborough.* Peterborough.

PALMER, S. (1829). Plants collected by the Rev. S. Palmer of Chigwell, Essex. *Loudon's Mag. Nat. Hist.,* **2,** 386.

PERRING, F. H., SELL, P. D. & WALTERS, S. M. (1955). Notes on the Flora of Cambridgeshire. *Proc. Bot. Soc. Brit. Isles,* **1,** 471–481.

PERRING, F. H. (1956). An MS. in Wisbech Museum. *Proc. Bot. Soc. Brit. Isles,* **2,** 133. (Manuscript by an unknown hand, headed 'This book contains the Catalogue of Plants contained in Mr Skrimshire's Hortus Siccus—sold some years since to Lord Milton—and also of Plants contained in a Hortus Siccus bought by me and collected by Mr Skrimshire after the sale of the former'.)

PERRING, F. H. (1958). Additions to the vascular flora of Cambridgeshire. *Nature in Cambridgeshire,* **1,** 27–29.

PERRING, F. H. & SELL, P. D. (1959). Further notes on the flora of Cambridgeshire. *Proc. Bot. Soc. Brit. Isles,* **3,** 165–171.

PERRING, F. H. (1959). Flowering plant records. *Nature in Cambridgeshire,* **2,** 36.

BIBLIOGRAPHY

PERRING, F. H. (1959). Field meeting, April 14th, 1957, Cambridge. *Proc. Bot. Soc. Brit. Isles*, **3**, 230–232.

PERRING, F. H. (1960). Vascular plant records in 1959. *Nature in Cambridgeshire*, **3**, 36.

PORRITT, G. T. (1879). A fortnight in the Fens. *The Naturalist*, **4**, 116–120, 129–133.

PRATT, J. (pre 1663). *Catalogus Plantarum Angliae*. (Manuscript in Brit. Mus. (Sloane no. 591).)

PROCTOR, M. C. F. (1956). A Bryophyte Flora of Cambridgeshire. *Trans. Brit. Bryol. Society*, **3**, 1–49.

PRYOR, R. A. (1873). *Hypericum dubium* in Cambridgeshire. *J. Bot., Lond.*, **11**, 274.

PRYOR, R. A. (1874). Plants of Kirtling, Cambridgeshire. *J. Bot., Lond.*, **12**, 22–23.

PRYOR, R. A. (1881). *Osmunda regalis* in Cambridgeshire. *J. Bot., Lond.*, **19**, 54.

PRYOR, R. A. [A. R.] (1887). *A Flora of Hertfordshire*. London.

PUGSLEY, H. W. (1929). A revision of the British Euphrasiae. *J. Linn. Soc., Lond.*, **48**, 467–544.

PUGSLEY, H. W. (1933). Notes on British Euphrasiae. III. *J. Bot., Lond.*, **71**, 83–90.

RAVEN, C. E. (1942). *John Ray, Naturalist, his Life and Works*. Cambridge.

RAVEN, C. E. (1947). *English Naturalists from Neckam to Ray*. Cambridge. (For page references to Cambs plants see Simpson, N. D. (1960).)

RAY, J. (1660). *Catalogus Plantarum circa Cantabrigiam nascentium*. Cambridge. Another issue (1660) contains separately paged, *Index Plantarum Agro Cantabrigiensis*. Appendix 1 (1663). Appendix 2 (1685). (There is a copy of the main work in the Cambridge University Botany School Library annotated by J. Martyn.)

RAY, J. (1670). *Catalogus Plantarum Angliae*. London. Ed. ii (1677). (Plants which occur in Cambs are marked 'C'.) (Copy in British Museum annotated by J. Newton.)

RAY, J. *Synopsis Methodica Stirpium Britannicarum*. London. Ed. ii (1696). (Annotated copies in libraries of the British Museum and J. E. Raven.) Ed. iii (1724). (Annotated copy by J. Lightfoot and J. Hill in Library, Bot. Dept. Oxford.)

RELHAN, R. (1785). *Flora Cantabrigiensis*. Cambridge. Supplements 1786 and 1793. Appendix to 1793 Supplement. Ed. ii, 1802; Ed. iii, 1820. (There are annotated copies in the Cambridge University Botany School Library of a second edition by L. Jenyns and third editions by C. C. Babington and J. S. Henslow.)

REYNOLDS, B. (1908). *Lathraea clandestina* L. near Cambridge. *J. Bot., Lond.*, **46**, 123. (But cf. (1910) *J. Bot., Lond.*, **48**, 79 and (1920) **58**, 30.)

RICHARDS, P. W. (1949). Rushes in East Anglia. *New Nat. (East Anglia)*, **6**, 41–44.

BIBLIOGRAPHY

RICHARDS, P. W. & WALLACE, E. C. (1950). An annotated list of British Mosses. *Trans. Brit. Bryol. Soc.*, **1**, part 4, App. i–xxi.

RICHENS, R. H. (1958). Studies on *Ulmus*. II. The village elms of southern Cambridgeshire. *Forestry*, **31**, 132–146.

RICHENS, R. H. (1960). Cambridgeshire elms. *Nature in Cambridgeshire*, **3**, 18–22.

RIDDELSDELL, H. J. (1931). Field notes, chiefly 1930. *J. Bot., Lond.*, **69**, 240–244, 309–313.

RISHBETH, J. (1948). The Flora of Cambridge walls. *J. Ecol.*, **36**, 136–148.

SALMON, C. E. (1910). Tragopogon hybrid. *J. Bot., Lond.*, **48**, 284.

SCHOTSMAN, H. D. (1954). A taxonomic spectrum of the section Eu-Callitriche in the Netherlands. *Acta Bot. Neerl.*, **3**, 313–384.

SIMPSON, N. D. (1960). *A Bibliographical Index of the British Flora*. Privately printed.

SMITH, J. E. (1799). *English Botany*. **9**. London.

STUART, H. C. (1853). A botanical stroll from Cambridge to the Cherry Hinton chalk-pits on the 1st of November. *The Naturalist* (conducted by B. R. Morris), **3**, 53–54.

STYLES, B. T. (1962). The taxonomy of *Polygonum aviculare* and its allies in Britain. *Watsonia*, **5**, 177–214.

TANSLEY, A. G. (1911). *Types of British Vegetation*. Cambridge. (Especially pages 178–181 on the Fleam Dyke.)

TIMM, E. W. & CLAPHAM, A. R. (1940). Jointed rushes of the Oxford district. *New Phytologist*, **39**, 1–16.

TURNER, D. & DILLWYN, L. W. (1805). *The Botanist's Guide*, **1**, 41–71. London.

TUTIN, T. G. (1931). *Bromus britannicus*. *J. Bot., Lond.*, **69**, 316.

WALLIS, A. In MARR, J. E. & SHIPLEY, A. E. (1904). *Handbook to the Natural History of Cambridgeshire*, 209–237. Cambridge.

WALTERS, S. M. (1961). Wild Juniper on the Fleam Dyke. *Nature in Cambridgeshire*, **4**, 40–41.

WATSON, H. C. (1835). *The New Botanist's Guide*, **1**, 143–155 and Suppl. 598–601. London.

WATSON, H. C. for COLEMAN, W. H. (1840). Note on exhibit of *Carum bulbocastanum* and *Seseli libanotis*. *Proc. Linn. Soc., Lond.*, session **1**, 51.

WATSON, W. C. R. (1958). *Handbook of the Rubi of Great Britain and Ireland*. Cambridge.

WATT, A. S. (1936). Studies in the Ecology of Breckland. I. Climate, Soil and Vegetation. *J. Ecol.*, **24**, 117–138.

WATT, A. S. (1940). Studies in the Ecology of Breckland. IV. The Grass Heath. *J. Ecol.*, **28**, 42–70.

WELCH, D. (1961). Water Forget-me-nots in Cambridgeshire. *Nature in Cambridgeshire*, **4**, 18.

WEST, W. Jnr. (1898). Notes on Cambridgeshire plants. *J. Bot., Lond.*, **36**, 246–259, 491–492.

BIBLIOGRAPHY

WHITEHOUSE, H. L. K. (1958). Additions to the Bryophyte Flora of Cambridgeshire. *Nature in Cambridgeshire*, **1**, 25–27.

WHITEHOUSE, H. L. K. (1959). *Riccia* in Cambridgeshire. *Nature in Cambridgeshire*, **2**, 37.

WHITEHOUSE, H. L. K. (1961) Bryophytes added to the County list during 1960. *Nature in Cambridgeshire*, **4**, 43–44.

Wicken Fen, A Guide to. Published by the National Trust, London. Ed. i (1932); ed. ii (1939); ed. iii (1949); reprinted (1950); reprinted with addendum (1959).

Wicken Fen, Card index of species occurring there. (In the Herbarium, Botany School, Cambridge.)

WILLIS, J. C. & BURKILL, I. H. (1893). Observations on the Flora of pollard willows near Cambridge. *Proc. Camb. Phil. Soc.*, **8**, 82–91.

WOLLEY-DOD, A. H. (1930–31). A revision of the British Roses. *J. Bot., Lond.*, Suppl., **68** and **69**.

YAPP, R. H. (1908). Sketches of vegetation at home and abroad. IV. Wicken Fen. *New Phytol.*, **7**, 61–81.

INDEX

Accepted names are in Roman type, synonyms in italic. Page numbers given in bold type are those on which the full account of the species is given, numbers in Roman type are those on which an accepted name is mentioned, and numbers in italic are where a synonym is mentioned.

INDEX

Amaranthus (*cont.*)
quitensis, **81**
retroflexus, **81**
Amaryllidaceae, **234**
Amblystegium, **312**
juratzkanum, **312**
serpens, **312**
varium, **312**
Anacamptis, **243**
pyramidalis, **25, 243**
Anacharis alsinastrum,
220
Anagallis, **160**
arvensis, **160**
minima, 30, **160**
tenella, **160**
Anchusa sempervirens, 164
Anemone, **42**
nemorosa, 27, **42**
pulsatilla, 42
Anemone, Wood, **42**
Angelica, **139**
sylvestris, 20, 21, **139**
Angiospermae, xi, **41**
Anisantha
gussonii, 269
sterilis, 269
Anomodon, **310**
viticulosus, **310**
Antennaria, **202**
dioica, **202**
Anthemis, **203**
arvensis, **204**
cotula, 23, **203**
nobilis, 204
tinctoria, **203**
Anthocerotaceae, **321**
Anthoceros, **321**
punctatus, **321**
Anthocerotales, **321**
Anthoxanthemum, **279**
odoratum, **279**
puelii, **279**
Anthriscus, **132**
caucalis, **132**
neglecta, 132
sylvestris, **132, 133**
vulgaris, 132
Anthyllis, **102**
vulneraria, **102**

Antirrhinum, **171**
majus, **171**
orontium, 171
Apargia
autumnalis, 212
hispida, 212
Apera, **277**
interrupta, 12, **277**
spica-venti, **277**
Aphanes, 32, **114**
arvensis, **114**
microcarpa, **114**
Apium, **134**
dulce, **134**
graveolens, **134**
inundatum, **134**
nodiflorum, **134**, 137
repens, **134**
Apocynaceae, **161**
Apple
Crab, **121**
Cultivated, **121**
Aquifoliaceae, **93**
Aquilegia, **47**
vulgaris, 22, **47**
Arabidopsis, **64**
thaliana, 29, **64**
Arabis, **61**
glabra, 61
hirsuta, **61**
turrita, **61**
Araceae, **243**
Araliaceae, **131**
Archangel, Yellow, **185**
Archidiaceae, **286**
Archidium, **286**
alternifolium, **286**
Arctium, **207**
lappa, **207**
majus, 207
minus, **207**
pubens, **207**
tomentosum, 207
vulgare, 207
Arenaria, 3, **78**
leptoclados, **78**
serpyllifolia, **78**
tenuifolia, 78
trinervia, 78
Aristolochia, **141**

clematitis, 3, **141**
Aristolochiaceae, **141**
Armeria, **158**
maritima, 30, **158**
Armoracia, **59**
amphibia, 63
rusticana, **59**
Arnoseris, ix, **211**
minima, 30, **211**
Arrhenatherum, **274**
avenaceum, 274
elatius, **274**
Arrow-grass, **220**
Arrow-head, **219**
Artemisia, **206**
absinthium, **206**
biennis, **206**
campestris, **206**
maritima, **206**
verlotorum, **206**
vulgaris, **206**
Arum, **244**
maculatum, 27, **244**
Ash, 26, 27, 131, **161**
Mountain, **120**
Asparagus, 8, **228**
Bath, **229**
officinalis, **228**
Aspen, **152**
Asperugo, **163**
procumbens, 3, **163**
Asperula, **191**
arvensis, **191**
cynanchica, 24, **191**
odorata, 191
Asphodel, Bog, **227**
Asphodelus, **227**
fistulosus, **227**
Aspidiaceae, **37**
Aspleniaceae, **36**
Asplenium, **36**
adiantum-nigrum, **36**
ruta-muraria, **36**
trichomanes, **36**
Aster, **202**
laevis, **202**
lanceolatus, **202**
novae-angliae, **202**
novi-belgii, **202**
salignus, **203**

345

INDEX

INDEX

INDEX

349

INDEX

353

INDEX

INDEX

INDEX

358

INDEX

INDEX

INDEX

INDEX

INDEX

INDEX

Printed in the United States
By Bookmasters